ANATOMY AND ANATOMISTS IN EARLY MODERN SPAIN

The History of Medicine in Context

Series Editors: Andrew Cunningham and Ole Peter Grell

Department of History and Philosophy of Science
University of Cambridge

Department of History
Open University

Titles in the series include

The Fate of Anatomical Collections
Rina Knoeff and Robert Zwijnenberg

Sudden Death: Medicine and Religion in Eighteenth-Century Rome
Maria Pia Donato

*Suzanne Noël: Cosmetic Surgery, Feminism and Beauty in
Early Twentieth-Century France*
Paula J. Martin

Wounds in the Middle Ages
Edited by Anne Kirkham and Cordelia Warr

The One-Sex Body on Trial: The Classical and Early Modern Evidence
Helen King

Anatomy and Anatomists in Early Modern Spain

Bjørn Okholm Skaarup

Routledge
Taylor & Francis Group

LONDON AND NEW YORK

First published in paperback 2024

First published 2015 by Ashgate Publishing

Published 2016 by Routledge
4 Park Square, Milton Park, Abingdon, Oxon OX14 4RN

and by Routledge
605 Third Avenue, New York, NY 10158

Routledge is an imprint of the Taylor & Francis Group, an informa business

British Library Cataloguing in Publication Data
A catalogue record for this book is available from the British Library

The Library of Congress has cataloged the printed edition as follows:
Skaarup, Bjxrn Okholm, author.
Anatomy and anatomists in early modern Spain / by Bjxrn Okholm Skaarup Ashgate.
    p. ; cm. – (History of medicine in context)
    Includes bibliographical references and index.
    ISBN 978-1-4724-4826-2 (hardcover)

    I. Title. II. Series: History of medicine in context.
    [DNLM: 1. Anatomy–history–Spain. 2. Anatomists–history–Spain. 3. Anatomy–education–Spain. 4. History, 16th Century–Spain. QS 11 GS6]
    QM11
    611.00946–dc23

                                                                    2014026174

    ISBN: 978-1-4724-4826-2 (hbk)
    ISBN: 978-1-03-292305-5 (pbk)
    ISBN: 978-1-315-56697-9 (ebk)

    DOI: 10.4324/9781315566979

To my gorgeous wife, best friend, colleague,
(and fellow Ashgate author),

Joanna Milstein

Contents

List of Figures

Acknowledgements

I wish to thank everyone who has contributed to this work. Many thanks go to my supervisor, Professor Antonella Romano (European University Institute), who provided guidance and advice during the development of this project. Thanks also go to the jury and examining board of my doctoral thesis, Professor Andrea Carlino (Institut d'Histoire de la Médicine et de la Santé, Geneva), Professor Rafael Mandressi (Centre Alexandre-Koyré, Paris) and Professor Bartolomé Yun-Casalilla (European University Institute). A large number of Spanish scholars, universities, libraries and archives have been consulted during the writing of this monograph, and have been extremely helpful. I would especially like to thank Professor José Pardo Tomás (CSIC, Barcelona) and Professor Àlvar Martínez-Vidal (IHMC, Valencia), whose scholarship on Spanish anatomy theatres first opened my eyes to the exciting topic of Spanish Renaissance anatomy and who have offered friendly assistance ever since. Thanks also to Professor Victor Navarro-Brotóns and Professor José Luis Fresquet Febrer (IHMC, Valencia), Professor Anastasio Rojo Vega, (University of Valladolid), Professor Fernando Bouza (Universidad Complutense de Madrid), and Dr Asunción Fernández (University of Zaragoza). I am furthermore indebted to Professor Cynthia Klestinec, (Miami University, Ohio) for generously sharing her knowledge of Renaissance anatomy and surgery with me and to Professor Michele L. Clouse (Ohio University) for her contributions to the study of sixteenth-century Spanish medical history. I would like to acknowledge Niels Lynnerup, Professor of Forensic Anthropology, for many years of anatomical training used in our joint attempts to reconstruct the faces of the past, and Jørgen Tranum-Jensen, Professor of Anatomy (Panuminstituttet, University of Copenhagen), for allowing me to access original textbooks from the anatomical institute library. Thanks also to my father, Dr Jørgen Skaarup, my wife, Dr Joanna Milstein, Dr Gojko Barjamovic (Harvard University), Dr Ivan Boserup (Royal Library, Copenhagen), Professor Ian Maclean (University of Oxford) and Professor Pamela H. Smith (Columbia University) for reading through one or more chapters of the text. Dr Alessio Assonitis and Dr Sheila Barker from the Medici Archive Project in Florence have provided many useful elucidations. For financial support during my doctoral and postdoctoral studies at the EUI, the Warburg Institute and Columbia University, I would like to thank the Danish Rectors Conference and The Carlsberg Foundation. I also wish to thank Lucy Morris for thoroughly editing the text. Finally, I am grateful to Tom Gray, Emily Yates, Katie McDonald, Jon Lloyd, Matthew Irving and the series editors of *The History of Medicine in Context* at Ashgate for facilitating the publication of this monograph. Thanks at last to my parents Jørgen Skaarup and Lulu Okholm (and my sister Grith Okholm Skaarup) for a lifelong encouragement of my interests in history and the arts.

Chapter 1

Early Modern Spanish Anatomy and 'La polémica de la ciencia española'

Paradoxically, studies of early modern Iberian science have only been carried out coherently and collaboratively during the last few decades, even though fierce debates on the subject have dominated Spanish historiography for more than two centuries. Since the late 1700s, Spain has been the stage of an often bitter controversy known as 'la polémica de la ciencia española'. The polemic has been played out between *criticos*, lamenting the absence of Spanish contributions to modernity, and *apologistas*, presenting Spain as a leader rather than a failure in the intellectual and scientific history of Europe. The former have mourned the role of the Inquisition, censorship, and religious orthodoxy, and the intellectual isolation of early modern Spain cemented by Philip II's 1559 decree, which prohibited Spanish students and scholars from teaching and studying abroad. According to these critics, the mid-sixteenth century represented the opening stage of subsequent centuries of cultural and intellectual backwardness, and their conclusions often advocated the needs of radical and immediate reforms of Spanish society. In strong opposition to this pessimist view, the apologists, regularly referred to by their counterparts as *nationalistas* or *triunfilistas*, have presented the period in question as 'a golden century' (*Siglo de Oro*), the glorious culmination of Spain as an unequalled European superpower and creator of the first global empire.

The long polemic on Spanish science has raged since the 1782 article 'Espagne' by the French polymath Masson de Morvilliers, in the *Encyclopédie méthodique*, which presented Spain as the only European nation that had never contributed to the progress of science.[1] Since then, the controversy has not only taken the shape of an academic dispute, but has also acted as a forum for different political agendas and reactions to dramatic events and ruptures in

---

[1]   Nicolas Masson de Morvilliers, 'Espagne' in *Encyclopedie méthodique: Géographie moderne: tome premier, Vol. 1, Part 2* (Paris: Chez Panckoucke, 1782), p. 565: 'Aujourd'hui le Danemark, la Suede, la Russie, la Pologne même, l'Allemagne, l'Italie, l'Angleterre et la France, tous ces peuples ennemis, amis, rivaux, tous brûlent d'une généreuse émulation pour le progrès des sciences et des arts! Chacun médite des conquêtes qu'il doit partager avec les autres nations; chacun d'eux, jusqu'ici, a fait quelque découverte utile, qui a tourné au profit de l'humanité! Mais que doit on à l'Espagne? Et depuis deux siècles, depuis quatre, depuis dix, qu'a-t-elle fait pour l'Europe?'

contemporary Spanish history. Initially, strong anti-French reactions followed the article by Masson de Morvilliers and peaked during the subsequent *Guerra de la independencia* (1808–14). The apologists held the upper hand during the later period of restoration and reaction.[2] Revisionist history writing from that period incorrectly attributed several anatomical discoveries to Spanish renaissance physicians; Andrés Laguna was credited as the first explorer of the appendix and the ileocecal valve in Antonio Hernández Morejón's *Historia bibliográfica de la medicina española*, published after the author's death in 1836.[3] Anastasio Chinchilla's subsequent *Historia bibliografica de la medicina Española* (1842–52) claimed that the significant theory of the general circulation of the blood had been introduced by Francisco de la Reina half a century before William Harvey, who is normally credited with this discovery.[4]

The nationalist self-sufficiency enabling such publications was drastically challenged by the end of the century in Marcellino Menéndez Pelayo's *Esplendor y decadencia de la cultura científica española* (1894) and by the admission the following year of the renowned anatomist/neurolologist Santiago Ramon y Cajal to the Spanish Academy of Science. Cajal was the first Spaniard to be awarded the Nobel Prize in Medicine for his research in neurons and the structure of the nervous system. In 1897 he outlined a critical survey of the deficiencies in Spanish science and medicine caused by centuries of isolation and intellectual segregation. It offered a scientific programme entitled *Reglas y consejos sobre investigación científica*, which was reprinted and translated in numerous later editions. This manifesto called for a renewed Spanish engagement in the natural sciences, from which the country had been long largely secluded: 'We have been living in a shell for centuries, turning the wheels of Aristotelianism and scholasticism, uninterested and scornful (except for a few parenthesises) in the powerful critical and revisionist movements in the sciences and the arts.'[5] Cajal's

---

[2]   Henry Kamen, 'The Decline of Spain: A Historical Myth?', *Past and Present*, 81 (1978), p. 25.

[3]   Antonio Hernández Morejón, *Historia bibliográfica de la medicina española*, vol. II (Madrid: Imprenta de la viuda de Jordan e hijos, 1845), p. 256. The appendix was first described in 1521 by the Bolognese anatomist Jacopo Berengario da Carpi, and the discovery of the ileocecal valve (also known as Tulp's valve) was published in 1641 by the Dutch anatomist Nicolaes Tulp.

[4]   Anastasio Chinchilla, *Anales historicos de la medicina general, Tomo primero* (Valencia: Lopez y compañia, 1841), p. 155–60. Chinchilla and later Spanish historians have on very loose foundations attributed this discovery to Francisco de la Reina and his veterinary treatise *Libro de Albeyteria en el qual se veran todas quantas enfermedades y desastres suelen acaecer a todo genero de bestias* (Zaragoza, 1562).

[5]   'Hemos vivido, pues, durante siglos recluidos en nuestra concha, dando vueltas a la noria del aristotetelismo y del escolasticismo, y desinteresados y desdeñosos (con excepcion de pocos paréntisis) del poderosos movimiento crítico y revisionista que impulsó a Europa a las ciencias y artes. Fuera empero, injusticia olvidar que algunos de nuestros sabios y filósofos conocieron y profesaron

plea for renewed scientific initiatives was followed by the imminent 'year of disaster', 1898, which marked the total defeat in the Spanish-American war and led to the combined loss of Spain's formerly unmatched navy and last overseas colonies, Cuba and the Philippines, Puerto Rico, Guam and the Mariana Islands. This watershed event was a severe blow to the apologists and the last imperial legacy of the 'Siglo de Oro', and signalled a call for modernisation and 'regeneración', which gained rapid momentum.[6] The following years marked a new and critical phase in 'la polémica de la ciencia española', in which 'the generation of '98 often presented the historical development of their country as a blind alley of religious orthodoxy and political suppression.

A 50-page book entitled *La anatomía y los anatomistas del siglo XVI* (Granada, 1902) written by the anatomist Victor Escribano García clearly reflected the changing sentiments in medical history studies of early modern Spain. Escribano García openly distanced himself from the writings of prior medical historians, such as Morejón and Chinchilla, and lamented their excessive insistence on important Spanish contributions in science and medicine. As an example, his booklet refuted Morejón's claim of Andrés Laguna's anatomical discoveries, which was disregarded as a case of 'false patriotism': 'I am sorry to dispute this pleasant legend, which, if true, would have allowed us to add a Spanish name to the numerous others presented by the professor of anatomy during his lectures. But our primary concern here is respect for the truth.'[7] Escribano García furthermore mourned the hardship and lack of recognition suffered by Andreas Vesalius during his 1559–64 employment at the Madrid court of Philip II. He even suggested that a different reception of the famous anatomist might have changed the entire course of the Spanish history of science and medicine: 'What vast areas of investigation could have been offered to this creator of human anatomy studies! What a different future the natural sciences in Spain could have had!'[8]

---

las novísimas verdades matemáticas, astronómicas, físicas y biológicas, conquistadas por Copérnico, Galileo, Torricelli, Newton, Descartes, Vesalio, Harveo, Lavoisier, pero poquísimos de ellos tuvieron el arranque necesario para trasladarse a los grandes centros culturales y adquirir el contagio tonificante de la genialidad creadora.' Santiago Ramon y Cajal, *Los tónicos de la voluntad. Reglas y consejos sobre investigación científica* (Madrid: Edición de Leoncio López-Ocón, 2006), p. 227.

[6]   Martin Blinkhorn, 'Spain: The "Spanish Problem" and the Imperial Myth', *Journal of Contemporary History*, 15 (1980), p. 13.

[7]   'Me da pena deshacer tan grata leyenda, que de ser cierta nos permitiría añadir un nombre español á los muy contados que el catedrático de Anatomía puede citar en sus explicaciones, pero lo primero es el respeto á la verdad.' Victor Escribano García, *La anatomía y los anatomistas del siglo XVI* (Granada: Traveset, 1902), p. 10.

[8]   Ibid., p. 41: 'La consideración es la siguiente: si á la manera que Isabel la Católica y su esposo podian ver en el humilde genovés Christobal Colón, descubridor del Nuevo Mundo, hubieran apreciado aquellos monarcas la divina intelligencia del modesto cirujano de sus tercios,

The national awe which followed Santiago Ramon y Cajal's Nobel Prize in 1906 gave the reformers a further popular and political backing, and led to widespread scientific renovation programmes later referred to as 'la cajalización de España'.[9] These initiatives involved intensified public investments in research, institutions and laboratories throughout Spain, overseen by the newly established *Junta para Ampliación de Estudios e Investigaciones Científicas* (JAE). From its creation in 1907, the JAE directed the course of Spanish science under Cajal's leadership until his death in 1934, shortly before its abolition in 1939. Cajal's own writings called for a complete break with an ill-fated course of history, which had reduced Spain from a global empire to an underdeveloped backdrop of Europe: 'For centuries Spain has been in debt to civilization, and if she persists in such shameful neglect of duty, Europe will lose patience and eventually ostracize her.'[10]

The historian Julián Judérias brushed aside the increasingly negative conception of Spanish history as *la leyenda negra* and thus coined a term that has reappeared in almost any historical account of early modern Spain since its introduction in 1913.[11] The 'black legend' consequently presented itself as a self-imposed image – an attempt to reject what was conceived of as not only an unreasonable perception of Spanish history among its native critics, but also a dismissal of centuries of real and imaginary foreign claims of Spanish backwardness.[12] In 1920 the Spanish philosopher and historian José Ortega y Gasset added to the black legend with his claim of Spain's isolation or *tibetización* from the most important intellectual currents of previous centuries. This conception appeared repeatedly in the historical narratives from the first three decades of the twentieth century. It was emphasised in *Compendio de historia de la anatomía*, published in 1929 by Dr R. Alcalá Santaella, *Profesor auxiliar de anatomía en la facultad de medicina de Madrid*, who regretted the decline of Spanish anatomy studies since the late sixteenth century: 'Spain seized to carry out anatomical studies of dead bodies, and abandoning this practice left the anatomy studies in this nation in a deplorable state, to the extreme that in the following century there were no anatomists left,

---

luego medico de SS. MM. que varía suerte la del infortunado belga! Que campo de investigación tan vasto podrían haber ofrecido al creador de la anatomía humana! Que diverso porvenir el de las ciencias naturales en España!'

[9]   Leoncio López-Ocón Cabrera, *Breve historia de la ciencia española* (Madrid: Alianza, 2003), pp. 343–78.

[10]   Ramon y Cajal, *Los tónicos de la voluntad. Reglas y consejos sobre investigación científica*, p. 191.

[11]   Victor Navarro Brotons, *Espanya i la revolució científica: Aspectes historiogràfics, reflexions i perpectives*. in *Actes de la VII trobada d'història de la ciencia i de la tècnica* (Barcelona: SCHCT, 2003), p. 16.

[12]   Cited in Victor Navarro Brotons and William Eamon (eds), *Más allá de la Leyenda Negra. España y la Revolución Científica* (Valencia: CSIC, 2007), p. 29.

and those who taught anatomy here had to come from other countries.'[13] The Professor of Anatomy at the University of Salamanca, Casto Prieto Carrasco, wrote another critical historical account entitled *La enseñanza de la anatomía en la Universidad de Salamanca*, which was published in 1935, only a year before his arrest and execution by Francoist forces.

Opposing efforts were made to replace the black legend with a 'leyenda rosa', which saw Spain's historical contributions as at least equal or even superior to those of other European countries. The adherents of 'la leyenda rosa' finally and effectively silenced their opponents with the fascist takeover of Spain in 1939. In that year the still-dominant Spanish state council for scientific research, the *Consejo Superior de Investigaciones Científicas* (CSIC), replaced the JAE with an explicit programmatic 'resolve, first of all, to lay the foundation of a restoration of the unity of the classical and Christian sciences, which was destroyed in the eighteenth century'.[14] The imperial past was once again embraced and sanctified by the new regime, which sought inspiration and legitimacy in the 'Siglo de Oro' for its own cultural policy. As a surviving legacy, visitors to El Escorial and the neighbouring Valle de los Caídos can still visit the mausoleums of both Philip II and Francisco Franco on a combined ticket.

In spite of the hostility towards foreign tendencies to exclude or ignore Spanish contributions to European culture and history, the Francoist regime in many ways confirmed the idea of Spanish distinctiveness, which was literally presented in its tourist slogan 'Spain is different'. Post-war studies of Spanish science celebrated this unique status, as was the case in the only book-length study to this day on Spanish renaissance anatomy – Luis López Alberti's *La anatomía y los anatomistas españoles del renacimiento*, which was published by CSIC in 1948 and presented Spain as a leading contributor to anatomical insight in sixteenth-century Europe. The frequent use of vernacular language in early Spanish publications on anatomy was seen by the author as a sign of this precursory status and the true renaissance spirit of Spanish anatomy studies compared to those of other European nations: 'This does not mean that this characteristic renaissance trait never appeared in the rest Europe, but evidently its development was much slower, and never had the same impact as in Spain.'[15]

---

[13] 'En España se deja de hacer anatomía sobre el cadaver, y este abandono condujo a un deplorable estado de la anatomía en nuestro país, hasta el extremo, que en el siglo siguiente no hubo ningún anatómico y los que exlicaban la anatomía tenían que venir de otros países.' Dr Rafael Alcalá Santaella, *Compendio de historia de la anatomía* (Madrid: Javier Morata, 1929), p. 85.

[14] 'Tal empeño ha de cimentarse, ante todo la restauración de la clásica y cristiana unidad de las ciencias destruída en el siglo XVIII.' Luis Enrique Otero Carvajal, *La destrucción de la ciencia en España: las consecuencias del triunfo militar de la España franquista* (Madrid: Complutense, 2006), p. 139.

[15] Luis López Alberti, *La anatomía y los anatomistas españoles del renacimiento* (Madrid: CSIC, 1948), p. 199: 'Esto no quiere decir que en el resto de Europa no se manifestase esta nota

In the last lines of Alberti's work, the eminence of Spanish renaissance anatomy was even claimed to 'precede the ideas later launched by Vesalius'.[16]

In the wake of the Spanish Civil War and the political and cultural isolation of Spain of the early post-war years, a rather subdued and one-sided *polemica* replaced the formerly fierce debates, but publications from that period were not of consequence outside Spain. An example of the more moderate writings of contemporary scientific culture within Spain is provided by Gregorio Marañon's *La literatura científica en los siglos XVI y XVII*, which was published in a multi-volume history of Spanish literature in 1953. Marañon occasionally distanced himself from the more extreme partisans among nineteenth-century *apologistas*, but nevertheless wrote of 'Spanish science, full of brilliant episodes, and sometimes of pure genius'.[17] Similar attempts were made to refute the 'black legend' and to revise the history writing on early modern science, but these initiatives failed almost completely. While the dictatorship isolated Spain from its European neighbours both politically and intellectually, its scattered efforts to place the nation in a prominent role in the scientific history of Europe were largely ignored outside Spain. The leading Falangist scholar and medical historian Pedro Laín Entralgo later acknowledged the disastrous influence that this policy had on the post-war scientific milieu in his *Descargo de conciencia, 1930–1960* (Madrid, 1976). Recent studies have even presented the aftermath of the civil war as 'the destruction of Spanish science', with significant and enduring restraints, censorship, and the exile – both external and internal – of scientists and scholars.[18]

A breakthrough in history writing on science in early modern Spain coincided with the change of political system and was achieved with José Maria

---

renacentista, pero evidentemente su evolución fué mucho más lenta y no tuvo nunca la amplitud que en España.'

[16]   Ibid., p. 233: 'en este sentido preludían las ideas que poco después lanzaría Vesalio'.

[17]   *'La ciencia española, llena de episodios brilliantes y a veces geniales.'* Gregorio Marañon, 'La literatura cientifica en los siglos XVI y XVII' in *Historia general de las literaturas hispanicas. Vol.3, Renacimiento y Barrocco* (Madrid: Vergara, 1953), p. 936.

[18]   The alleged Francoist destruction of a flourishing scientific culture is described in detail in Luis Enrique Otero Carvajal's *La destrucción de la ciencia en España: las consecuencias del triunfo militar de la España franquista* (Madrid: Complutense, 2006). Henry Kamen has presented a directly opposing view: 'The vision of a developing scientific community in Spain that has been destroyed through the expulsion of its best members was comforting as a belief in Spain's capacities, but is certainly baseless. A number of personnel with technical and medical qualifications – a high proportion of them state employees whose jobs had been taken away by the new regime – certainly left Spain as a result of the Civil War, but there was no significant emigration of "scientists", nor any identifiable impact for good or ill on Spanish science, least of all a "blood-letting".' Henry Kamen, *The Disinherited: Exile and the Making of Spanish Culture, 1492–1975* (New York: HarperCollins, 2007), p. 415.

López Piñero's as yet unequalled *Ciencia y técnica en la sociedad española de los siglos XVI y XVII*, which was published in 1979. Originally a medical historian, Piñero showed a remarkable academic scope and range of interests in numerous fields of early Spanish science. Besides *Ciencia y tecnica*, his 1983 co-edition of the two-volume *Diccionario histórico de la ciencia moderna en España* dealt with a broad range of subjects, including medicine, botany, astronomy, optics and metallurgy. Piñero, the driving national force in the promotion and publication of early Spanish science, went on to establish the Valencia-based *Instituto de Historia de la Ciencia y Documentación*, which later changed its name to *Instituto de Historia de la Medicina y de la Ciencia López Piñero*. One must, however, question to what extent the institution succeeded where its predecessors failed in terms of presenting Spain as an important contributor to the scientific and medical history of early modern Europe. With rare exceptions, such as David Goodman's *Power and Penury – Government, Technology and Science in Philip II's Spain*, few scholars outside Spain seem to have paid much attention to their Spanish peers' many attempts to 'resurrect' the history of Iberian science.

The English-Spanish language barrier has probably been a major cause of this lack of awareness beyond Iberia of the many important Spanish contributions to the history of early European science. Piñero himself was fully aware of this problem in his 1999 contribution to a publication commemorating the quatercentenary of Philip II's death. In *Felipe II, la Ciencia y la Tecnica* (Madrid, 1999) Piñero blamed both his foreign colleagues for their ignorance of non-English publications and his fellow Spaniards for their tendency to often publish in inferior vernacular journals for a very limited readership.[19] Ironically, Piñero and his colleagues were blamed for the same tendency of engaging only in minor studies and the gathering of data rather than producing an overall vision or synthesis on Spanish Renaissance science; in his polemical article 'Iberian Science in the Renaissance: Ignored How Much Longer?', Jorge Ganizares-Esguerra scorned history writing on Spanish science for its inability to force through a revision of scientific writing about the sixteenth and seventeenth centuries. He criticised the lack of persuasive argument put forth by historians who preferred 'sorting, weighing (credibility), counting and cataloguing over grandiose speculation' and blamed his Spanish colleagues for their lack of international impact. However, Ganizares-Esguerra foresaw a drastic future development in the understudied field of early modern Iberian science:

---

[19]   'La mayoría de los trabajos de investigación aparece en serias monográficas en que J. Puerto ha calificado recientemente con humor negro como 'clandestinas'. José Maria López Piñero, 'Actividad Científica y Sociedad de la España de Felipe II' in Enrique Martinez Ruiz (ed.), *Felipe II, la Ciencia y la Tecnica*. (Madrid: Actas, 1999), p. 19.

It is extraordinary that decades after the publication of Maravall's 'Ancients and Moderns' and Lopez Piñero's 'Ciencia y Técnica' north-Atlantic historians' narratives still can so easily afford not to study this literature. Fortunately, as demographic patterns in the USA change, the self-satisfying north-Atlantic narratives of the origins of modernity are in the rude awakening. It is just a matter of time before books in English begin donning dust jackets with the frontispiece in García de Céspedes' 'Regimento de Navegación' instead of Bacon's 'Great Instauration'.[20]

Francis Bacon's *Instauratio magna* (London, 1620) borrowed its title page (depicting ships leaving the well-known Mediterranean for the unknown ocean beyond the Pillars of Hercules) from Andres García de Céspedes' earlier *Regimento de Navegación* (Madrid, 1606). The focus on such Spanish contributions to the development of early modern science has consolidated a third legend, this time 'green', in which Imperial Spain is seen as a pioneering figure in numerous scientific disciplines of the sixteenth and seventeenth centuries. *La leyenda verde* was the title given by Javier Puerto to his 2003 publication, which was subtitled *Naturaleza, sanidad y ciencia en la corte de Felipe II*, and presented Philip II as a leading benefactor and instigator of several braches of natural science and medical learning. The same attitude was promoted by an international conference entitled *Más allá de la leyenda negra. España y la Revolución Científica*, which was held in Valencia in 2005. The ensuing publication, written in both Spanish and English (Valencia, 2007), appealed to a wider readership beyond the traditional domestic audience and outlined the international scope and objective of the contributors. In the following year, Jorge Ganizares-Esguerra's introduction to *Science and Medicine in the Spanish and Portuguese Empires* (Stanford, 2008) maintained that '"science" was the handmaiden of the Iberian empires. Cosmography and natural history were the backbone upon which the Spanish and Portuguese Crown built their mighty Christian monarchies'.[21] The polemic regarding early Spanish science evidently still provokes conflict, whose most extreme participants have been divided between a patriotic or triumphalist stance, and an equally bizarre bias excluding Spain from standard historical narratives on early European science. The revisionist agenda appears strident at times and yet is right to pinpoint a 'provincial' tendency in traditional studies on the history of science, which is at least very evident when dealing with the 'anatomical revolution' of the sixteenth and seventeenth centuries.

---

[20] Jorge Ganizares-Esguerra, 'Iberian Science in the Renaissance: Ignored How Much Longer?', *Perspectives on Science*. 12 (2004), pp. 106 and 117.

[21] Daniela Bleichmar (ed.), *Science in the Spanish and Portuguese Empires* (Stanford: Stanford University Press, 2008), p. 1.

In a canonised and almost Hegelian narrative, this significant development in the history of science and medicine still focuses on Italy, which eventually hands its torch of insight into the human body over to Northern Europe. This 'canon' has resulted in a series of rather repetitive accounts outlining the same anatomical precursors and protagonists in Italy, France, the Netherlands and the UK, which are almost devoid of glimpses at anatomy and anatomists in other countries. Such omissions ignore several noteworthy initiatives from the latter half of the sixteenth century during which Spanish medicine embraced new practices and theories from the European centres of anatomical research. Indeed, throughout the Iberian Peninsula, several books on anatomy and surgery, anatomical art treatises, reformed statutory orders, and new anatomy chairs and theatres appeared during the sixteenth century. However, this national context has been largely ignored in most non-Spanish scholarship, and within Spain has been dealt with by only a handful of scholars in the last century. *La polémica de la ciencia española* does not provoke similar strength of sentiment as in earlier periods, but the perceived irrelevance and almost non-existence of early modern Spanish science still discourages many historians from exploring this unknown field and expanding their own narratives and horizons. The present study focuses on a neglected national context and a number of individuals traditionally ignored in the history of early modern anatomy. This work will describe a number of these regional and institutional settings and agents, and some of their differing responses to one of the most significant developments in mid-sixteenth-century medicine and science – the so-called 'Vesalian Revolution'.

The Vesalian Revolution in a Spanish Context

The 'Vesalian Revolution' is given a prominent position in accounts of the scientific revolutions of the sixteenth and seventeenth centuries. Although increasingly questioned and rejected by some historians of science, scholars do still actively support the notion of paradigmatic revolutions in science and have occasionally even dared to pinpoint the initial year of this innovative break with a continuity dating back to Greco-Roman antiquity. The historical coincidence that 1543 marks the publication year of both Andreas Vesalius' *De corporis humani fabrica* and Nicolaus Copernicus' *De revolutionibus orbium coelestium* has made that year a tempting and convenient starting point for the so-called 'scientific revolution'.[22] Attempts to date sudden ruptures and precursory events are often easy to deconstruct. Still, there are good reasons for regarding

---

[22]   Among the scholars and books which favour a 1543 dating of the scientific revolution are: Eduard Jan Dijksterhuis, *The Mechanization of the World Picture* (Princeton: Princeton University Press, 1986); Richard S. Westfall, *The Construction of Modern Science* (Cambridge: Cambridge

the two 1543 publications as important contributions to new chapters and perspectives in the history of science, and to the study of the smallest and largest phenomena of the natural world. Vesalius' work represented the, until that time, most systematic questioning of previously infallible medical truths of the Aristotelian-Galenic *Microcosmos*: the structure and functions of the human body. Copernicus instead envisioned an enhancement of the cosmology of the Aristotelian-Ptolomaic *Macrocosmos*: the structure of the universe and the movement of its heavenly bodies.

Throughout its long existence, the Aristotelian worldview and its structure and order of the natural world had developed into a closed, finitely coherent discourse; it was a qualitative, hierarchical system of linked and mutually dependent natural phenomena. The complex cosmology of correlation and dependence within the natural world presented an ideal interaction of the heavenly spheres and the bodily parts, which imperceptibly linked the order and chaos of the universe to the health and disease of the human body. It furthermore advocated a complicated system attaching different qualities to different aspects of the physical world – celestial bodies as well as bodily organs. The Vesalian and Copernican corrections of this worldview came in the wake of new hands-on investigations of human anatomy and mathematically based theories and models of the structure of the universe. This development represented the gradual introduction of interests in autonomous functions and mechanisms, and a quantitative rather than qualitative approach to natural science. The new insights into the fabric of the human body and the introduction of an infinite heliocentric universe have since been perceived as radical intellectual challenges to a system of formerly unquestioned truths. There were, however, significant differences between the practices and theories of the two 'revolutionaries' of 1543, and in the respective receptions of their parallel attempts to give priority to observation over authority. Copernicus' vision of a solar system was still idealistic and speculative, and his proposed alternative cosmology would not be partially confirmed or even tested until the following century. Vesalius' numerous revisions of Galenic anatomy had a more immediate impact and were soon examined by numerous physicians skilled in human dissection. Vesalius did not introduce the study of human anatomy to European universities, where such training had been carried out since the late thirteenth and early fourteenth centuries. Nonetheless, while earlier public dissections had taken the shape of ritual practices and endeavoured to prove the unquestioned authority of ancient medicine, Vesalius' more systematic approach used dissection as a method for producing and presenting new anatomical evidence.

University Press, 1971); Paolo Rossi, *The Birth of Modern Science* (Oxford: John Wiley & Sons, 2001); John Gribbin, *Science* (London: Penguin, 2009).

Before Vesalius was employed by the University of Padua in 1537, studies and lessons in human anatomy were most often carried out according to a tripartite model: a professor (lector) read aloud from Galenic texts on medicine, while a surgeon (sector/incisor) opened the dead body and a demonstrator (ostensor) pointed out the visible organs. Vesalius' break with this practice has been presented as a cornerstone of the 'Vesalian Revolution'. Yet even in the first decades of the sixteenth century, anatomists such as Berengario da Carpi and Niccolò Massa had reduced the three teachers traditionally involved to one medically trained anatomist who carried out the dissections alone.[23] This gradual break with the tripartite model was instrumental in establishing anatomy as a field of research and an academic discipline rather than a ritualised didactic tool. It furthermore represented the beginnings of a union between previous divisions of manual and intellectual labour, and the separation of hand and mind, which had been strongly advocated since classical antiquity. This division was explicitly stressed in the Hippocratic Oath, which prohibited the doctor from ever using the scalpel, an instrument suited only for the uneducated barber and surgeon.[24]

Both Aristotle and – to a much larger extent – Galen had been among the exceptions to disobey this rule, and both carried out numerous dissections and vivisections on pigs, dogs, and monkeys, but seemingly never on human bodies. The combined role as dissectors of both humans and animals enabled Vesalius and other contemporary anatomists to compare the claims of the Aristotelian and Galenic canons with their own empirical studies of human anatomy. Their research corrected and dismissed several small but significant assertions of ancient medical authority, and Vesalius even reached the challenging conclusion that Galen had apparently only dissected animals and not humans, since he was plainly unaware of several obvious traits of human anatomy. Among these corrections was a rejection of the *rete mirabile* or 'wonderful network' of blood vessels found in the lower brain of cows and sheep, but which had already been questioned as an extant trait of human anatomy by Berengario da Carpi in 1522. The reality of this imaginary bridge between the human body and soul was nonetheless still maintained in Vesalius' 1538 Tabulae anatomicae sex and during his public anatomies presented in Bologna in early 1540.[25] Vesalius' *Fabrica*, completed a few years later, disproved this and other notions and claims of Galenic anatomy, such as the horned uterus, and the four- or five-lobed liver, as well as the idea that the human mandible and sternum were divided into two and seven parts respectively – instead of one and three as Vesalius'

---

[23]   Rafael Mandressi, *Le regard de l'anatomiste. Dissections et invention du corps en Occident* (Paris: Seuil, 2003), p. 90.

[24]   Ole Peter Grell and Peter Elmer, *The Healing Arts: Health, Disease and Society in Europe 1500–1800* (Manchester: Manchester University Press, 2004), p. 13.

[25]   Mandressi, *Le regard de l'anatomiste*, pp. 87–88.

own dissections had shown. These discoveries and, according to Vesalius' own allegations, another 200 corrections of Galen may seem insignificant, but nonetheless represented a challenge to 1,300 years of consolidated medical truths which had been repeatedly affirmed in studies and treatises on human anatomy.[26] Vesalius not only advocated the abandonment of prior methods of dissection and an increased scepticism towards ancient medical authority, but also sought new ways to make recent anatomical knowledge and insights accessible to a broader public. The detailed woodcuts of his *Fabrica* were crucial in establishing the book as the most famous work on human anatomy published to date – and an often-imitated matrix for anatomical illustration. Its deliberate use of visual aids would be imitated only a few years later in the first two 'post-Vesalian' anatomy books written in vernacular Spanish: Bernardino Montaña de Montserrate's *Libro de la anothomia del hombre* (Valladolid, 1551) and Juan Valverde de Amusco's *Historia de la composición del cuerpo humano* (Rome, 1556). Vesalius' richly illustrated deluxe binding was a noticeable promotion of the academic study of anatomy and was thus instrumental in establishing the practice as a university discipline in several prominent medical faculties throughout Europe. The 'Vesalian synthesis' of reforms in anatomical practice, research and representation did not escape the attention of its contemporaries, but was met with immediate approval as well as fierce resistance by reformers and defenders of ancient medical authority. Most radical in the latter group was Vesalius' former Parisian teacher Jacobus Sylvius, who insisted on the infallibility of Galen's anatomy and later produced a bitter written attack on 'Vesanus', his unmannerly and 'insane' student.[27]

Discussions about Vesalius and his *Fabrica* have since divided scholars, who still disagree on aspects of his life and work, and about his status as a 'scientific revolutionary'. A thorough examination of the seemingly endless number of publications with different views on Vesalius is beyond the scope of this book, but among the titles that deserve attention is Andrea Carlino's *Books of the Body. Anatomical Ritual and Renaissance Learning* (Chicago, 1999). Carlino's study stands out as a refreshing and multifaceted analysis of changes in anatomical practice, teaching and visual representation from Greco-Roman antiquity to the late Renaissance. In this work the author presents Vesalius as the reformer of 'an epistemological norm that had constricted anatomy for over a millennium'.[28] The publication of Fabrica is understood by Carlino as the crucial event in the break with the tripartite anatomical practice which had been dominant in Europe since Mondino di Luzzi carried out and published his pioneering

---

[26]   Carlo Pedretti, *L'anatomia di Leonardo da Vinci* (Florence: Grantour, 2000), p. 175; Andrew Cunningham, *The Anatomical Renaissance* (Aldershot: Sholar Press, 1997), p. 77.

[27]   Rossi, *The Birth of Modern Science*, p. 46.

[28]   Andrea Carlino, *Books of the Body* (Chicago: Chicago University Press, 1999), p. 213.

studies of human dissections in Bologna in 1315–16. Carlino's work elaborated upon Vesalius' almost all-encompassing significance as a reformer of anatomical practice, knowledge, and representation, which all came together in his *Fabrica*: 'By directly observing the cadaver and through his own possession of a profound knowledge of earlier anatomical literature, Vesalius was able to confirm, discuss, and correct everything that had been said previously about the different parts of the body. He often entered into open controversy with those who fiercely defended a still vigorous Galenic tradition. His repeated references to dissections he had performed provide what the author himself suggested was the major feature of the Vesalian revolution as later defined by historians.'[29] Since presenting Vesalius as the instigator of a new anatomy, Carlino has become increasingly lukewarm towards the idea of a 'Vesalian Revolution', which he openly dismissed in a recent article entitled 'Vesalio e la cultura delle anatomia e le stampe del rinacimento' (Bologna, 2004).[30] His reservations towards the hagiographical tendencies so common in the biographies of scientific pioneers are both fitting and justified. Similar caution should be exercised when referring to 'scientific revolutions' and 'revolutionaries' in early modern Europe, as emphasised by I.B. Cohen and Steven Shapin.[31] Wherever the present account refers to this 'revolution' or to 'pre- and post-Vesalian anatomy', it is with these reservations in mind, and only as well-known chronological cornerstones for the analysis of subsequent studies in the field of anatomy. Vesalius' own career as an 'anatomical revolutionary' culminated, and effectively ended, with the publication of his *Fabrica* in 1543, even though he devoted some time to important improvements of its second 1555 edition. During the following two decades, he served as a court physician to Charles V and as imperial surgeon during military campaigns in Germany, Italy and Flanders. He finally settled in Spain at Philip II's Madrid court, where he lived from 1559 until shortly before his death in 1564.

Paradoxically, the host country of Vesalius' later years has generally escaped the interest of historians investigating Renaissance anatomy. Spain has been

---

[29]   Ibid., p. 39.

[30]   Andrea Carlino, 'Vesalio e la cultura delle anatomia e le stampe del rinacimento' in *Rappprasentare il corpo – Arte e anatomia da Leonardo all' illuminismo* (Bologna: Bononia University Press, 2004), p. 76: 'Ho, infatti qualche riserva sulla tendenza all'agiografismo di cui pecca una certa storia della scienza e da cui non è asente neppure la storia dell'arte: storie scandite da grandi opere e da uomini geniali che, quasi piovuti dal cielo, 'rivoluzionano' i saperi, le practiche, le discipline, i modi di agire e di pensare. La Fabrica, sembrerà banale dirlo, è semplicemente un prodotto del suo tempo. La rivoluzione vesaliana non è altro che una rivoluzione delle forme e dei modi di usare e sfruttare una serie di tecniche, di strumenti, di saperi già in buona parte acquisiti e consolidati nell'uoso: soprattutto la dissezione, la stampa, l'illustrazione anatomica.'

[31]   I.B. Cohen, *Revolutions in Science* (Cambridge: Belknap Press of Harvard University Press, 1985); Steven Shapin, *The Scientific Revolution* (Chicago: University of Chicago Press, 1996).

excluded from most existing historical narratives of the anatomical revolution and, if included, has often been defined only negatively as the unfitting home of Vesalius' final bitter years. Vesalius' only published works written in Spain were a short 50-folio *Examen* of *Observationes anatomicae* written by his successor at the University of Padua, Gabrielle Fallopio, and a letter to Giovanni Filippo Ingrassia from Madrid dated Christmas 1562, which was later included in Ingrassia's *Quaestio de purgatione per medicamentum* (Venice, 1568). Another work produced in manuscript form during Vesalius' stay in Spain is a recently discovered translation from Italian to Castilian from December 1562 of Alessio Piemontese's book of natural secrets, *I secreti del reverendo donno Alessio Piemontese* (Venice, 1555).[32] The *Examen*, finished in late 1561 and published in Venice three years later, included a passage that seems to justify the poor reputation of Spanish Renaissance anatomy: 'It is not without pleasure that I have read through your *Observationes* to the end, and with delightful recollections of the very happy life I enjoyed while teaching anatomy in Italy, true nurse of talents. I would that I might readily reply to those things you have communicated to me with so much profit so that, when you succeed in exposing further mysteries of nature, you will inform me of such things as arise. I meanwhile can foresee no possible opportunity for performing dissection – here I cannot even obtain a skull.'[33]

Two cases which emphasised Spain as the very antithesis of the anatomical revolutions elsewhere in Europe were the first in-depth biographies of Vesalius' life and work – Moritz Roth's *Andreas Vesalius Bruxellensis* (Berlin, 1890) and Charles D. O'Malley's *Andreas Vesalius of Brussels* (Berkeley, 1964). Both biographies offer an unconditional appraisal of the famous anatomist. The major parts of these works are dedicated to his early years and triumphs at the University of Padua, and the making of 'De corporis humani fabrica'. Sad appendices to these biographies describe Vesalius' final and supposedly harsh years at the Spanish court in sharp contrast to his early career. O'Malley formally distanced himself from Roth's 'too enthusiastic appreciation of Vesalius' achievement without sufficient credit to his predecessors and contemporaries',[34] but nonetheless presented an equally biased account of Vesalius' stay in Spain in his own 1964 publication. This biography included repeated references to the 'incompetence of Spanish physicians' and 'his lesser Spanish colleagues', and pointed out that 'the general atmosphere in Spain was hardly scientific'.[35]

---

[32]   Mar Rey Bueno, 'Primeras ediciones en castellano de los libros secretos de Alejo Piamontes', *Boletín de la Biblioteca histórica de la Universidad Complutense de Madrid*, 2 (2005), p. 32.

[33]   Charles D. O'Malley, *Andreas Vesalius of Brussels* (Berkeley: University of California Press, 1964), p. 29.

[34]   Ibid., *preface*, n.p.

[35]   Ibid., p. 454.

O'Malley introduced his first chapter on Vesalius' final years with a description of the ominous atmosphere of superstition and backwardness awaiting the great anatomist in Spain: 'The Spanish king ... gave instructions that he would leave for Spain on 8 August. Nevertheless, when all had been prepared, a counterorder was given, which was, according to the English ambassador Chaloner, the result of neither politics or religious scruples, "but the foolish Nostradamus with his threats of tempests and shipwrecks this month did put these sailors in a great fear". What must Vesalius, who had already indicated his lack of sympathy with the superstitions of that age have thought of such an excuse for delay?'[36]

Ironically, O'Malley's books and articles on sixteenth-century anatomy represented not only the culmination of hostility towards the ill-fated Spanish reception of Vesalius, but also the earliest English-language studies on anatomy in late Renaissance Spain. While O'Malley's 1964 biography of Vesalius clearly continued earlier inclinations to scorn or ignore early Iberian anatomy studies, his articles from the following decade on Spanish anatomists such as Andres Laguna, Pedro Jimeno and Bernardino Montaña were sincere attempts to establish Spain as a part of the history of sixteenth-century anatomy. In his 1972 article 'Pedro Jimeno: Valencian Anatomist of the Mid-sixteenth Century', O'Malley commented with admirable academic flexibility on his own earlier citation of Vesalius' problems in obtaining a skull during his tenure at the court of Philip II: 'Although this may have been true in Madrid it was certainly not the case elsewhere since Spanish universities were among the first outside Italy to accept Vesalian anatomy, and to carry on active programs of human dissection and research.'[37] Still, this increased sensibility and openness towards an inclusion of Spain in the history writing of early modern anatomy has not struck a general chord among medical historians. Even though numerous books on Renaissance anatomy have appeared in the last few decades, they are most often structured around the same well-known precursors and protagonists, and focus on the universities of Italy, France, Holland and England, with only a few sidelong glimpses at happenings elsewhere. Reviews of some of these works have suggested that Spain be included in future research, yet this proposal has been largely ignored.

Benjamin A. Rifkin's *Human Anatomy*, a recent popular narrative on the history of anatomical representation, presented a near-caricature of Spanish Renaissance anatomy in its description of the Spanish anatomist Juan Valverde as an uncompromising Galenist and fierce critic of Vesalius: 'Like the new vision of the place of the earth in the universe that was slowly gaining adherents, the new vision of man and woman enshrined in Vesalius was controversial in

---

[36]   Ibid., p. 288.
[37]   Charles D. O'Malley, 'Pedro Jimeno: Valencian Anatomist of the Mid-sixteenth Century' in A.G. Debus (ed.), *Science, Medicine, and Society* (London: Heinemann, 1972), p. 69.

conservative religious circles. Furthermore, anatomists who wanted to make their mark in the immediate post-Vesalian world needed a position from which to attack him, and that was provided by Vesalius' mild attack on the stature of Galen. It is not surprising that one of these controversialists should have been Spanish, given that Spain was the bastion of catholic traditionalism and the most dangerous place in Europe to espouse new ideas on controversial topics.'[38] Rifkin's account demonstrated not only a very superficial idea of contemporary Spanish history, but also a complete ignorance of the actual contents of Valverde's *Historia de la composicion del cuerpo humano*, which included numerous written passages and a daring iconography of outspoken criticism of Galenic anatomy. The fact that such notions continue to appear in publications on the history of anatomy confirms the continued seclusion or misrepresentation of Spain in standard narratives on this subject. In recognition of this fact, the Wellcome Trust historian of medicine Vivian Nutton has repeatedly advocated the need for new national perspectives and approaches to the history of anatomy, and highlighted Andrea Carlino's *Books of the Body* as a study worthy of imitation: 'Its methodology could with profit be adopted for other geographical areas such as Germany, England, or the much neglected Spain.'[39] Nutton repeated this idea in his review of Andrew Cunningham's *The Anatomical Renaissance: The Resurrection of the Anatomical Projects of the Ancients* and suggested that new narratives would facilitate a richer and fuller understanding of the history of early modern anatomy, 'but for that a different book is wanted, one that would leave Italy for Vienna, Oxford, or Salamanca.'[40] In spite of this encouragement, one must still turn almost exclusively to Spanish scholarship in order to piece together an idea of the anatomical studies carried out on the Iberian Peninsula before, during and after Vesalius' Spanish residence between 1559 and 1564.

Following on from Nutton's suggestion to look at post-Vesalian anatomy at the University of Salamanca, it is striking that the 1561 statutes from this renowned institution not only embraced the new anatomical practices, but explicitly – and uniquely in a European context – called on teachers and students to familiarise themselves with Vesalian writings and images: 'The professorship of anatomy is obliged to obtain human bodies for the dissections, and when he cannot procure them, he should demonstrate what he is teaching with Vesalius' prints and illustrations.'[41] No less noteworthy is the appearance in

---

[38] Benjamin A. Rifkin and Michael J. Ackermann, *Human Anatomy* (London: Harry N. Abrams, 2007), p. 95.

[39] Vivian Nutton, 'Renaissance Anatomy', *Medical History*, 44 (2000), p. 547.

[40] Vivian Nutton, 'Review of *The Anatomical Renaissance: The Resurrection of the Anatomical Projects of the Ancients*', *Medical History*, 42 (1998), p. 258.

[41] 'Sea obligado el dicho catedratico a poner diligencia para auer cuerpos humanos do se hagan las dichas dissectiones, y no pudiendo auerse, lo que fuere leyendo en su lection y cathedra lo vaya mostrando en las stampas y figuras de Besalio, para que se entienda lo que se va leyendo.'

the University of Salamanca's statutes of another of the 'revolutionaries of 1543', Nicolaus Copernicus, whose heliocentric theories were placed next to those of Ptolemy in the astronomy curriculum. This has not escaped the attention of two of the founders of the Valencian *Instituto de Historia de la Ciencia y Documentación*, José Maria Lopéz Piñero and Victor Navarro Brotóns, who have pioneered studies of the Spanish reception of Vesalius and Copernicus.[42] Their studies showed that these controversial works on anatomy and astronomy were seemingly never banned by the Spanish Inquisition, which prohibited a number of arguably more harmless books, such as Leon Battista Alberti's treatise on architecture and Andres Laguna's Castilian translation of Dioscorides' medical botany.[43] It is dubious, however, whether Vesalius or Copernicus had any lasting effects on the intellectual life and practices at this foremost Spanish university. The next generation of university statutes at Salamanca from 1594 no longer contained references to Vesalius, and Galenic medicine once again stood unchallenged in the medical curriculum. In these revised statutes Galen's anatomical treatise *De usu partium* seemed to have eliminated all competition: 'The professor of anatomy is obliged to read the books of 'De usu partium' for two years. In the first year the first eight books, and in the second the last nine, and the professor of anatomy should not read anything else, because these books contain all the material that is worth knowing.'[44]

Enrique Esparabé Arteaga, *Historia de la Universidad de Salamanca* (Salamanca: Francisco Nuñez, 1914), p. 261.

[42]   José Maria Lopéz Piñero has produced a large number of articles on anatomy in early modern Spain, listed here chronologically: *Antologia de la Escuela Valenciana del siglo XVI* (Valencia: Cátedra e Instituto de Historia de la Medicina, 1962); 'La obra de Juan Tomas Porcell', *Medicina Española*, 52 (1965); 'El saber anatómico y la dissección de cadaveres humanos en la España en la primera mitad del siglo XVI', *Cuadernos de la historia de la medicina Española*, 13 (1974); *Medicina moderna y sociedad Española (siglos XVI–XIX)* (Valencia: Cátedra e instituto de la historia de medicina, 1976); 'The Vesalian Movement in Sixteenth Century Spain', *Journal of the History of Biology*, 12 (1979); *Ciencia y técnica en la sociedad Española de los siglos XVI y XVII* (Barcelona: Labor, 1979); *Diccionario historico de la ciencia moderna en España*, 2 vols (Barcelona: Península, 1983); *Los saberes medicos y su enseñanza en el siglo XVI. Historia de la medicina Valenciana* (Valencia: Vicent Garcia, 1988); *Los temas polémicas de la medicina renacentista – Los controversias de Frensisco Valles* (Madrid: CSIC 1988); *La medicina* in *Historia de la Ciencia y de la Tecnica en la Corona de Castilla*, vol. III (Madrid: Junta de Castilla y León, Consejería de Educación y Cultura, 2002). On Victor Navarro Brotón's studies of the Spanish reception of Copernicus, see Victor Navarro Brotóns, 'The Reception of Copernicus in Sixteenth-Century Spain – The Case of Diego de Zuñiga', *Isis*, 86(1) (1995) and 'La actividad astronomica en la España del siglo XVI: perspectivas historiograficas', *Arbor*, 142 (1992).

[43]   López Piñero, 'Actividad Científica y Sociedad de la España de Felipe II', p. 143.

[44]   'En la cátedra de anatomía se han de leer los libros "De usu partium" en dos años. En el primero se han de leer los ocho libros primeros, y en el segundo los otros nueve, y no se ha de leer otra cosa en cátedra de anatomía, porque en estos libros se contiene toda la material que

How should we interpret these ruptures in the statutes? The study of Spanish Renaissance anatomy contains similar evidence of discontinuity throughout the late sixteenth century, not only in Salamanca, but also at the large Castilian universities in Valladolid and Alcalá de Henares. Yet, only a few decades earlier, ambitious programmes of anatomy and later surgery had been established at these and other universities across the Iberian Peninsula. Some of these initiatives even anticipated the 1559 arrival of Vesalius himself by almost a decade. While only the University of Valencia held a chair in anatomy by the time of the first Spanish publication on post-Vesalian anatomy, Pedro Jimeno's *Dialogus de re medica* (1549), the situation was transformed less than two decades later. In 1563 five universities within the Iberian Peninsula had established a university chair in anatomy, numerous anatomy books were published and the first permanent anatomical theatre had been built. During the early seventeenth century, however, the situation changed again and contemporary sources seemed ignorant of the fact that any anatomy studies had been carried out in Spain during the previous century. In 1611 the leading medical authority of the University of Valladolid, Luis Mercado, seemed to be unaware that his own university had been the first within the Crown of Castile to practise public dissections – which had been performed by Alonzo Rodríguez de Guevara between 1548 and 1550 and were recorded shortly after in Bernardino Montaña's 1551 *Libro de la anothomia del hombre*. Mercado rejected a proposal by the university senate to establish a chair in anatomy and indicated a surprising ignorance of the recent anatomical reforms of mid-sixteenth-century Spain: 'In Spain good medicine has been practised for more than 200 years without use of that discipline, and we do not have any persons in the country who are sufficiently trained to practise it.'[45]

This development shows how an academic discipline that had been introduced and institutionalised over a period of decades was nonetheless often unable to leave a lasting legacy or to prevent Spanish medicine from entering into a renewed consolidation of ancient authority. In Kuhnian terms, the introduction of anatomy at the universities in Castile and Aragon represented a late stage of integration and the development of a novel scientific paradigm. It was no longer carried out by a handful of isolated individuals, but had become 'normal science' integrated into the statutes and medical curricula of the most prominent Iberian universities – yet without causing a revision and eventual collapse of existing medical authority. As a consequence, the 'Vesalian revolution' in Spain did not lead to generations of sustained and significant anatomy studies; indeed, in some

---

es menester saber.' Fransisco Javier Alejo Montes, *La reforma de la Universidad de Salamanca a finales del siglo XVI: Los estatutos de 1594* (Salamanca: Universidad de Salamanca, 1990), p. 139.

[45] 'En España se ha hecho Buena medicina durante más de doscientos años sin necessidad de tal disciplina y no existen en el país personas lo sufficiente preparadas para su practica.' Àlvar Martínez-Vidal and José Pardo-Tomás, 'Anatomical Theatres in Early Modern Spain', *Medical History*, 49 (2005), p. 254.

university contexts the new medical practice enjoyed only a relatively short-lived integration. Pedro Lain Entralgo's article 'Spanish Contributions to World Science' admitted that even the University of Valencia – which spearheaded the study of anatomy throughout Spain during the mid-sixteenth century – once again embraced Galenic orthodoxy in the next century (advocated by its renowned anatomist, Matías Garcìa) and then became 'a centre for the most uncompromising support of the theories of Galen (...) emerging as the avowed enemy of such important novelties as the discovery of the circulation of the blood. We simply have to accept the cold, historical fact that a man of the stature of Matías Garcìa, an intelligent and experienced anatomist, as far from dry-and-dust pedantry as could well be imagined, nevertheless devoted his efforts, his brains and his experience to combating the evidence in support of Harvey's discovery.[46]

It is the objective here to chart the apparent 'rise and fall' of anatomy studies in sixteenth-century Spain, as indicated in primary sources that date from roughly 1550 to 1600, and to present a synthesis of the highly complex and previously often oversimplified circumstances surrounding anatomical research in this national context. The empirical evidence accumulated for this purpose suggests a far more multifaceted narrative of Iberian Renaissance anatomy than has been presented before, and shows both an early susceptibility and an increasing animosity towards reforms in anatomical education and practice. The sources necessary for a one-sided narrative are readily available and one can easily choose to adhere to a rigid stance, as several scholars have done during the long and often fruitless 'polémica de la ciencia española'. It is not the aim here to present Spain as either an obscurantist or a precursor in the history of sixteenth-century anatomy; rather, the intention is to look somewhat dispassionately at the history of early modern anatomy in this particular national context, where anatomical studies and teaching programmes were frequently instigated and supported by the highest authorities, but also often collided with a society embedded in religious, scientific and medical orthodoxy.

A number of Spanish anatomy books referring directly or indirectly to Vesalius' new anatomy were published prior to the anatomist's own arrival in Spain.[47] These publications are crucial to a comprehensive understanding of contemporary Spanish anatomy, despite the fact that they provide very

---

[46]   Pedro Lain Entralgo, 'The Spanish Contributions to World Science', *Cahiers d'historie mondiale*, 6 (1961), p. 962.

[47]   Pedro Jimeno, *Dialogus de re medica compendia ratione, prater quaedem alia, universam Anatomen humani corporis perstringens, summe necessarius omnibus Medicinae candidatis* (Valencia: Typis Ioannis Mey Flandri, 1549); Bernardino Montaña, *Libro de la anothomia del hombre* (Valladolid: Sebastian Martinez, 1551); Luis Collado, *Cl. Galeni Pergameni liber de ossibus ad tyrones, interprete Ferdinando Balamio, Enarratore Ludovico Collado medico* (Valencia: Typis Ioannis Mey Flandri, 1555); Juan Valverde de Amusco, *Historia de la composición del cuerpo humano* (Rome: Antonio Salamanca and Antonio Lafrery, 1556); and Alonso Rodríguez de

different accounts of the spread of anatomy throughout the Iberian Peninsula. Pedro Jimeno and Luis Collado – Valencian anatomists, colleagues and alleged disciples of Vesalius – both credited the University of Padua as the place of origin for this new anatomy and named Vesalius its pioneer. They also claimed to have personally introduced the study of Vesalian anatomy at the University of Valencia. As Charles D. O'Malley has shown in his article 'Pedro Jimeno: Valencian Anatomist of the Mid-sixteenth Century, a significant part of Jimemo's book, which dealt with human osteology, was copied verbatim from Vesalius' *Fabrica*.[48] Jimeno praised 'my great master, Andreas Vesalius' and in his final dialogue between the citizen Caspar and the physician Andreas, the name of the latter was an obvious allusion to Vesalius.[49] A decade after this publication, Jimeno himself was credited posthumously by Fransisco Valles, one of the leading medical authorities in Spain, for having introduced anatomy studies at the Castilian University of Alcalá de Henares, where he had allegedly held a short-lived lectureship in anatomy at an unknown period in the 1550s. Luis Collado succeeded Jimeno after the latter left Valencia, but offered a slightly different account, crediting himself with spearheading Vesalian anatomy outside Aragon through the employment of his pupil, Cosme Medina, at the leading Castilian university in Salamanca.[50] There was nonetheless complete agreement between the two Valencian colleagues regarding the origins of their training and experience in this field in the works of 'Vesalius, the restorer of anatomy.'[51] Collado's *Cl. Galeni Pergameni Liber de ossibus...* was one of the first published defences of Vesalius following the attacks launched by Jacobus Sylvius against his 'mad' pupil's writings on osteology. Sylvius claimed that any error found in Galen's osteology could be explained by changes in the human skeleton which had occurred since classical antiquity. Collado not only presented this claim as ridiculous, but reduced Sylvius to the status of *singularis imitator Galenis*[52] and instead offered one of the most flattering contemporary appraisals of Vesalius.

In opposition to the notions of Valencia as a gateway for Vesalian anatomy to spread throughout Spain, and that Jimeno and Collado were at the very forefront of this development, Bernardino Montaña and Alonzo Rodríguez de Guevara instead asserted that Guevara's public dissections at the University of Valladolid between 1548 and 1550 were the first anatomical lectures to be given in Spain. Neither Montaña nor Guevara made any mention of the contemporary activities in Valencia in their two only known publications,

---

Guevara, *In pluribus ex iis quibus Galenus impugnatur ab Andrea Vesalio Bruxelensi in constructione et usu partium corporis humani, defensio* (Coimbra: Ioan Barrerium, 1559).

[48]   O'Malley: *Andreas Vesalius of Brussels*, p. 70.

[49]   Jimeno, *Dialogus de re medica*, Epistola, n.p.

[50]   Luis Collado, *Cl. Galeni Pergameni liber de ossibus ad tyrones*, f. 30v.

[51]   Ibid., f. 4v.

[52]   Ibid.

Montaña's *Libro de la anothomía del hombre* (1551) and Guevara's *In pluribus ex iis quibus Galenus impugnatur ab Andrea Vesalio Bruxelensi in constructione et usu partium corporis humani, defensio* (1559). Montaña's book was based on his attendance of Guevara's lectures in anatomy during a 20-month course attended by medical students and professors at the University of Valladolid.[53] This course marked the short-lived introduction of anatomy studies at the University of Valladolid, which remained the only leading Castilian university not to establish a permanent chair in anatomy during the sixteenth and seventeenth centuries. At the time of Montaña's publication, Guevara had left Valladolid for Sigüenza, from which he graduated in medicine before leaving for Portugal, where he was appointed physician to the royal family in 1556.[54] In spite of the seemingly passing interest in anatomy at the University of Valladolid, Montaña's textbook still presented Valladolid as the only Spanish centre of European anatomy studies: 'The surgeon who wishes to study it well (division or dissection) should go and practice at the universities where it is carried out regularly, such as Montpellier in France, Bologna in Italy, and Valladolid in Spain, where once again the discipline is practised skilfully, and with royal authority and support, by Bachelor Rodriguez, a surgeon, and an excellent man experienced in this art.'[55] While the title and contents of Guevara's *Defensio*, which was printed in Coimbra in 1559, explicitly referred to Vesalius – and indeed presented itself as a defence of Galen, in opposition to Vesalian anatomy – Montaña did not mention Vesalius or the University of Padua, but instead credited the University of Bologna, a far less significant institution by the mid-1500s, as the Italian centre for contemporary anatomy studies. The publications from Valencia and Valladolid/Coimbra were in direct opposition to one another and indicate a rivalry between the medical faculties and practitioners of the Iberian university towns.

Juan Valverde de Amusco's *Historia de la composicion del cuerpo humano*, published in Rome in 1556, provided yet another model for the understanding of contemporary Spanish anatomy and ultimately confused the conflicting accounts of anatomists from Aragon, Castile and Portugal rather than providing any sort of settlement. While other Spanish anatomists disagreed about the

---

[53]   M.B. de Barbosa Sueiro, 'Sumula de vida interlope de Alonzo Rodriguez de Guevara', *Archivio de anatomía y antropología*, 29 (1956), p. 260.

[54]   Juan Riera, *Cirujanos, urulogos y algebristas del renacimiento y barocco* (Secretariado de Publicaciones, University of Valladolid, 1990), p. 15.

[55]   'El cirujano que quiera bien hacella (la división o la dissección) vaya á prender este exercicio á las universidades donde se acustumbra de hacer ordinariamente, como en Francia á Mompellier, en Italia á Bolonia, en España á Valladolid, donde agora nuevamente se comienza a hazer muy artificiozamente, con auctoridad del consejo de su Ma. por el Bachiller Rodriguez cirujano, muy excelente hombre y experimentado en este arte.' Bernardino Montaña, *Libro de la anothomia del hombre*, f. 3r.

Iberian centres and routes of anatomical research and teaching, Valverde's Roman publication disregarded the very idea that activities in the field of anatomy were being undertaken in Spain. In his prologue and dedication, Valverde lamented the de facto non-existence of Spanish anatomy studies and was emphatic in his assurances to his patron and readers that only a very low level of medical research was being carried out in mid-sixteenth-century Spain. He bemoaned 'the great lack of men of this nation who understand anatomy, partly because it is considered an ugly thing among Spaniards to dissect bodies, and partly because only few of them have come to Italy where they could have learned it'.[56] The author presented not only a deep-rooted discomfort among his fellow Spaniards in dealing with the 'ugly' hands-on empirical investigation of human anatomy, but also a widespread lack of ability among Spanish scholars to read and write Latin, the scientific lingua franca of the era. By communicating in vernacular Spanish, the author intended to lead the Spaniards out of their ignorance of new anatomical practices and discoveries then being produced at the most prominent medical centres of late Renaissance Italy.

The opposing accounts of mid-sixteenth-century Spanish textbooks on anatomy make it difficult to trace the spread of post-Vesalian anatomy in an Iberian context. Moreover, these books provide only limited information about the everyday practice and institutionalisation of anatomy at universities and hospitals. Surgical textbooks often provide additional information on practical dealings with human anatomy in medical faculties, hospitals and military battlefields. The Spanish source material in this field is surprisingly rich and varied, and often in direct opposition to contemporary foreign sources, which tell of very poor working conditions for the Spanish surgeons. Roth's and O'Malley's biographies of Vesalius both incorrectly referred to a letter from May 1562 in which the Tuscan ambassador in Spain, Leonardo di Antonio de' Nobili, informed Duke Cosimo I that Vesalius and his incompetent Spanish colleagues were at odds about how to save the life of the wounded Spanish crown prince Don Carlos: 'Those who have not witnessed it themselves would not believe the poor skills of these surgeons ... And these Spanish physicians have until

---

[56]   Juan Valverde, *Historia de la Composicion del Corpo Humano, Al Illustriss. y Reverenvdiss. S. Don Fray Ioan de Toledo Cardenal, y Arzobispo de Santiago, el Dotor Ioan de Valverde su Medico*, n.p.: 'Considerando Illusrtiss. Señor la gran falta, que la nación nuestra tiene de hombres, que entienda la anathomia: assi por ser cosa fea entre los Españoles despedacar los cuerpos muertos, como por auer pocos, que venidos á Italia, donde la podrian deprender. no huelgan antes de occuparse en otros excercícios que en este, por no estar accostumbrados a semejantes cosas: y visto el daño que desto se sigue a toda la nación Española, parte por los Cirujanos (a quien falta más haze no entenderla) saber poco latín, parte por auer escrito el Vesalio tan escuramente, que con difficultad puede ser entendido, sino de aquellos que primero algunas vezes an tenido el cuerpo delante de sus ojos, y muy bien maestro que se le declare: pareciame cosa muy conueniente, escribir esta historia en nuestra lengua.'

now obstructed the opinions of Vesalius.'[57] A later travel account by the Swiss medical student Thomas Platter the Younger commented on the undeveloped state of Iberian surgery as witnessed in late-sixteenth-century Barcelona: 'As to the surgeons and barbers, their shops look out on the streets with no more than a simple curtain over the door.'[58] Juan Valverde explicitly offered his *Historia* as an aid to Spanish surgeons, whose poor knowledge of anatomy and complete ignorance of Latin kept them unaware of the latest anatomical reforms and discoveries of contemporary Italy. Valverde blamed Vesalius, whose lack of pedagogical skills had failed to mitigate this ignorance, grounded in the rugged style and inaccessibility of *Fabrica* for non-Latinist readers, and warned his fellow Spaniards of 'the damage, which this causes to the entire Spanish nation, partly because the surgeons (who miss the most by not understanding it) know only a little Latin, and partly because Vesalius has written in such a complicated style.'[59]

In spite of this alleged backwardness in the field of surgery, the Spanish universities were the first outside the Italian Peninsula to offer academic studies and university degrees in surgery during the sixteenth century. While several of the books, statutes and decrees on anatomy stressed the need for a new generation of surgeons familiar with anatomy, many Spanish books on surgery included detailed anatomical treatises and introductions.[60] One of the results of the establishment of a new anatomical practice among medical doctors was the increasingly close relationship between the work of the anatomist and the surgeon, which occasionally made the two professions difficult to distinguish

---

[57]   'Chi non vede non può creder la poca prattica di questi cerujici ... E questi medici spagnoli hanno indugiato sino a ora a volere il Vesalio.' This quotation was used by Vesalius' two primary biographers of the nineteenth and twentieth centuries, Moritz Roth and Charles D. O'Malley, who both elaborated upon its content to support their claims of Spanish backwardness in the fields of surgery and anatomy. Moritz Roth, *Andreas Vesalius Bruxellensis* (Berlin: G. Reimer 1890); p. 269, O'Malley, *Andreas Vesalius of Brussels*, p. 420. The reference to Nobili is incorrect, however, and is seemingly rooted in Roth's uncritical use of Luis Prosper Gachard's *Don Carlos et Philippe II* (Brussels: Emm. Devroye 1863), which quotes Nobili's letter twice (pp. 75 and 78), but without any archival reference. A closer look reveals that Leonardo Nobili was not in Spain as Cosimo I's ambassador until 1565 and therefore could not have written the letter from May 1562. The real author turns out to be Duke Cosimo's agent in Spain, Bernardetto Minerbetti. Thanks to Director of the Medici Archive Project, Alessio Assonitis, for his invaluable efforts and assistance in solving this small mystery. ASF. MdP, vol 5040, fols. 206 r. and v.

[58]   Thomas Platter, *Journal of a Younger Brother: The Life of Thomas Platter as a Medical Student in Montpellier at the Close of the Sixteenth Century* (London: F. Muller, 1963), p. 227.

[59]   'El daño que desto se sigue a toda la nación Española, parte por los Cirujanos (a quien falta más haze no entenderla) saber poco latín, parte por auer escrito el Vesalio tan escuramente.' Juan Valverde de Amusco, *Historia de la composicion del cuerpo humano* (Rome, 1556), p. 3.

[60]   Pedro García Barreno, *La medicina en El Quijote y en su enterno* in *La Ciencia y El Quijote* (Barcelona: Crítica, 2005), p. 162.

from one another, as seen in a number of Spanish universities, where the chairs in surgery and anatomy eventually melted together into one professorship.[61]

The first 'post-Vesalian' book of anatomy to be written in vernacular Spanish, Bernardino Montaña's *Libro de la anothomia del hombre*, did not mention Vesalius in its text, but nonetheless revealed its dependence on his work in a number of crude woodcuts which were plagiarised from *Fabrica*. Juan Valverde's *Historia de la composición del cuerpo humano*, published in Rome in 1556, made more deliberate use of the Vesalian images, which were reworked on copper and supplemented by a only a few original illustrations arguably carried out by the Spanish artists Pedro de Rubiales and Caspar Becerra.[62] It is a significant indication of the contemporary Spanish dominion over Italy that the printing of Valverde's book in 1556 coincided with the preparations for another Spanish invasion of Rome during the peak of the short-lived 'Caraffa War'. This 1555–57 struggle between the Caraffa Pope Paul IV and Philip II was the closest the Eternal City ever came to a repetition of the 1527 'Sacco di Roma' and was prevented only by the final papal submission to the young Spanish monarch.[63] The patron of Valverde's work was the exceedingly powerful Cardinal Juan Álvarez de Toledo, son of the second Duke of Alba and General Inquisitor at the curia of Paul IV. Toledo was at that time the mightiest and most influential of the Spanish settlers in Rome, who by the end of the century made up almost one-third of its inhabitants. The far-reaching Spanish sphere of influence in contemporary Europe was seen elsewhere in Philip II's 1555 grant of a privilege to the surgeons' guild in Amsterdam, permitting the dissection of an executed criminal once a year.[64] The guild's annual dissection was later celebrated in Rembrandt's famous 1632 *Anatomy Lesson of Dr. Nicolaes Tulp*, which has since become an icon of the enlightened spirit of Protestant Europe, ironically referring to a practice instigated in Amsterdam by the Most Catholic King. Antwerp, and particularly the Plantin publishing house, became another

---

[61]   Teresa Santander Rodriguez, *La creación de la cátedra de cirugía en la Universidad de Salamanca. Cuadernos de historia de la medicina Española*, vol. 4 (Madrid, 1965), p. 196.

[62]   Andrea Carlino, 'Tre piste per la anatomia di Juan di Valverde. Mélanges de l'ecole francaise de Rome', *Italie et Méditerranée*, 114 (2002), pp. 525–41

[63]   The Spanish dominance and cultural influence in Renaissance Italy has in recent years been the subject of several interesting studies, such as Aurelio Musi, *Nel sistema imperiale. L'Italia Spagnola* (Napoli: Edizioni Scientifiche Italiane 1994); Manuel Rivero, *Felipe II y el gobierno de Italia* (Madrid, Sociedad Estatal para la Conmemoración de los Centenarios de Felipe II y Carlos V, 1998); and Thomas J. Dandelet's *Spanish Rome, 1500–1700* (New Haven: Yale University Press, 2001). The latter work presented mid-sixteenth-century Rome as a fully integrated part of the Spanish empire, an idea further explored by T. Dandelet (ed.) in *Spain in Italy, Politics, Society and Religion, 1500–1700* (Leiden: Brill, 2007).

[64]   Tim Huisman, *The Finger of God. Anatomical Practice in 17th-Century Leiden* (Leiden: Primavera Pers, 2009), p. 166.

important centre for the dispersal of Spanish medical publications; both a Latin and a Dutch translation of Valverde's anatomy were produced and printed by Plantin in 1566 and 1568 respectively. The complex geography and influence of Imperial Spain during the 'Siglo de Oro' extended far beyond the Iberian Peninsula and even to recently discovered American colonies, where the first books on anatomy and surgery appeared as early as the 1570s – Alonzo López de Hinojosos' *Summa y recopulacion de cirurgia* (Mexico City, 1578) and Augustin Farfán's *Tractado breve de anothomia y chirugia* (Mexico City, 1579).[65]

This study focuses primarily on the Iberian universities with chairs of anatomy during the latter sixteenth century, and on the differences and general patterns in the support and institutionalisation of anatomy as an academic discipline. While there are no records of anatomical practice in Castile until the mid-1500s, such studies had been carried out since the fourteenth century at the medical school of Montpellier, which was at that time a part of the Aragonese Empire.[66] Due to this continuity and experience, anatomists from the Crown of Aragon, and especially from Valencia, were often in the forefront during the subsequent establishment of anatomy studies throughout the Spanish realms. The universities of Aragon and Castile were structured very differently, with the former under the control and protection of their municipal governments, and the latter subjected to the direct supervision and intervention of the Crown. This may previously have been an obstacle in the attempts to introduce anatomy studies into the curriculums of the universities of Castile. Yet the late reign of Charles V facilitated crown-mandated requests between 1550 and 1552 for the introduction of anatomy studies at the Universities of Salamanca, Valladolid and Alcalá de Henares. As David Goodman's *Power and Penury* and (in much more detail) Michele L. Clouse's *Medicine, Government and Public Health in Philip II's Spain* have shown, the medical renovation programmes of the Spanish crown ranged far beyond its request for anatomical education at the three leading universities of Castile. Initiatives for the welfare of the proto-absolutist state also included systematic searches for medical plants in the American and Asian territories, as well as constructions of hospitals for the poor, and new regulations of the Tribunal del Protomedicato, instigated by *los Reyes Católicos* in 1477 to secure a high and standardised level of medical practice throughout the Spanish realms.[67] As a consequence of this direct involvement, the initial support for anatomy at the Castilian Universities of Salamanca, Alcalá and Valladolid was not forced through *within* the universities, but often *upon* the universities by outside pressure from the crown

---

[65]   T.N.V. Persaud, *A History of Anatomy – The Post-Vesalian Era* (Springfield: Charles C. Thomas Publisher, 1997), p. 283.

[66]   López Piñero, 'Actividad Científica y Sociedad de la España de Felipe II', p. 309.

[67]   David C. Goodman, *Power and Penury* (Cambridge: Cambridge University Press, 1988), p. 216; Michele L. Clouse, *Medicine, Government and Public Health in Philip II's Spain: Shared Interests, Competing Authorities* (Farnham: Ashgate, 2011).

and regional municipalities. These extensive reforms were in direct opposition to the traditional images of Imperial Spain as a reactionary and static monolith opposed to intellectual and scientific reforms. The medical faculties of the mid-1500s were in fact often quick to amend their university statutes on anatomy and surgery – in an ironic contrast to the medical faculty of Padua (often celebrated as the very birthplace of the 'Anatomical Revolution') whose Galenic statutes remained unaltered well into the seventeenth century.[68]

Considering the dispersal of professorships of anatomy in numerous universities of mid-sixteenth-century Spain, it is surprising that Vesalius never referred to any of his Spanish disciples and followers during his five-year stay at the Madrid court of Philip II. Only circumstantial evidence suggests that he was aware of the research carried out by his alleged apprentices and disciples in Valencia, Pedro Jimeno and Luis Collado. Pedro Jimeno's *Dialogus de re medica* (1549) referred to Jimeno's own recent discovery of the third auricular bone – the stapes – which was the only one of the three auditory ossicles that was not mentioned in the 1543 and 1555 editions of Vesalius' *Fabrica*.[69] Luis Collado instead credited himself and his pupil Cosme de Medina with the same discovery six years later in his *Liber de ossibus* (1555).[70] In his *Examen*, written in Spain to Gabrielle Fallopio in 1561, Vesalius referred to his recent acknowledgement of the existence of this third auricular bone, but did not reveal the source of his new insight.[71] In the same text he complained about his difficulties in obtaining a single skull to examine, which makes it unlikely that the discovery could have been made on his own. O'Malley suggested that Vesalius owed his awareness of this osteological feature to Juan Valverde and that he may have heard of it from the Italian anatomist Giovanni Ingrassia, who discovered the ossicle around the same time as Jimeno, but published his study much later.[72] Valverde, like

---

[68]   Carlino, *Books of the Body*, p. 228.

[69]   Jimeno, *Dialogus de re medica*, f. 20r: 'Duo, Andreae Vesalio praeceptori nostro plurimum observando, Caesareo medico, viro incomparabili: et nobis tertium idque minimi momenti, quae a Galeno neque per somnium excogitata fuere, quod ex nostri amitissimi praeceptoris sententia, dixeram olim in Gal. in nostris illis publicis disputationibus, tertium illud ossiculum, repertum est a me frequenter, in caluaris quae passim occurunt exsiccatis, postmodum in omnibus recentibus quas priuatim saepe aggressus sum, eius rei gratia id sedulo animaduerti.'

[70]   Collado, *Cl. Galeni liber de ossibus*, f. 30r: 'Reperiuntur etiam duae cavitates, una meatus auditorii, in qua Andreas Vesalius duo ossicula in gratiam instrumenti auditus a natura fabricata invenit: quorum unum incudi, alterum malleo non inepte, propter similitudinem, comparavit. Ego autem una cum Cosmo Medina in inclyta academia Salmanticensi nunc publico Anatomes professore longe doctissimo discipulo meo mihi charissimo, aliud os repperi, qui, quod simile esset equitandi instrumento, quo pedes firmantur, stapedae nomen imposui.'

[71]   O'Malley, *Andreas Vesalius of Brussels*, p. 291.

[72]   'Possibly he became acquainted with the third ossicle from the reference to it – an unacknowledged borrowing from Ingrassia – by Valverde in 1556.' Ibid., p. 291.

O'Malley, was seemingly unaware of the writings of Jimeno and Collado when he wrote of 'the third ossicle, of which nobody before me has made any mention'.[73] The coincidence that Vesalius learned of this discovery during his stay in Spain, where it was first presented in print, at least indicates an indirect recognition of the anatomical discovery made and published by his Valencian disciples. Vesalius' only written allusion to a Spanish anatomist was a fierce attack on Valverde's Historia de la composicion del cuerpo humano, which Vesalius in his 1561 Examen accused of being a mere plagiarism of his own Fabrica: "Valverde, who never put his hand to a dissection and is ignorant of medicine as well as of the primary disciplines, undertook to expound our art in the Spanish language only for the sake of shameful profit." Footnote: Cited in Charles D. O'Malley, Andreas Vesalius of Brussels (Berkeley: University of California Press, 1964), p. 267. While only scant evidence suggests that Vesalius knew anything else of the activities of his Iberian colleagues, acclamations of Vesalius among his Spanish followers were frequent and numerous.

The earliest Spanish colleague of Vesalius was probably the ill-fated Miguel Servet, whose theories on the anatomy of the blood and the soul introduced the idea of the pulmonary system and the 'smaller circulation' of the blood, later improved upon by Juan Valverde and Realdo Colombo. Servet's combined theological and medical thoughts and observations were presented in *Christanismi Restitutio*, one of the rarest books in existence, with only three copies still surviving today. Servet's controversial ideas were immediately denounced by both Calvinist and Catholic authorities, and the text was burned at the stake with its author in 1553, with Servet having been sentenced to death for heresy in Calvinist Geneva. Before his tragic end, Miguel (Servet) Villanovano and Andreas Vesalius were praised together in the same lines in the *Institutiones anatomicae* (Paris, 1536), a text published in Paris by their medical teacher Günther de Andernach: 'In the task (of revising this book in the light of numerous dissections), which was not easy, I was assisted by Andreas Vesalius ... together with Miguel Villanovano, who assisted me in several dissections.'[74] Another Spanish physician with close ties to Vesalius was Fransisco Hernández,

---

[73]   Valverde, *Historia de la composicion del cuerpo humano, Declar. delas fig. del Lib. I.*: 'El tercer ossezuelo, del qual ninguno antes de mi ha hecho mención.'

[74]   Miguel Servet, *Restitución del Cristianismo* (Madrid: Fundación Universitaria Española, 1980), p. 25: 'En esta tarea (de revisar el libro a la luz de numorosas dissecciones), nada facil, me ha ayudado, ante todos Andrés Vesalio, un joven, por Hercules, muy diligente en anatomía y professor de medicina pura, por quien no hay por que preoccuparse: recientemente, al publicarse esta obra en Venecia, la ha corregido excelentemente. Junto a él Miguel Villanovano, quien me assistió en dissecciones amistosamente: un hombre que sería el honor de qualquier rama de las letras y que es segundo a nadie en doctrina galénica. Con ayuda de ambos examiné en muchos cuerpos humanos las partes interiores y las exteriores, los musculos, venas, arterias y nervios, y se los mostré a los estudiosos.'

the Royal Protomedico and so-called 'Pliny of the New World' who described Vesalius as 'a very good friend of mine, while he was still alive' in his commentary on Pliny, which was written between 1565 and 1569, but remained unpublished until recently.[75]

Hernández's accounts of his experience as an anatomist at the monastery of Guadalupe offer an interesting indication of the spread of contemporary anatomy beyond the university faculties of Castile. García de Talavera, a later chronicler of the monastery, confirmed Hernández's descriptions of active anatomy studies at the monastery hospital in his *Historia de nuestra Señora de Guadalupe* (Toledo, 1597). In Hernandez's manuscript on Pliny, *Historia natural de Cayo Plinio Segundo – traslada y anotada por el Doctor Francisco Hernández*, he praised the attempts in Guadalupe to elevate contemporary Spanish anatomy to a higher level and 'the experience, which these men in Guadalupe had in cutting bodies open with their own hands, and through the grace of God put anatomy up to date'.[76] Among his associates in Guadalupe, Hernández mentioned Francesc Micó, who was later appointed to the first joint chair of anatomy and *materia medica* at the University of Barcelona. Ten years later, Hernández made use of his anatomical knowledge again, this time in a New World context, as the instructor to the surgeon Alonzo López de Hinojosos during their autopsies of victims of the epidemic *cocolitztli*, which raged throughout Mexico. This epidemic was first described by Hernandez in 1577 and then in much more detail in Hinojosos' *Summa y recopulacion de cirurgia* the following year. Like other Spanish anatomists such as Andres Laguna, Francesc Micó, and Juan Fragoso, Hernández was first of all credited later for his botanical writings. Anatomical studies seemingly comprised only a relatively small facet of the broader education of these naturalists; all produced treatises or commentaries on human anatomy, but were primarily engaged in the foundations of the Royal gardens of Escorial and Aranjuez, comments on Dioscorides and Pliny, and the gathering and descriptions of plants from the Iberian Peninsula, East India and the Americas.[77] In parallel to the works of these polymaths, an increased

---

[75]   Germán Somolinos d'Ardois, *Vida y obra de Francisco Hernández* (Mexico City: Universidad Nacional de Mexico, 1960), p. 132: 'Andreas Vesalio, varón excellente en anatomía y mientras vivía amigo nuestro.'

[76]   Ibid., p. 122: 'exercicio que en cortar por mano ajena hombres tuve en Guadalupe, donde dexamos puesta por la bondad de dios el anatomía de su punto como hasta allí no se hubiesen cortado todo sino los miembros interiores solo'.

[77]   These medical publications include Andrés Laguna's *Pedacio Dioscorides Anazarbeo, Acerca de la material medicinal, y de los venenos mortiferos* (Madrid: Mathias Gast, 1552); Fransisco Micó's *Libro del regalo y utilidad de beber frío y refrescado con nieve* (Barcelona: Casa de Diego Galván, 1576); Juan Fragoso's *Discurso de las cosas aromaticas, arboles y frutales, y de otras muchas medicina simples que se traen de la India Oriental, y sirven al uso de la medicina* (Madrid: Casa de Fransisco Sánchez, 1572); and Fransisco Hernández's unpublished manuscripts on New World

specialisation and professionalisation was developing among Spanish anatomists during the latter sixteenth century. Following a 1564 plague epidemic in Zaragoza, the Sardinian physician Juan Tomas Porcell carried out one of the first known pathological studies of plague victims. In the following decades the surgeon Fransisco Díaz authored general treatises on both surgery and anatomy, and also more specialised publications on the urinary system, which were among the earliest Renaissance treatises on urology and acknowledged a legacy from both foreign and Spanish predecessors. Most notable were Díaz' references to his Spanish colleagues Pedro Jimeno, Bernardino Montaña and Juan Valverde, and their studies in Valencia, Valladolid and Rome – thus uniting this first Spanish generation of 'post-Vesalian' anatomists, who in their own writings never acknowledged any contributions or even the existence of one another: 'The curious reader, if he is a physician or a surgeon, should read Galen, Vesalius, Charles Estienne, and Valverde, and Jacobus Sylvius, and Realdo Colombo, and Montaña, who are the most learned in this field, and among them Jimeno should be placed as well, the erudite Valencian, and the first who elegantly and skilfully began to dissect and carry out anatomy in the city of Valencia, where medicine and anatomy is presently flourishing.'[78]

---

flora and fauna, parts of which were later published in a printed version in *Cuartro libros de la naturaleza y virtudes de las plantas, y animales que estan recividos en uso de la medicina en la Nueva España* (Mexico City, Diego Lopez Dávolos, 1615).

[78] Fransisco Diaz, *Tratado nuevemente impresso, de todas las enfermedades de los Riñones, Vexiga, y Carnosidades de la verga, y Urina, diuidido en tres libros* (Madrid: Francisco Sanchez, 1588), f. 19r and 19v: 'advertiendo primero, que el curioso lector, si es medico o cirujano, lea a Galeno, Vesalio, Carolo Stéfano y a Valverde, y a Jacopo Sylvio, y a Realdo Colombo, y a Montaña, que son los que deste negocio trataron eruditíssamente y de proposito, y tambien entre estos puede poner a Ximeno, doctissimo valenciano, y el primero que con elegancia, y gran destreza comenzó a poner la execución de cortar y a hacer anatomía en la ciudad de Valencia, donde tanto resplandece la medicina y la anatomía al presente.'

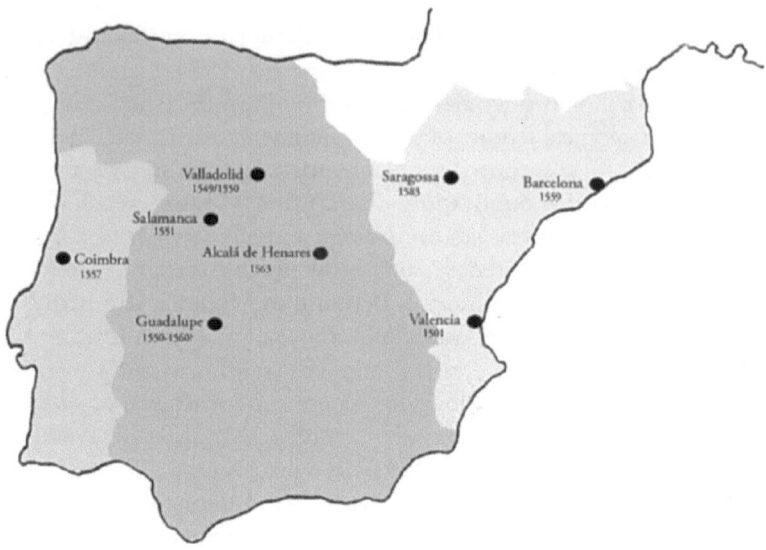

Figure 1.1 Map of Iberian centres of anatomy, 1500–1600. There is evidence of human dissection having taken place in the Crown of Aragon as early as the fourteenth and fifteenth centuries, when royal privileges enabling this practice were granted to guilds of surgeons in Lleida, Valencia, Zaragoza and Barcelona. The Crown of Castile, however, had no known tradition in this field when it was first introduced in the mid-sixteenth century. A rapid development of anatomy studies can be traced back to the latter sixteenth century in many of the largest university towns of the Iberian Peninsula. At the beginning of the century, only the University of Valencia had a permanent chair in anatomy, but by the mid-1500s, similar permanent professorships had been established in Salamanca, Alcalá de Henares, Coimbra and Barcelona. *Credit*: Author

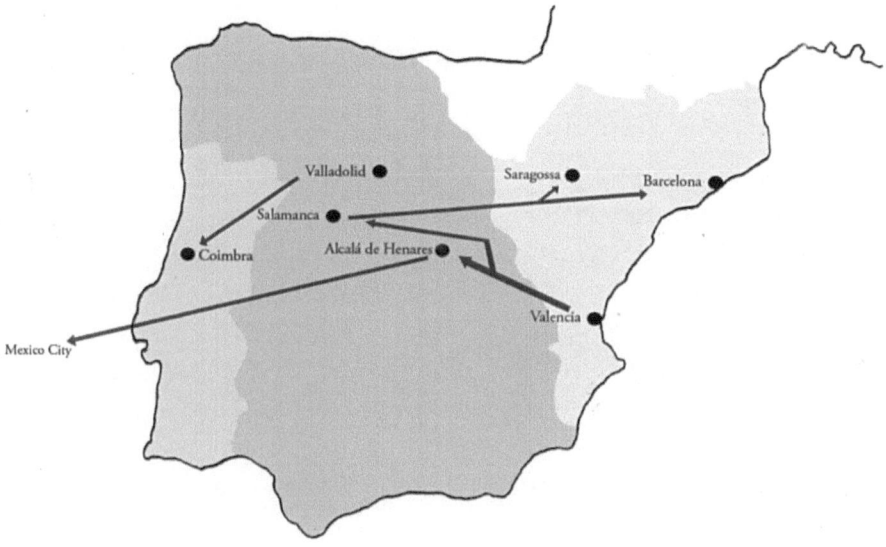

Figure 1.2   Map of the circulation of Iberian anatomists, 1550–1600. The influx of anatomists from the *Estudi General* of Valencia into Castile led to the first formally appointed university chairs of anatomy in Salamanca (1552) and Alcalá de Henares (1563). The University of Salamanca appeared as a later counterpart of this flow of anatomists between Castile and Aragon, and as the educational base for the first anatomists active in Barcelona and Zaragoza. The University of Alcalá de Henares stood out as the largest contributor of New World anatomists, with Francisco Hernández, Augustin Farfán and Juan de Barrios as the most prominent representatives of this exchange between New Castile and New Spain. While no permanent chair of anatomy was created at the University of Valladolid during the sixteenth century, the surgeon Alonso Rodríguez de Guevara brought the practice to the University of Coimbra after his short 20-month course of anatomy in Valladolid between 1549 and 1550. *Credit*: Author

Chapter 2

Valencia

The anatomical studies established at the University of Valencia in 1502 were significant and unique in a number of different ways, and were crucial to the later development of this new practice elsewhere in the Iberian Peninsula. Consequently, the timeframe from 1550 to 1600, which represents the chronological span of later case studies, is here extrapolated back to the early 1500s. University anatomy was not only first practised and formally approved in Valencia half a century before the leading universities of Castile, but numerous Valencian anatomists were also dispatched to universities beyond the Aragonese borders. The first anatomists to establish this practice at the Castilian Universities of Salamanca and Alcalá de Henares were all born and educated in Valencia, a fact celebrated by the Valencian chronicler Caspar Escolano in an early seventeenth-century account of the history of his native city: 'From our University and medical doctors derive the knowledge of anatomy throughout the rest of Spain. It was unknown in Castile until Valencian Doctors came to teach it in Salamanca and Alcalá.'[1]

One of these doctors, Cosme de Medina, left Valencia in the autumn of 1551 to fill the first permanent chair of anatomy at the University of Salamanca and later became the primary professor of its medical faculty. The Valencian influence also spread to Alcalá de Henares, where the university appointed no fewer than five consecutive anatomists of Valencian origin during the latter half of the sixteenth century. Within the Crown of Aragon, the Universities of Barcelona and Zaragoza were organised according to the Valencian model, with chairs of anatomy established from the very foundations of the medical faculties in 1559 and 1583 respectively.

To understand and explain this unique position, a variety of reasons and preconditions must be taken into account: the uniquely advanced development of the medical school of Valencia was the product of close historical ties between the Crown of Aragon and the Italian Peninsula, and ties between the school in Valencia and the medical school of Montpellier, which was placed

---

[1]   Cited in José Maria López Piñero, 'El saber anatomico y la dissección de cadaveres humanos en la España en la primera mitad del siglo XVI', *Cuadernos de la historia de la medicina Española*, 13 (1974), p. 83: 'De nuestra Universidad y medicos ha tenido su origin la noticia que generalmente se tiene en España de anathomia; ni la conocieron en Castilla hasta que fueron valencianos a leerla en Salamanca y Alcalá.'

within Aragonese borders during the Crown of Aragon's imperial zenith in the 1200s and 1300s. In 1340 the Valencian Professor of Medicine Fransisco Conill consolidated the practice of anatomical dissection begun at this renowned medical school a few decades earlier by the French surgeon Henri de Mondeville. In Montpellier, the privilege of dissecting a condemned criminal every second year was later extended, with students being permitted to carry out annual anatomy studies. The Montpellier model was later imitated in Barcelona, Valencia and Zaragoza, where medical schools and guilds of surgeons gained similar rights in 1402, 1478 and 1488, respectively. Although the students in Barcelona were granted dissection privileges long before their Valencian counterparts, it was in Valencia that the most systematic practice in this field developed during the late fifteenth century.

None of the largest cities of Aragon could boast fully developed universities until the following century, when Valencia, Barcelona and Zaragoza each established an *Estudi General* in 1502, 1559 and 1583, respectively. It should be noted that Valencia had established a progressive school of surgery, or *Estudi de Cirurgía*, as early as 1433. From 1463, this institution was regulated with official support from the city council and began to enjoy an increase in academic stature with the nomination of a claustro of medical professors in 1480.[2] In 1478 the *Collegi de Barbers i Cirurgians* was granted the privilege of carrying out anatomical dissection with the official approval of King Juan II.[3] In 1486 the *Estudi di Cirurgía* further consolidated its increasing prestige and monopoly over surgical education when a five-year training period at this institution became compulsory for Valencian students of surgery.[4] Following the introduction of these privileges, surgeons in Valencia made a request for a dissection room in the precincts of the local hospital general.[5] Uncertainty surrounds the immediate

---

[2]   Mercedes Gallent Marco, *El colegio de cirujanos de Valencia: Aportación documental* (Valencia: *Saitabi, Revista de la Facultat de Geografia i Història*), p. 150.

[3]   'Per que de aquell se puixa fer notomia obrint tallant e administrant aquell per totes aquelles parts membres e lochs dela persona de tal sentenciat que volran e conexeran que fer se deia la dita anotomia.' Luis Garcia Ballester, 'La cirugía en la Valencia del siglo XV. El privilegio para disecar cadaveres en 1478', *Cuadernos de historia de la medicina española*, 7 (1967), p. 169.

[4]   The incorporation of surgical practice into the newly established Estudi General in 1502 has been analysed in an article by the medical historian Jose Luis Fresquet Febrer in *La práctica médica en los textos quirúrgicos españoles en el siglo XVI* (Granada: Dynamis, 2002), p. 255: 'El prestigio que llegó a alcanzar este centro dió lugar a que en 1478 obtuviera un privilegio real para disecar cadaveres y que desde 1486 se hiciera obligatorio haber cursado cinco años para poder ejercer como cirujano. Por tanto las constituciones de la Universidad de 1499, no hicieron más que incorporar a los planes de estudio de la futura Universidad esta formación quirúrgica.'

[5]   'per quant nosaltres no tenim ne loch destinat tal qual mester seria per fer exercir aquell arte de anotomia, del qual tenim previlegi de sa magestat real, seria e és cosa molt necesaria haver una casa o ort qui fos del nostre col.legi e mortizada per lo senyor rey, en la qual los dits actes

outcome of this petition, but references to anatomical studies carried out within the hospital general appear from the 1520s onwards and may indicate the successful granting of a 'casa o ort', as specified in the surgeons' request of 1488. Unlike other university towns of the Iberian Peninsula, Valencia possessed long-established expertise with surgery and anatomy – practices that had been known and performed in Aragon since the fourteenth and fifteenth centuries, and were incorporated into the Valencian university curriculum in the early 1500s. The three leading Aragonese universities in Valencia, Barcelona and Zaragoza were all founded during the 'Siglo de Oro' and included this novel practice in their statutes and medical faculties from the very beginning. These institutions furthermore continued their engagement with anatomical studies longer and more continuously than any of the leading universities of Castile. In Valencia, surgery and anatomy were essential areas of medical practice and as such were incorporated into the curriculum during the 1502 foundation of the *Estudi General*. Later Castilian universities had no such tradition to build upon when Valencian anatomists joined the Universities of Salamanca and Alcalá in the mid-sixteenth century. A professorship of surgery had already been created with the foundation of the Valencian University, but it was met with resistance when royal provisions in 1566 called for the establishment of surgery as a university discipline in Salamanca and later in 1593–94 at the Universities of Valladolid and Alcalá de Henares.

Further disparities emerge when considering the appearance and overall organisation of the *Estudi General* of Valencia and the leading universities of Castile. While the most prominent Castillian universities were subject to the direct governance and intervention of the Crown, the University of Valencia was ruled and financed by a local municipality and oligarchy in order to fulfil the most immediate needs of the city. The local powers judged the education of medical doctors to be one of the most urgent matters of the new university. Whereas the teaching of medicine at the Universities of Salamanca and Alcalá remained fairly rudimentary compared to the much larger institutional cornerstones of law and theology, the Valencian university faculty of medicine grew to become the most comprehensive medical school in mid- and late sixteenth-century Spain. By the end of the 1500s, the medical faculties of Valladolid, Salamanca and Alcalá retained between four and six university chairs, whereas Valencia boasted nine permanent professorships of medicine, one of which was the chair of anatomy.[6] While no permanent university chair of anatomy was ever established in Valladolid, and was only maintained with difficulty or with long periods of

---

se poguessen fer'. Cited in Àlvar Martínez-Vidal and José Pardo-Tomás, 'Anatomical Theatres in Early Modern Spain', *Medical History*, 49 (2005), p. 264.

[6] José Maria López Pinero, *Valencia y la medicina del renascimiento y barrocco* in *III Congresso Nacional de Historia de la Medicina. Volumen II* (Valencia, Actas, 1969), p. 96.

vacancy in Salamanca and Alcalá, the Valencian professorship was filled without interruption during the sixteenth and seventeenth centuries.

Another important facet to the specific prerequisites of Valencian medicine was the multicultural character of its medical practitioners until the 'ethnic cleansing' of Iberian medicine from the late fifteenth century onwards.[7] Juan II's approval of dissection at the Valencian *Estudi di Cirurgía* referred to the differing creeds of his subjects and emphasised that the condemned and anatomised subject should preferably be Muslim, 'catiu serrahi', or a suicide, 'persona desesperada'.[8] The multi-religious background of Valencian medical practitioners found increasing conflict at the dawn of the sixteenth century, as exemplified by the misfortune of Lluis Alcanayis, author of the first printed Valencian book on medicine, whose problems with 'limpieza de sangre' eventually placed him before the Inquisition. Alcanayis had originally been a professor of surgery at the *Estudi di Cirurgía* until his 1502 appointment to the first combined anatomical/botanical chair of the *Estudi General*. The position was originally entitled 'segona cadira de medicina' and was later named 'anatomía y simples' or 'anatomía y materia medica'.[9] Before the establishment of this university chair, a number of publications on surgery and anatomy had appeared in Valencia, including some of the earliest printed textbooks on plague and syphilis. The first of these, Lluis Alcanyis' *Regiment preservatiu e curatiu e la pestilencia*, was published in 1490, when the author still held his professorship at the *Estudi di Cirurgía*. The radicalisation of state politics towards Jewish descendants or *judeoconversos* did not initially prevent Alcanyis from obtaining the first prominent professorship of the 'segona cadira' at the *Estudi General* after its inauguration in 1502. He held this chair until 1504, when he was put

---

[7]   'Hemos visto hasta ahora la relación de la comunidad quirúrgica valenciana con el mundo de Montpellier, Barcelona e Italia y el peso decisivo de la influencia de ésta. Pero no hemos tenido en cuenta un hecho cierto, y es la íntima relación existente y comprobada entre las otras dos comunidades que convivían con los christianos, la judía y la musulmana ... Concretamente en Valencia tenemos numerosas noticias de permiso de ejercicio quirúrgico, previo examen, concedido a musulmanes (serrahins) y a judíos y del lugar íntimo y importante que estos últimos ocuparon cerca de la corte.' Luis Garcia Ballester, 'La cirugía en la Valencia del siglo XV. El privilegio para disecar cadáveres de 1478',*Cuadernos de Historia de la Medicina Española*, 7 (1967), p. 160. A more detailed study in a broader national context can be found in Luis Garcia Ballester, *Los moriscos y la medicina. Un capítulo de la medicina y la ciencia marginada de la España del siglo XVI* (Barcelona: Labor, 1976).

[8]   Ibid., p. 169.

[9]   This university chair continued as a twin professorship until 1560, when it was split into two separate positions within the medical faculty. Until then, the professor of the 'segona cadira' taught anatomy in the cold months of the year and *simples* or *materia medica* during the summer months, as described by Jose Maria López Piñero in *La Facultad de Medicina de Valencia (1502–2002)* (Valencia: University of Valencia, 2002), p. 42.

before the Inquisition and subsequently imprisoned. In 1506 he received a death sentence and was burned alive with other 'judaizantes', allegedly after informing on his own wife. Alcanyis' horrific death has not attracted the same attention as the similarly ill-fated Spanish anatomist Miguel Servetus in Calvinist Geneva half a century later. Nonetheless, the incident presents a clear example of the cleansing of a medical discipline that had previously included several members of the religious minorities residing within the Crown of Aragon.

Later holders of this second chair published textbooks on anatomy during their university appointments, and Valencia is once again unique as the only Iberian university town in which publications on anatomy were produced throughout the sixteenth century. It was furthermore the first Iberian city with known physical facilities for anatomical dissections, as presented in documentary evidence from 1524, which mentioned the presence of an irrigation ditch and a house for dissections within the *hospital general*. In an Iberian context, this is the earliest known reference to a place where anatomical dissections were carried out, but unfortunately it provides no descriptions of the shape and size of the location.[10] During the early sixteenth century, the medical faculty in Valencia had already been closely linked to the humanist medical tradition initiated in Italy and France, as opposed to the 'galenismo arabizado', which still hindered the development of humanist medicine at the medical faculties in Castile. The hostility towards medieval authority and the humanist renovation programmes in Valencian medicine from c. 1500 onwards is further indicated by the lack of a university chair in 'Avicenna' at the *Estudi General*, such as that found in both Valladolid and Salamanca. This tendency was further manifested in the establishment at the University of Valencia of a chair of 'Aforismos de Hippocrates'. The close scholarly relationship with Renaissance Italy was strengthened by the nomination of the Valencian cardinal Roderic de Borja as Pope Alexander VI in 1492; some of the first medical university publications from the early sixteenth century were dedicated to his son, Cèsar Borja who at that time filled positions as both Italian warlord and Archbishop of Valencia.

Italian influence also appeared in the repeated appraisals of the Bolognese anatomist Berengario da Carpi, such as that found in Miguel Juan Pascual's surgical/anatomical treatise *Libro o practica de cirugía* (Valencia, 1537). Pascual's book was a commented translation of a surgical text by Pope Julius II's physician Giovanni di Vigo, 'con ciertas addiciones marginales provechosas para

---

10   Martínez-Vidal and Pardo-Tomás, 'Anatomical Theatres', p. 167: 'It is clear that the place where the practice of dissection and the teaching of anatomy were carried out in Valencia was the General Hospital. However, information on the architectural features of the theatre, as well as its exact location within the hospital area, varies a great deal and is very vague. The oldest reference, provided by a document dated 1524, alludes to an irrigation ditch that flowed through the hospital grounds, next to the place where the surgeons had built a house to carry out anatomical dissections.'

los cirurgianos'. Its countless references to Rhazes and Avicenna expressed the enduring 'galenismo arabizado' of previous centuries, but were supplemented with contemporary notations on human anatomy from Berengario de Carpi's *Isagogae breves* (Bologna, 1522). Pascual credited the Bolognese anatomist repeatedly and even quoted Carpi's erroneous description of the multi-lobed liver as a recent and important anatomical insight: 'Carpi says that the liver sometimes has five, four, three, and two lobes.'[11] Pascual's work began with a *Libro dela nothomia* and an insistence that this scientific discipline was crucial for any practising surgeon: 'the surgeon who practices without any knowledge of anatomy is like a blind man working as a carpenter'.[12]

Pascual's textbook did not refer to Galen's personal experience with human dissection, which was later questioned by Vesalius, but was steadfastly defended by devoted Galenists throughout Europe. His text nevertheless encouraged the teaching of anatomy through dissection, and not merely by using textbooks and images, in accordance with views advocated by Galen himself in the second century when he carried out his renowned animal dissections and vivisections.[13] Three Valencian professors of anatomy during the mid- and late sixteenth century – Pedro Jimeno, Luis Collado and Vicente Garcia Salat – all wrote anatomical treatises. Jimeno's *Dialogus de re medica* from 1549 was the first anatomical text in an Iberian context to be based predominantly on Vesalius' teachings. Luis Collado's *Liber de ossibus* from 1555 was an early written defence of Vesalius following the attacks on his osteology by Jacobus Sylvius in 1551. Vicente García Salat, who held the Valencian professorship of anatomy from 1578 to 1611, is known to have produced an unfortunately unpublished and now lost manuscript entitled *De anathomia*, allegedly written around 1590. The close ties

---

[11]   Miguel Juan Pascual, *Libro y practica de cirurgia...Compuesto por el muy famoso doctor Juan de Vigo, medico y cirurgiano del Sumo pontifice Julio Segundo. Traducido nuevamente de latin en lengua castellana. Por el Doctor Miguel Juan Pascual. Con ciertas addiciones marginales provechosas para los cirurgianos* (Valencia: 1537), f. 3r: 'Dize Carpo que el higado tiene alguna vez cinco lobos o penulas alguna vez quarto y alguna tres y alguna dos.'

[12]   'Como el ciego obra en el leño: asi obra el cirugiano ignorante la anothomia. El ciego cortando el madero muchas vezes yerra tomando mas o menos delo que es necessario. Y assi yerra el cirurgiano que ignora la anothomia. El anothomia es sciencia recta con la qual los miebros del cuerpo humano singularmente con incision son diudidos': ibid., f. 3r. This metaphor of a blind carpenter was originally introduced in Guy de Chauliac's *Chirugia magna* and was copied almost literally by Pascual, and also by Juan Calvo in his later surgical/anatomical treatise, *La chirurgia universal y particular del cuerpo humano* from 1580: 'si (el cirujano) no sabe anatomía será como el carpintero ciego de su natividad, que corta el madero por donde no lo ha de cortar' (p. 168).

[13]   Pascual, *Libro y practica de cirurgia*, f. 5r: 'Por la sciencia dela anothomia puede el medico pronosticar y curar. Y por tanto el buen Gal. vino a noticia dela nothomia en los cuerpos delas simias puercos y otros animales: y no ha hecho como algunos que con pinturas se han esforçado a mostrar la nothomia.'

between Valencian scholars and the medical schools of Italy were also present in works by Jimeno and Collado. Both made references to their education at the University of Padua, where they claimed to have been direct apprentices of Vesalius. At the time of the publication of Pedro Jimeno's textbook *Dialogus de re medica* in 1549, a number of appraisals and rejections of Vesalian anatomy had appeared in other medical textbooks and were published in Valencia from the 1540s onwards, as in Miguel Jerónimo Ledesma's *De pleuritide commentariolus* from 1546 and Pedro Jaime Esteve's *Hippocrates Coi Medicorum omnium principis epidemion liber secundus* (Valencia, 1551).

While both these authors condemned the Arabic translations and misinterpretations of Greco-Roman purity and authority, they were more ambivalent in their views of the anatomical reforms spearheaded by Andreas Vesalius. In what is possibly the earliest mention in Iberia of Vesalius' *Fabrica* (1543), the medical humanist Miguel Jeronimo Ledesma referred in 1546 to 'Andreas autem Vuesalius, uir quidem de re medica bene meritus' and emphasised his own familiarity with dissection and anatomy. Like Ledesma, Pedro Jaime Esteve later insisted that anatomy be taught using human dissection, yet remained critical of Vesalius' Galenic scepticism and the claim that Galen had never dissected a human body: 'To suspect that Galen never dissected a human body is nonsense, and to dare to affirm it is an act of great madness.'[14] Ledesma and Esteve's contradictory attitude towards the anatomical reformer contrasts starkly with the unconditional praise offered by Pedro Jimeno and Luis Collado.

Pedro Jimeno: First Vesalian Anatomist at the *Estudi General*

While previous Valencian textbooks had dealt only fleetingly with the study of anatomy, Pedro Jimeno's book focused chiefly on this subject and comprised a tribute to Andreas Vesalius and the new anatomy taught at the University of Padua. Jimeno succeeded Esteve in the *cátedra de anatomía y materia médica* in 1547 and held this position until 1549, when he was promoted to the *cátedra de practica* and published his only book, which was entitled *Dialogus de re medica, compendiaria ratione, praeter quaedam alia, vniversam anatomen humani corporis perstringens, summe necessarius omnibus Medicinae candidatis*. Almost nothing is known about Jimeno's alleged apprenticeship under Andreas Vesalius, whose lectures he claimed to have followed during an unspecified period between 1537 and 1543, when Vesalius held his famous and short-lived position as anatomy lecturer at the University of Padua. Jimeno made repeated reference to his mentor and even copied long verbatim passages from *Fabrica*. This has led some

---

[14]   Pedro Jaime Esteve, *Hippocrates Coi Medicorum omnium principis epidemion liber secundus* (Valencia: Typis Ioannis Mey Flandri, 1551), f. 147v.

Spanish historians of medicine to mistakenly conclude that his educational background was similar to Vesalius' own, since some of the borrowed excerpts in Jimeno's text feature both the cities and universities of Paris and Louvain. In a short article on Jimeno published in 1972, Charles D. O'Malley corrected some such misconceptions in the existing accounts of his life and work, such as the erroneous idea that Jimeno had been involved in a grave robbery of a cremated and decomposed body in Louvain similar to Vesalius' own experience related in Book One of *Fabrica*.[15] While Jimeno's claims regarding his educational background at the University of Padua have not been supported by evidence beyond his own textbook, the author insisted that he had received his schooling under Andreas Vesalius at this renowned institution – even if he admitted not to have attended every single lecture of the revered master: 'My great mentor Andreas Vesalius, whose lessons, if not all of them, I attended in Padua where he was there frequently teaching anatomy in a magnificent way before a numerous and happy audience.'[16]

A search for references to Jimeno and Collado at the archive of the University of Padua did not reveal any evidence of their entry or graduation as

---

[15] Charles D. O'Malley, 'Pedro Jimeno, Valencian Anatomist of the Mid-sixteenth Century', *Science, Medicine and Society* (1972), p. 70: 'Vesalius is frequently mentioned and the text … depends very heavily on Vesalius' Fabrica. In fact the Dialogus is a severe abridgement of that work with occasional sentences borrowed verbatim. The most striking instance is to be found on folios 26–38, which offer verbatim the text of the Fabrica, pages 155-162. This represents the entire chapter XXXIX of book I entitled "How the bones and cartilages of the human body may be prepared for observation", that is Vesalius' celebrated account of the method of preparing the human skeleton. Although the Andreas of Jimeno's Dialogus introduces this extensive borrowing with the remark that he is going to present what Vesalius wrote 'at the end of book one of his Fabrica humana on the preperation of the skeleton', the fact that this is Vesalius' account and not original with Jimeno seems not to have been realized by either Morejon or Chinchilla. Indeed, they have based their statements that Jimeno studied in Louvain and Paris on what are in fact Vesalius' reminiscenses.' Claims of Jimeno's training in Louvain and Paris have been repeated frequently ever since the appearance of Morejon's and Chinchila's works on the history of Spanish medicine, *Historia bibliográfica de la medicina Espanola* and *Anales históricos de la medicina en general*, both published in Madrid in 1841, and as late as in Jose Vazquez Vicente's 1935 article 'Los anatomicos de la epoca del renacimiento in Trabajos de la cátedra de historia critica de medicina'. The misconception presented Jimeno's involvement in an attempt to smuggle the cremated bones of a condemned criminal into the University of Louvain in 1536 – in the company of Vesalius' friend from the student years, Frisius Gemma, who assisted Vesalius in this endeavour (as described in detail in Book One of *Fabrica*, p. 161–62).

[16] 'Meo potissimum institutore Andrea Vesalio, ex cuius prelectionibus admodum multa si non omnia Patavij quam ibi ultimo anatomen numerosissimo frequentissimoq; auditorio felicissime de more exerceret, obseruaueram.' Pedro Jimeno, *Dialogus de re medica compendia ratione, prater quaedem alia, universam Anatomen humani corporis perstringens, summe necessarius omnibus Medicinae candidatis* (Valencia: Typis Ioannis Mey Flandri, 1549), f. 2r.

medical students at this institution. However, the matriculation lists of medical laureati from the University of Bologna proved very useful in revealing hitherto unnoticed facts regarding Jimeno's Italian university education. A list of graduate students of arts and medicine from the University of Bologna included in a *Liber secretus* from 1505–75 not only locates Jimeno at the institution but also gives us his year and date of graduation: '1544. marz 8. D. Petrus Ximenez Valentinus Hispanus'.[17] Strangely, this reference has escaped the attention of Vincenzo Busacchi's thorough study on *Gli studenti spagnoli di medicina e di arte in Bologna dal 1504 al 1575* (Bologna, 1956), which is based on some of the same material, but does not mention Jimeno in the otherwise detailed list of Spanish medical students at Bologna. Busacchi did find the following reference from 8 February 1544, a month before Jimeno's date of graduation, as evidenced above: 'Die octavo Petrus hyspanus asumpsit gradus medicine.'[18] It is reasonable to assume that this also refers to Pedro Jimeno, even though the two graduation dates are conflicting.

The surprising discovery of Jimeno's graduation from Bologna and not Padua does not altogether exclude the possibility of an affiliation with Vesalius, who occasionally gave anatomical lectures in Bologna. These events were recorded by the German medical student Baldassar Heseler, who documented Vesalius' first anatomical lecture there in January 1540, and by Francesco dal Pozzo, who witnessed one of Vesalius' last anatomical demonstrations, which was given in Bologna in January 1544, only two months before Jimeno's graduation.[19] It could also be that Jimeno may have attended some of Vesalius' lectures in the neighbouring University of Padua. Yet his alleged anatomical education at the side of this mentor and his claims of having attended almost all his lessons in Padua are cast into doubt by the new documentary evidence. Jimeno himself gave no exact date of his purported education under Vesalius, whereas the dedicatory epistle of *Dialogus de re medica* presented the winter of 1548–49 as the initial phase of the publication, coinciding with a critical study of Galen's *De usu partium*, which he compared to his own experiences under Vesalius. The book was dated 31 July 1549 and was dedicated to Mencía de Mendoza, Duchess of Calabria, and her personal physician, Pedro Lozano *viro literis & nobilitate ornatissimo*, who was presented by Jimeno as a close friend and colleague. It is unclear whether the apparent affiliation to the exceedingly powerful Mendoza family was already secured by the time *Dialogus* was published or if the

---

[17] *Universitatis bononiensis monumenta. Noticia doctorum sive catalogus doctorum qui in collegiis philosophae et medicinae bononiae laureati fuerunt*, p. 35.

[18] Vincenzo Busacchi, 'Gli studenti spagnoli di medicina e di arte in Bologna dal 1504 al 1575', *Bulletin Hispanique* (1956), p. 196.

[19] Baldasar Heseler, *Andreas Vesalius' First Public Anatomy at Bologna, 1540*, translation and commentary by Ruben Erikson (Uppsala: Ruben Erikson, Almqvist & Wiksells boktrykkeri, 1959); and O'Malley, 'Pedro Jimeno, Valencian Anatomist of the Mid-sixteenth Century', p. 468.

dedication to the Duchess was rather an attempt to establish such a connection to a potential benefactor.

The *Dialogus* consists of a series of questions and answers on human anatomy played out between *Caspar civis* and *Andreas medicus*, with the name of the latter an obvious allusion to Jimeno's acclaimed mentor. The author's relationship with Vesalius was emphasised on several levels in this work and was not limited to the use of his name in the dialogue and praise for his teachings in Padua. The structure of the book, which comprised chapters on systematic anatomy, was borrowed from the alleged master, as well as the long verbatim passages of *Fabrica*, which covered several pages. Finally, the detailed description of new anatomical findings and corrections was fully in line with Vesalian scepticism and recommended direct manual approach to anatomical teaching and research. Still, very few certainties surround the known life and career of of Pedro Jimeno. After *Dialogus,* his first and only publication, his name reappears only a few times, as in Fransisco Valles' textbook *De locis patientibus* (1559), which referred to Jimeno's arrival at the University of Alcalá, shortly before he died: 'Jimeno, a very good friend of mine, who had come from Valencia to Alcalá de Henares to teach the art of dissection, in which he was very skilled, and who died not long after having begun his work here.'[20] Another prominent anatomist from Alcalá, Fransisco Diaz, claimed to have studied under Jimeno in Valencia around the same period, and lauded *Ximeno, doctissimo valenciano* as: 'The first who elegantly and skillfully began to dissect and carry out anatomy in the city of Valencia, where the study of medicine and anatomy is presently flourishing.'[21]

The reverence with which Jimeno regarded Vesalius' teachings and texts, and the intellectual debt he confessed, was very explicit in his writings, in stark contrast to the reservations and ambivalence found in his Valencian colleagues' earlier portrayals of Vesalius. In fact, criticism of Vesalius is completely absent from the publications of both Pedro Jimeno and Luis Collado, who succeeded Jimeno as Professor of Anatomy in 1549. The respective introductions of their publications from 1549 and 1555 in fact represented some of the most unconditional vindications of Vesalius amongst contemporary writings. Both publications furthermore followed the Vesalian programme of corrections and additions of anatomical knowledge through empirical manual dissections of the human body. This approach supposedly led to Jimeno's discovery of the third

---

[20]   'Ximenii amicissimi mei qui nuper e Valentia Complutum, ut dissecandi artem cuius erat peritissimus profiteretur, venerat, neque multo post hic agens vita defunctus est.' Fransisco Valles, *Claudii. Gal. Pergameni de Locis Patientibus Libri Sex, cum Scholiis* (Lyon: Luis Gutierrez, 1559), p. 5.

[21]   Fransisco Diaz, *Tratado nuevemente impresso, de todas las enfermedades*, f. 19: 'Ximeno, doctissimo valenciano, y el primero que con elegancia, y gran destreza comenzó a poner la execución de cortar y a hacer anatomía en la ciudad de Valencia, donde tanto resplandece la medicina y la anatomía al presente...y no tengo yo poca jactancia de haber gastado en esta ciudad algún tiempo, y tener por maestro al peritissimo doctor Collado y al doctor Ximeno.'

auricular ossicle, the stapes, a small but significant anatomical finding which appeared in print for the first time in *Dialogus de re medica* and was given its present name in Collado's *Cl. Galeni Pergameni liber de ossibus*.[22] The two treatises took the shape of detailed osteological textbooks rather than all-encompassing mappings of human anatomy and thus exhibited the advancements within anatomy by the mid-1500s, when this novel science was already moving towards specialisation. The idea of dividing anatomy into specialised sub-branches continued with the emergence of treatises on urology and anatomical pathology only a few decades later, authored by two other anatomists active in Valencia, Fransisco Diaz and Juan Calvo. However, the fact that the mapping of the exact number of bones in the human skeleton was still incomplete and continued to be be supplemented and corrected until the mid-sixteenth century is an indication that contemporary anatomical practice and knowledge was still in its infancy.

To overcome such uncertainties, Jimeno's text on osteology emphasised the use of an assembled and mounted skeleton as an indispensable part of anatomical teaching, and copied large sections of Vesalius' original text on the proper preparation of a skeleton for this purpose. The use of human frames, rather than scattered bones, had been strongly advocated by Vesalius for the teaching of anatomy, and in some cases he had even assembled them himself. This was the case after the dissection of a certain Jakob Karrer von Gebweller, who had been condemned to death as a criminal in Basel in 1543 and whose bones were later assembled by Vesalius and donated to the local University, where they are still on display in the anatomical museum.[23]

The use of this pedagogical tool in Valencia is one of the earliest known cases in Europe and certainly is the first known occurrence of this practice in an Iberian context. Similarly, the first and most detailed part of Jimeno's anatomy book was almost entirely based on Vesalius' practices as described in Book I of *Fabrica*, which was dedicated to the description of the human skeleton. The recommendations were radically different from the tradition passed down since Mondino di Luzzi, where studies of human anatomy were carried out on regional sections of the body rather than through systematic dissections of anatomical categories such as bones, muscles, veins, etc. Only a decade before Jimeno's *Dialogus*, Andres Laguna's anatomical treatise *Anatomica methodus* (Rome, 1535) and Miguel Juan

---

[22] While this ossicle was first described and published in Jimeno's *Dialogus*, it was discovered simultaneously, or possibly even earlier, by the Neopolitan anatomist and Sicilian protomedico Felipe Ingrassia. His account, however, did not appear in print until much later, in the posthumous publication *In Galeni librum de ossibus doctissima et expectatissima commentaria* (Palermo, 1603). Ingrassia's study is described in detail in Charles D. O'Malley, 'Los saberes morfológicos en el Renacimiento. La anatomía' in Pedro Lain Entralgo (ed.), *Historia universal de la medicina* (Barcelona: Salvat, 1973), p. 67.

[23] Charles D. O'Malley, *Andreas Vesalius of Brussels, 1514–1564* (Berkeley: University of California Press, 1964), p. 144.

Pascual's *Libro y practica...* (Valencia, 1537) still focused on the descriptions of the 'softer' organs of human anatomy and left little or no room for the osteology and myology which formed the focal points of Jimeno's later anatomical studies. Organised as a dialogue between the two 'interlocutores', *Caspar civis* and *Andreas medicus*, Jimeno's text openly admitted to only a superficial focus on the inner organs, which had attracted most attention in previous anatomy books, due to their recognised contribution to the production of blood and humours, which were essential functions and components in Galenic physiology. Jimeno's book had a different purpose and, as a direct consequence, the answer by the medical mentor Andreas on the digestion and nutritive system was interrupted when Caspar invited his counterpart for dinner.[24] According to Jimeno's *Dialogus*, the form and function of the inner organs were supposedly better known and less difficult to explain than the secrets of bones and muscles, and consequently deserved only minor attention.[25] Jimeno concentrated his criticism of ancient osteology on newly found anatomical features, such as the third auricular bone, 'which would not even have been imagined by Galen during his sleep', and recent corrections to the descriptions of the abdominal muscles, which had been similarly misrepresented in ancient explanations of human anatomy.[26] As Jimeno himself emphasised repeatedly, his account was an appreciation of the empiricist approach to anatomy as spearheaded by Vesalius, prioritising direct observation over written authority.

Despite the lack of documentary evidence to support Jimeno's claims to direct training under Vesalius in Padua, he often referred to his unique educational background. Jimeno endeavoured to correct a series of contemporary misunderstandings of the structure and morphology of the human body, and concentrated chiefly on features of osteology and myology, similar to Vesalius' own anatomical research. Jimeno began his treatise on the osteology of the skull by insisting upon the undivided nature of the frontal bone. He rejected the idea of a bipartite frontal bone 'against popular and vulgar superstition' (and thereby revealed his own unawareness of the relatively common frontal or metopic suture), though he moderated his dismissal with the claim that the anatomical trait was at least unknown in female skulls.[27] In his attempts to number the bones in the human cranium, Jimeno described his own discovery of the third auricular bone, noteworthy as the first new anatomical finding explained in any detail in contemporary Spanish anatomy.

---

[24] Jimeno, *Dialogus de re medica*, f. 64r.

[25] Ibid., f. 68r.

[26] Ibid., f. 53v.

[27] Ibid., f. 19. While this assertion is most often correct, it is also an indication of Jimeno's unfamiliarity with the division of the frontal bone in two halves, as found in infant skulls, and occasionally seen in the craniums of adults where the two parts have not fused properly together.

This important contribution stands as a rather isolated case, but nonetheless provides important evidence that anatomy in this regional context had moved from being a didactic tool to a scientific method for accumulating new data for the chart of human anatomy. The importance of this anatomical find was further indicated by the fact that two other Spanish anatomists, Luis Collado and Juan Valverde, later claimed the discovery as their own in 1555 and 1556. Their boastful claims contrasted starkly with Jimeno's own modest description of the find as a mere supplement to Vesalius' far superior discoveries of the two other auricular bones, the hammer (malleus) and the anvil (incus): 'Within the cavity devoted to the organ of hearing we have recently found three ossicles, two after thorough observation discovered by my master Andreas Vesalius, imperial physician, and an extraordinary man; and the third – of little consequence – by myself.'[28] As previously mentioned, Vesalius was made aware of this finding – overlooked by himself in both the original 1543 and revised 1555 versions of *Fabrica* – during his time at the Madrid court of Philip II. His mention of this third auricular ossicle was recorded for the first time in his 1561 *Examen* of Falloppio's anatomy published in 1564, in which Vesalius also complained about his inability to acquire and study a human skull. Vesalius made no claims to knowledge of Jimeno's *Dialogus*, where this ossicle had first been described, but was aware of Valverde's *Historia de la composicion del cuerpo humano*, where the author declared himself to be its discoverer. Valverde's textbook depicted the ossicle visually for the first time next to the two known ossicles on a pedestal, which was flanked by a contemplating skeleton copied from *Fabrica*. In the explanatory text, Valverde deviously referred to 'the third ossicle, of which no one before me has made any mention.'[29] It is a remarkable testament to the limited circulation and audience of Jimeno's *Dialogus* that no contemporary sources credited Jimeno as the true discoverer of this osteological novelty. Juan Valverde and Realdo Colombo each claimed ownership of the discovery in 1556 and 1559, respectively, and Jimeno's successor at the chair of anatomy in Valencia, Luis Collado, did the same in his *Liber de ossibus* (1555). While Collado disregarded Jimeno's previous contributions to the new discovery, he instead admitted to have been assisted by his apprentice Cosme de Medina, who later became the first anatomical professor at the University of Salamanca: 'I, together with Cosme Medina, now a distinguished professor at the renowned University of Salamanca and before that my skilled disciple, found the ossicle. Because of its similarity with an instrument of horsemanship, which supports the feet, we named it the stapes.'[30]

---

28 Ibid., f. 19–20.

29 Juan Valverde, *Historia de la composicion de cuerpo humano*, Declar. delas fig. Del Lib. I.

30 'Ego autem una cum Cosmo Medina in inclyta academia Salmanticensi nunc publico Anatomes professore longe doctissimo discipulo meo mihi charissimo, aliud os repperi, qui, quod

The discovery was attributed to Collado half a century later by the Valencian chronicler Caspar Escolano in his *Década primera de la historia de la insigne y coronada Ciudad y Reino de Valencia* (1610) and clearly reflects Jimeno's waning influence following his 1549 publication. After his brief appointment in Alcalá, as related by Fransisco Valles in 1559, Jimeno's name disappears almost completely from the documentary sources. Escolano's account paid a brief tribute to this skilful but little-known anatomist.[31] The chronicler acknowledged Jimeno as the author of a textbook on anatomy, but was seemingly unaware of his discovery of the stapes, instead crediting Luis Collado with the find. Despite the lack of contemporary and later recognition, Jimeno's *Dialogus* undeniably included one of the only anatomical discoveries made in Iberia during the sixteenth century. It was quickly incorporated into the ever-expanding and increasingly detailed European atlas of the human body and was given its present classification in Collado's *Liber de ossibus* from 1555. While Jimeno may have been a pioneer in his search for unknown features of the human skeleton, he was far from an approximate understanding of the pulmonary system which would soon be described by Spanish anatomists practising and publishing outside Spain, and his cordial anatomy was completely in line with the Galenic conception of invisible pores between the two heart ventricles.

Luis Collado: Consolidation of Vesalian Anatomy at the *Estudi General*

Whereas Pedro Jimeno's lack of recognition has left little behind in the historical record, his successor Luis Collado's life and work is very well documented. Collado's career continued through the latter part of the sixteenth century and his dedication was eventually rewarded with the position of *Protomédico de Valencia* during the last decades of the century. After some years of rotation between the three medical chairs of *anatomía y simples*, *practica* and *principiis*, he held the chair of *materia medica* or *simples*. This chair was separated from anatomy following a university reform of 1561, which appointed the anatomical professorship to a certain Luis Arcis y de Soler. Almost nothing is known of this first Valencian professorship entirely dedicated to anatomy, but a short reference in the university's reformed statutes stated that both systematic and

---

simile esset equitandi instrumento, quo pedes firmantur, stapedae nomen imposui.' Luis Collado, *Cl. Galeni Pergameni liber de ossibus ad tyrones, interprete Ferdinando Balamio, Enarratore Ludovico Collado medico* (Valencia:, Typis Ioannis Mey Flandri, 1555), f. 30.

[31] 'Consumado en todas buenas letras y lenguas y solo nos dejo por uñas de león, para conocerle, sus 'Dialogus de anatomía.' Caspar Escolano, *Década primera de la historia de la insigne y coronada Ciudad y Reino de Valencia* (Valencia: Pedro Patricio May, 1610), vol. I, p. 102.

sectional anatomy were to be taught hereafter.[32] This division of teaching into *anatomía particular* and *anatomía universal* was explained in far more detail in the contemporary 1561 statutes of Salamanca, which specified that the former category should include dissections of the head, eyes, kidney, heart, arms and legs, and that dissections of bones, muscles, veins and nerves would form the latter category.[33] Solar was replaced in 1563 by the equally unknown Luis Almenara, who was succeeded the year after by Miguel Juan Pascual, whose *Libro o practica de cirugía* (1537) was mentioned earlier. This ageing professor retired after two years and was succeeded by Vicente García Salat, who held the chair of anatomy for almost half a century until his death in 1612. In 1574, a professorship of *Practica particular* was created for Collado by the University and City Council of Valencia. The new university chair was established with high expectations of Collado's advanced skills in many medical fields.[34] This appointment coincided with Collado's promotion to the position of *Protomédico de la ciudad y reino de Valencia* in 1576, which he held until his death in 1589.[35] His osteological textbook from 1555 seems to have been read and appreciated throughout the century, and Collado's reputation as an anatomical pioneer was later acknowledged in Escolano's Valencian chronicle: 'He was the first among the Spaniards who taught and shared the roots and secrets of anatomy, and who with his own hands cut through the most invisible small parts of the human body. In the study of the anatomy of the ear, he was the first to discover an ossicle inside the ear called the stapes, a thing never sighted by the ancients.'[36]

Collado's *Cl. Galeni Pergameni liber de ossibus* (Valencia, 1555) was dedicated to the municipal magistrate of Valencia, Bernat Lluis Vidal, whose father,

---

[32] 'Amb pacte que haja de fer lo exercici de anatomies universales e particulars', cited in Jose Maria López Piñero, *Antologia de la Escuela Valenciana del siglo XVI* (Valencia: Cátedra e Instituto de Historia de la Medicina, 1962), p. 62.

[33] Enrique Esparabé Arteaga, *Historia de la Universidad de Salamanca* (Salamanca: Francisco Nuñez, 1917), p. 260–61.

[34] 'Que los estudiantes que han seguido los cursos de medicina pueden perfeccionarse y conocer el modo cómo han de curar las enfermedades y qué clase de medicina han de applicar an cada caso.' Cited in Luis Garcia Ballester, *Las obras medicas de Luis Collado. Nota a proposito de un manuscrito del British Museum (MS Sloane, 2489)* (Madrid: Notas y Ensayos, Gráficas Orbe, 1971), p. 262.

[35] Ibid., p. 263. A reference to Collado's death and Philip II's nomination of a new 'Protomedico de la ciudad y reino de Valencia' is dated 21 October 1589: 'Cumber obitum dilecti nostri ludivici Collado medicinae doctoris vaccum in posse use regice curice officium Protomedici.'

[36] 'Fue el primero que entre los españoles aprendió y ahondó tan de raiz y delgadamente los secretos de la Anatomía que cortaba por sus manos las partecillas más invisibles del cuerpo humano: y fue el primero que en la anatomía de la oreja descubrió un huesecillo que es en el órgano de oír, llamado stapeda: pieza nunca atinada por los antiguos.' Escolano, *Década primera de la historia de la insigne y coronada Ciudad y Reino de Valencia*, vol. I, col. 1062.

Honorato Benito Vidal, was credited in one of Collado's later publications, *Ex Hippocratis et Galeni monumentis isagoge* (Valencia, 1561), which referred to his prior improvement of the standard of anatomical teaching at the *Estudi General*.[37] Collado's book was a systematic defence of Vesalius' writings on osteology in the wake of Jacobus (Dubois) Sylvius' written attack, entitled *Vaesani cuiusdam calumniarum in Hippocratis Galenique rem anatomicam depulsio* (Paris, 1551). In this work Sylvius refuted Vesalius' claim that Galen had never carried out any human dissections himself and that he was unaware of a number of obvious differences of human osteology compared to that of dogs, apes and pigs. According to the Parisian professor, any differences between recently dissected bodies and Galen's ancient descriptions were due to changes in human morphology which had taken place since classical antiquity and which affected the shape and number of certain bones, such as the sternum, and the phalanges of the hands and feet. Sylvius ended his scathing attack on Vesalius' revision of Galenic authority with a plea to Charles V to obstruct this troublemaker and curb the spread of his poisoning influence on established medical knowledge throughout the European continent.[38] Collado's *Liber de ossibus* delineated a detailed and outspoken condemnation of Sylvius' work 'in which he refers to Vesalius as arrogant, shameless, ignorant, ungodly, insolent, an obstructor of truth and nature, the most malcontent slanderer and villain; it is then unavoidable that in ardently defending my master Vesalius, I might be seen as too hard on Sylvius'.[39] Collado initially acknowledged that his blunt criticism of this renowned authority could seem unjust. He emphasised his own former respect and admiration for the old Parisian professor, whose writings he claimed to have introduced single-handedly into the medical faculty of the *Estudi General*: 'I was the first and only one who brought his books for my public lectures at the University of Valencia'.[40] Still, Collado made no attempt to conceal a strong disagreement with Sylvius and his erroneous understanding of human osteology, even if the discord was downplayed in the introduction to the account.

---

[37]　　The Vidals were among the most important families of sixteenth-century Valencia and were possible benefactors of Collado, and perhaps of previous Valencian anatomists as well, as indicated by the reference to Honorato Benito Vidal and his involvement in the organisation of the professorship of anatomy in the early 1530s. López Piñero, *Antologia de la Escuela Valenciana del siglo XVI*, p. 49.

[38]　　Iacobus Dubois Sylvius, *Vaesani cuiusdam calumniarum in Hippocratis Galenique rem anatomicam depulsio* (Paris: viduam Jacopo Gazelli, 1551), p. 116.

[39]　　'In quibus Andream Vesalium, arrogantem, imprudentem, ignorantem, impium, insolentem, assellum, veritati naturaeque obstrepentem, maledicentissimum, calumniatorem, momum, denique; uaesanum appellat, non potui me continere, quin pro praeceptore Vesalio excandescens aliquando in Iacobum Sylvium acerbior fuerim.' Ibid., f. 76v.

[40]　　'Sum enim Iacobi Sylvii tam admirator, quam qui maxime: quandoquidem eius libros unus ego in gymnasio Valentino publice interpretandos suscepi.' Ibid., f. 76v.

The moderate tone of the introduction progressed to vehement disapproval as the account numbered Sylvius' misunderstandings of obvious dissimilarities between human and animal osteology. As discussed previously, the Parisian professor had concluded that modern discoveries of hitherto unknown features in the human skeleton were the result of morphological changes occurring over the course of many centuries rather than any incorrect or insufficient observations made by Galen. Collado sought to disprove Sylvius' allegations by systematically showing how Galen had in several cases referred to the bones of animals rather than humans, mistakes already reported by Vesalius. Collado's book outlined a number of cases where human skeletal bones differed markedly from canine and primate osteology, not only in size and surface morphology, but also in solidity and inner structure, incontrovertible facts overlooked in the Galenic osteology steadfastly defended by Sylvius. The account furthermore scorned Sylvius' portrayal of his former student as a madman through use of the pun 'Vesanus/Vesalius'. Collado partly explained the pun by attributing it to the advanced age of the Parisian professor, who was 73 years old at the time of his anti-Vesalian publication in 1551, and whose death in 1555 coincided with Collado's *Liber de ossibus*.[41] Sylvius' reactionary insistence on Galenic osteology was refuted through a series of examples of the incompatibility between human and animal osteology, which Collado loyally based on the teachings of Vesalius.

Collado's description of different categories, sizes and shapes of skeletal parts presented the gracile and spongy characteristics of human bones, as compared to the more compact and robust features found in the skeletal structure of dogs, pigs, and apes. He exemplified previous misconceptions by comparing the vertebrae of the vertical human spinal column to the canine horizontal spine, which he claimed to be constructed and articulated differently: 'That would be true in the bones of dogs, but not in the human skeleton.'[42] Both dogs and monkeys were centrally represented in the title page and the osteological Book I of Vesalius' *Fabrica*, which depicted a canine skull next to its human counterpart to show their obvious differences in size, shape and structure. Vesalius had furthermore introduced a correction of the number of bones in the human sternum in Book I of the first edition of *Fabrica* (1543), and this literal attack on Galenic authority was supported wholeheartedly in Collado's *Liber de ossibus*. The reduction of the sternum from seven bones, as observed in dissected barbaby apes, to only three, as seen in the human skeleton, represented a substantial rejection of redundant anatomical evidence gathered from animal dissections during late classical antiquity. Leonardo da Vinci's unpublished anatomical atlas from the late fifteenth and early sixteenth centuries, Berengario da Carpi's *Isagogae breves* from 1522 and Bernardino Montaña's *Libro de la*

---

[41] José Barón Fernández, *Andres Vesalio, su vida y su obra* (Madrid: CSCI, 1970), p. 27.

[42] 'Hec in canibus vera sunt, non autem in hominum sceletis.' Ibid., f. 50v.

Anothomía from 1551 all erred in their descriptions of this tripartite bone and sustained the understanding of a septipartite sternum, which was later corrected by Vesalius and confirmed by Collado. These corrections were acknowledged by Fray Luis de Granada, whose theological/philosophical treatise *Introducción del símbolo de la Fe* (1583) included a chapter on human anatomy. The ageing theologian, an exile in Portugal, whose previous books were included in the Inquisitorial Index from 1559, was well aware of the differences between human and animal morphology observed by the anatomists of his own era and praised the recent corrections carried out in Spain and beyond by 'the new anatomists, who through experience have proven that our human body in some cases differs from those of the animals.'[43]

Among Collado's many osteological corrections, he disproved another erroneous description and rudiment of ancient anatomy, namely the idea that the heart was supported by an inner skeletal structure. In Collado's textbook, the imaginative yet non-existent 'ossicle of the heart' described in Book Seven of Galen's *De anatomicis administrationibus* was dealt with in the brief last chapter of *Liber de ossibus*, entitled *De cordis ossiculo*. This illusory anatomical sighting (occasionally found in the ox and other large mammals, but absent in humans) had appeared in numerous previous textbooks and was even described in Montaña's *Libro de la anothomia del hombre* as the 'huesso del corazon' only four years before Collado's own publication.[44] Collado's negation of the existence of this ossicle after repeated dissections of human hearts represented yet another case of the anatomical empiricism and scepticism which developed under Jimeno and Collado at the Valencian *Estudi General.*[45]

While earlier anatomical dissections at the university served mainly as teaching tools and tangible visual additions to existing written knowledge, the first years of post-Vesalian anatomy carried out in Valencia were far more significant. Jimeno and Collado disregarded numerous existing theories of human anatomy with their own hands-on investigation. Their work strictly

---

[43]   'Se debe notar que los antiguos médicos tenían por cosa de gran horror hacer esta experiencia en los cuerpos humanos, y por esto lo hacían en los animales que se hallaban más semejantes a ellos. Y para que se abaje la soberbia y vanidad de los gentiles hombres y mujeres, y vean de qué se vana glorian, sepan que los cuerpos que los antiguos hallaron más semejantes a los nuestros, aunque sea vergüenza decirlo, fueron los de las monas y puercos. Y así dice Galeno, que más divina y largamente trata esta matteria, se siguió en todo lo que concebió por la fabrica de los cuerpos de las monas. Y por esto es ahora corregido por los nuevos anatomistas, los quales hallan por experiencia que en algunas cosas se diferencian nuestros cuerpos de los de estos animales.' Fray Luis de Granada, *Introducción del símbolo de la Fe* (Salamanca, 1583) (Madrid: Fundación Universitaria Española, 1996), Chapter XXIV.

[44]   Bernardino Montaña, *Libro de la anothomia del Hombre*, f. 47v.

[45]   Collado, *Cl. Galeni Pergameni liber de ossibus ad tyrones, interprete Ferdinando Balamio, Enarratore Ludovico Collado medico*, f. 74v and f. 75r.

adhered to Vesalian practice and both offered unconditional praise for the renowned anatomist, whose reforms had been consolidated in Valencia well before his 1559 arrival in Spain. In recent studies on the history of science and medicine, the conceptions of Vesalius as a 'precursor' to scientific/medical methods and reforms have been repeatedly deconstructed. Yet the veneration of Vesalius in both Jimeno and Collado's writing reveals this comprehension to have been firmly rooted within the Valencian *Estudi General* as early as the mid-sixteenth century. Collado's most explicit appraisal of his mentor was found in a last addition to his textbook, included at the behest of his students. It shows the extent of the respect for Vesalius only a decade after the publication of the first edition of *Fabrica* and in the same year as its revised 1555 impression: 'Everyone knows that Andreas Vesalius caused admiration with the publication of his book on the structure of the human body. Who would not admire a youth of twenty-eight years, who in the tremendous obscurity of anatomical matters possessed such a rare skill in the dissection of bodies and so profound a knowledge of anatomy?'[46]

Collado's 1555 publication was followed the same year by another vindication of Vesalian osteology, published in Venice by the physician Renatus Henerus of Lindau and entitled *Aduersus Iacobi Sylvii depulsionum anatomicarum calumnias, pro Andrea Vesalio apologia: in qua praecipue totius negotij anatomici pene controuersiae breuiter explicantur Renato Henero Lindoense medico authore. Additus est & ipse Iacobi Siluij depulsionum libellus.* A contemporary defence of Sylvius was published by one of his followers, Francis Puteus, and was entitled *Apologia in anatome pro Galeno contra Andream Vessalium* (1562). The contested Parisian professor and former teacher of Henerus and Vesalius died in January 1555 and was therefore unable to follow the continuing quarrel among his adherents and opponents. Additional defences of Vesalian osteology were written by the Florentine surgeon Giudo Guidi and the Bolognese anatomist Giulio Cesare Arazzio, as well as Vesalius' successor in Padua, Gabrielle Fallopio, who even corrected a number of Vesalius' own osteological errors in his reduction of the bones of the sacrum from six to five. Henerus' *Apologia* was the only defence of Vesalius referred to by Charles D. O'Malley in his *Andreas Vesalius of Brussels* (1964). The author was seemingly unaware of Collado's contemporary publication and strong pro-Vesalian stance during the heated controversy. Collado was also completely absent from sixteenth-century writings published outside Spain and was not referred to until the eighteenth century, when he received a brief mention in Hermann Boorhaave's compilation of Vesalius' works, *Opera anatomica*

---

[46] 'Omnibus notum est, Andream Vesalium editis libris de fabrica corporis humani cunctis admirationi fuisse. Quis enim obsecro, non admiraretur, iuvenem octo et viginti annos natum, in tanta rerum anatomicarum caligine, tantam in dissecandis corporibus industriam, et negotii anatomici cognitionem sibi comparasse.'Ibid., f. 76r.

et chirurgica Andreæ Vesalii (Leiden, 1725). In contrast, Spanish support for Collado was extensive in sixteenth-century writings and also appeared much later in Nicolás Antonio's literary history of early modern Spain, *Biblioteca hispana nova...* (1672), which credited *medicus doctor Valentinae academiae, corporum sectionis anatomicaquae disciplinae omnis peritissimus.*[47]

Collado's contributions to the history of Renaissance anatomy were limited to an Iberian context and did not have much impact outside Spain, although López Piñero claimed differently in his article 'La dissección y el saber anatomico en la España de la primera mitad del siglo XVI'.[48] Collado produced only one publication on anatomy, despite continuing as rotating professor of the chair of *anatomia y materia médica*, together with that of *practica* and *principiis* until his appointments to the professorship of *practica particular* and as *Protomédico de la ciudad y reino de Valencia* in the 1570s. This last promotion to *Protomédico* elevated Collado to the prominent position of Philip II's Valencian overseer and representative in medical matters, and he retained this responsibility until his death in 1589. In a manuscript by his hand written the year before, Collado still showed an interest in anatomy and referred to his successor Vicente García Salat and his activities at the *caseta de notomies* of Valencia's General Hospital, where the dissections of the *Estudi General* were carried out and where Salat performed sectional anatomical demonstrations of the vena cava and the liver in February 1588.[49]

Among Collado's many apprentices, Cosme de Medina was the only one referred to in *Liber de ossibus* and was the first to introduce anatomical studies at the University of Salamanca, where he held the professorship of anatomy between 1552 and 1561. Parallel to Collado's own ascent to a chief medical position as *Protomédico* in his native Valencia, Medina eventually rose to the first chair of medicine at the renowned University of Salamanca. Vicente García Salat,

---

[47]   Cited in Luis Alberti Lopez, *La anatomia y los anatomistas del renacimiento* (Madrid: CSIC, 1948), p. 146.

[48]   'Habitualmente se cita la Apologia (1555) del medico de Lindau Renatus Henerus como la primera defensa abierta de Vesalio frente a los ataques de Silvio. El volumen de Collado, que apereció el mismo año, hay que alinearlo junto al de Henerus desde esto punto de vista. Sin embargo, su importancia es mucho mayor, en cuanto reflejo de una de las primeras escuelas europeas que cultivó la enseñanza y la indagación anatómicas de acuerdo con la línea encabezado por Vesalio.' López Piñero, 'El saber anatomico y la dissección de cadaveres humanos en la España en la primera mitad del siglo XVI', p. 82.

[49]   'Unde male arabici et Avicenna dicebant exprese deinde ex brachio contrario postea ex eadem nam illi sint intendebant revulsiones nostri simul evacuationes exercemus sed aliqua nostra variante nam vissus nobis fuit in hospicio pauperum Valentiae 3 die mensibus februari anni 1588 die martibus, in loco ad anatomes exercendam destinato directore doctore Vincentio García Salat venas cavas in vassi condii dividi in duos insignis Ramos quorum alter per dextras Jecoris partes alter per sinistras ascendebat.' Cited in Luis Garcia Ballester, 'Las obras médicas de Luis Collado', *Asclepio*, 23 (1971), p. 268.

another of Collado's students, was appointed to the Valencian professorship of anatomy in 1565 and held this post for a remarkable period spanning almost half a century until his death in 1612. Salat's anatomical treatise *De anathomia*, written around 1590, has since been lost, and only a posthumous publication on the diagnosis and cure of fever, *Utillisima disputatio de dignotione et curatione febrium* (Valencia, 1682), survives. Fransisco Diaz credited his mentor in a description of his apprenticeship in Valencia under Collado and Jimeno, who were mentioned together only in his *Tratado* (1588): 'I take no little pride in having spent some time in that city, and having had the erudite doctor Collado and Doctor Jimeno as my masters.'[50]

The absence of any appreciation from Collado himself for his predecessor and fellow Vesalian disciple, and his dubious claim of the discovery of the third auricular ossicle as his own, indicates a rivalry and enmity between the two first Vesalian anatomists of the *Estudi General*. Both brought their practice and knowledge in this field from Valencia into Castile, with Jimeno's own later move to Alcalá de Henares and the appointment of Collado's student Cosme de Medina to the first anatomical professorship in Salamanca. While no mutual recognition appears in the writings of Jimeno and Collado, they were united in enthusiastic accounts of their self-professed apprenticeships under Andreas Vesalius. The same inclination towards outspoken reverence for Vesalius, an apparent ignorance of Jimeno's contributions and the repeated support of Collado persisted in the only surviving publication on anatomy from late sixteenth-century Valencia, Juan Calvo's *Tratado primero de la Anotomia* (included in his *Chirurgia universal y particular del cuerpo humano* (Valencia, 1580). This text did not mention Jimeno, but was generous in its praise for Calvo's former teacher Luis Collado, who was lauded for his introduction of Vesalian anatomy to Castile and, once again, was credited with the discovery and classification of the third auricular ossicle.

Juan Calvo, *Tratado primero de la Anotomia verdadera del cuerpo humano*. Synthesis of Galenic and Vesalian Anatomy

Juan Calvo's *Tratado primero de la Anotomia verdadera del cuerpo humano* was brushed aside in López Piñero's 1979 publication *Ciencia y tecnica* and was only referred to in passing.[51] Calvo's publication is nonetheless a rare source of the

---

50 'No tengo yo poca jactancia de haber gastado en esta ciudad algún tiempo, y tener por maestro al peritissimo doctor Collado y al doctor Ximeno.' Fransisco Diaz, *Tratado nuevemente impresso, de todas las enfermedades de los Riñones, Vexiga, y Carnosidades de la verga, y Urina, diuidido en tres libros*, 1588, f. 19r and 19v.

51 'Se trata de una exposición de cierta amplitud – más de ochenta páginas – pero frio y escolar. Calvo cita a varias veces a Collado, de quien era discípulo, pero sigue una orientación

history of Iberian anatomy studies at the turn of the sixteenth century, during a period when similar publications are lacking from most other parts of the Spanish kingdoms. The introduction of anatomy to the leading universities of Aragon and Castile in the 1550s and 1560s had fallen into discontinuity and resignation by the time of Calvo's 1580 publication. In the same period, the chair of anatomy at Alcalá de Henares suffered almost a decade of vacancy between 1574 and 1583, and the revised university statutes of Salamanca from 1594 decreed that only Galen's *De usu Partium* would be tolerated in future teaching of anatomy after it had been sided with Vesalius' *Fabrica* in the previous 1561 statutes. Calvo's publication appeared at a time when several of the anatomical reforms launched in Valencia in the mid-sixteenth century were falling into decline at some of Spain's leading medical faculties.

Yet there was still hope for the continuing influence of the anatomical reforms and corrections of the previous decades in Calvo's *Tratado primero de la Anotomia*, which included several references to the accumulation of anatomical insight gathered by contemporary Italian and Spanish anatomists. In the former category, Realdo Colombo and Gabrielle Falloppio were frequently cited, and among the contemporary Spanish authorities referenced were Juan Valverde, Juan Fragoso, Christobal de Vega and Luis Collado. By the time of the publication of *Tratado primero de la Anotomia*, Calvo's former mentor Luis Collado continued to tower over his colleagues as the leading medical authority in Valencia and embodied the social advancement of Iberian anatomists during the mid-sixteenth century. There is no sign that Calvo enjoyed a similarly successful career. While he referred to his many years of surgical practice and teaching in Valencia, he seems never to have been appointed the chair of surgery at the *Estudi General*, even if this much is indicated in his later translation of Guy de Chauliac's *Chirurgia magna*, where Calvo was presented as *Lector en la misma facultad en la ciudad de Valencia*.[52] His aspirations to such a university appointment may have been the objective of the publication of his surgical/anatomical treatise from 1580, which was written as a textbook for students of surgery.

According to the bibliometric studies of Calvo's work made by Jose Luis Fresquet Febrer, Collado was repeatedly praised and only Galen's name appeared more often than Vesalius'.[53] Febrer's work is by far the most systematic study

titubeante y con escasas noticias de la anatomía posvesaliana.' Jose Maria López Piñero, *Ciencia y tecnica en la sociedad Española de los siglos XVI y XVII* (Barcelona: Labor, 1979), p. 330.

[52]   *Cirugia de Guido de Gauliaco con la glosa de Falco, agora nueuamente corregida, y emendada y muy anadida, y declarados los vocablos obscuros que en ella auia, con un tratado de los simples, por Iuan Caluo Doctos en Medicina, Lector en la misma facultad en la ciudad de Valencia* (Valencia: Pedro Patricio, 1596).

[53]   Fresquet Febrer, *La chirurgia universal y particular de Juan Calvo*, p. 191.

'A Galeno corresponde el 52,91 de las referencias, seguido de Vesalio (9,25), Realdo Colombo (8,45), Hipocrates (8,20) y Fragoso (5,29). Encontramos tambien las únicas 17

on this little-known Valencian surgeon from the late sixteenth century and his attempts to fuse the anatomy and physiology of Galen and Vesalius in a balanced synthesis. Calvo's treatise on anatomy was preceded by another longer dissertation on surgery and succeeded by a minor yet significant study of syphilis, in which Calvo acknowledged the recent nature of this disease and even specified its origins in the Americas: 'It is not an old, but a modern disease, which has been known for less than a century in our lands, but much longer in The Indies.'[54]

While Calvo's references to Collado were full of unconditional praise, this was not the case in several critical dealings with other contemporary Spanish surgeons/anatomists, such as Juan Fragoso, who was often the subject of scorn in Calvo's treatises on anatomy, surgery and syphilis. Educated at the University of Alcalá, Fragoso was practising at the Court in Madrid at the time of his publication *Erotemas chirurgicos* (Madrid, 1570). By the time of Calvo's publication, Fragoso was in the process of publishing his *Chirurgía universal* (Madrid, 1581), which may have been seen as a serious contender to Calvo's own surgical treatise. In one of a series of similar instances of medical rivalry between Spanish scholars and universities with regional differences and mutual antipathies, Fragoso's name had only negative associations in Calvo's account. For Calvo, the medical practitioners of Valencia towered far above other contenders in their unmatched standards of teaching and research. Indeed, the allocation of medical professors at the faculty of Valencia certainly outnumbered the other large universities of late sixteenth-century Spain.

Still around 1580, the field of osteology continued to be the centre of attention in studies of anatomy practised in Valencia. As a result, the skulls of newborn babies deformed by hydrocephalia were allegedly opened and drained without any mortal risks for the infant patients, unlike the alternative (and reportedly fatal) practices of Juan Fragoso.[55] Calvo's purported proficiency in the trepanation of infant skulls, and his chapter on osteology, which is by far

---

citas negativas hacia Galeno, y asi en el capítulo primero del Tratado de Anatomía se lee: Galeno en muchas cosas erró y assi en esta material, si se ha de dezir bien, no le podemos en todo seguir, como en los demás tratados y libros. Siguen en importancia las referencias a Vesalio, al que propina adjectives de gran admiración, y a Realdo Colombo, lo que demuestra la influencia que estos ejercieron durante un periodo de formación. Uno de sus maestros fue Luis Collado, al que nombra en los capítulos correspondientes a la osteología. Hemos de destacar finalmente la ausencia casi total de referencias a productores islámicos.'

[54] 'No es antigua sino moderna enfermedad, de menos de cien años a esta parte conocida, al de menos en nuestras partes, porque en las Indias más antigua es.' Calvo, *La chirurgia universal y particular del cuerpo humano*, p. 539.

[55] 'Si Fragoso dice que se mueren muchos de los abiertos, debe ser porque él les cura mal, o porque abre antes de tiempo, o porque debe sacar de una vez toda la serosidad allí contenida. Lo que yo sé dezir es que me he hallado y mandado abrir algunos, y ningunos ha muerto, porque en esta ciudad de Valencia, por la gracia de Dios se cura con gran método y orden.' Ibid., p. 520.

the most detailed part of his anatomical treatise, emphasised a loyalty to the studies initiated by Jimeno and Collado many decades earlier: 'We shall first describe the bones before the other parts, because they are the foundation of all the other parts of our bodies.'[56] Jose Luis Fresquet Febrer's studies supported this assertion and concluded that Calvo's chapter on osteology was the part of his anatomical treatise, which most clearly represented continuity with the Vesalian tradition at the medical faculty of Valencia.[57] Calvo saw himself as the heir of both Vesalius and his Valencian apprentice Luis Collado, whose calculations on the number of bones in the outer extremities (again erroneously decribed by Fragoso) he fully supported in his own treatise.[58] However, Calvo did not stray from traditional Galenic doctrines in his understanding of the function of the heart and the transmission of blood between invisible pores in the septum of the two ventricles. This was somewhat surprising, given his many references to Valverde and Colombo, who had triumphantly disproved this theory three decades earlier with their pioneering descriptions of the pulmonary system. Calvo's Galenic understanding of the anatomy of the heart represented a gradual return to such antiquated doctrines, even among medical scholars acquainted with new anatomical and physiological evidence: 'The heart contains two ventricles, one in the right side and one in the left, and in the middle of these is a septum with many holes, which enables the blood to pass from the right ventricle to the left.'[59]

---

[56] 'Trataremos primero de los huesos, que de las otras partes, porque ellos son fundamento de todas las demás partes que ay en nuestro cuerpo.' Juan Calvo, *Tratado primero de la Anotomia verdadera del cuerpo humano* (Barcelona: J. Cendrat, 1591), p. 498.

[57] 'El análisis del texto de la obra anatómica de Juan Calvo no depara elementos que demuestran, claramente una renovación de los saberes anatómicos del momento. Descubrimos, en cambio, como decíamos, la diferencia existente entre unos capítulos y otros, y con ello nos referimos, claro está, a los dedicados a la ostología, donde se hacen patentes las influencias de la escuela vesaliana. En el resto, es evidente una adhesion incondicional a los conocimientos tradicionales de raigambre galénica.' José Luis Fresquet Febrer, *El tratado de anatomía de Juan Calvo* in *Estudios dedicados a Juan Peset Alexandre* (Valencia: University of Valencia, 1982), p. 24.

[58] 'Acerca del numero de estos huessos ha auido diferentes pareceres: empero el major de todos es el de Vesalio, varon no solo en Anotomia, mas aun en medicina muy singular, al qual sigue Collado en el capitulo que haze destos huessos.' Ibid., p. 509.

[59] 'Hallanse en el corazón dos ventrecillos, el uno en la parte derecha, y el otro en la izquierda, en medio de los quales se halla un septo con muchos horados, para que pudiese pasar la sangre que entra por el derecho ventrecillo al izquierdo ... En el corazón se hallan cuatro vasos, dos en cada ventrecillo: en el derecho se halla la vena cava y la vena arterial y en el izquierdo la arteria magna y la arteria venal ... La vena cava entra ... lleva gran copia de sangre ... sale por la vena arterial a sustentar los pulmones ... Dizese vena porque lleva sangre venal ... Al ventrecillo izquierdo llega la arteria venal por la que entra aire preparado por los pulmones, en el corazón: y de este ayre, y de los vapores de la sangre que de alli se alzan se hacen los espíritus vitales ... El vaso que sale es la arteria

In accordance with Galenic theory, Calvo furthermore saw the liver as the first fully developed organ of the human body.[60] He asserted that the liver was formed by coagulated blood covered by a membrane within the embryo, and later took charge of production and supply of blood to the rest of the body. Galenic physiology viewed this organ as the originator of both the humours and the venal blood, which was created when digested food from the stomach was carried to the liver. The blood was then transferred through the heart and the vena cava, from which it finally reached the brain after having been refined with the 'animal spirit' of the arteries. In this process, blood was continuously expended and produced anew in an infinite system of 'ebb and flow'. This notion met with strong opposition and eventually exploded following William Harvey's theory of the circulation of the blood, which was presented in the following century (and fiercely rejected by Matias García who held the Valencian chair of anatomy from 1663 to 1687). Calvo's treatise on syphilis stood out as the most innovative part of his work, as it downplayed previously accepted ideas that external influences were largely responsible for the development of the new disease. Exterior factors were seen as crucial in many previous treatises on syphilis, which judged ominous preconditions such as bad air and ill-timed planetary positions to be instrumental in its appearance and development. The closeness of the connection between Valencia and Italy is evident here: the term *mal de simiente* was used for syphilis only in Italy and the Crown of Aragon, even though Calvo used its Latin name, *morbo galico*, in his own treatise. As Calvo underlined in Chapter 2, 'En el qual se enseña a ser el morbo gallico enfermedad nueva', the disease had not been known in Europe until the discovery of the Americas.[61] This new menace was given a series of different names, of which the most common in Spain were *bubas* or *mal francés*, which was translated from *morbo gallico* and based on the erroneous idea of its French origin and entrance into the Italian Peninsula with the invading French troops of the late fifteenth and early sixteenth centuries.

Calvo's scepticism towards ancient authority when faced with new evidence was evident in this case, and he disregarded the idea that unstable elements and humours were responsible for its development and diffusion. His descriptions and diagnosis were rather advanced in their understanding of the sexual transmission of the disease and exact account of the pathological changes of both the skin and bone tissue of infected patients.[62] Once again, Valencia was

---

magna, por la qual sale sangre arterial, y espíritu vital, para restaurar el calor natural que cada una de las partes pierda.' Ibid., p. 485.

[60]   'El hígado que es el primero que se engendra embroiologicamente para crear sangre venal, alimento para que se forman las demás.' Ibid., p. 458.

[61]   Ibid., p. 538.

[62]   'Se llaman también "bubas", mal muerto, por la vida miserable que pasan, llenos de úlceras, pustules, dolores óseos, con mal olor, con la nariz roma, principalmente cuando el huesso dellas se consume.' Ibid., p. 539.

placed above other Iberian cities with an alleged and unlikely medical expertise in the treatment of this horrific venereal disease, which remained incurable until the introduction of penicillin in the twentieth century.[63] While his treatise on syphilis included original contributions to the study of the new disease, Calvo acted mainly as a commentator on existing anatomical knowledge rather than revealing new evidence that he had gathered. The most interesting part of his work, and rather unique in an Iberian context, was a long reflection on the many virtues of the anatomical science and its different applications and connotations of a medical, ethical and empirical character.[64]

Restoration of Galenic Anatomy at the *Estudi General*

The progressive development and outspoken criticism of Galenic doctrines initiated in the mid-sixteenth century had fallen into decline by the dawn of the seventeenth century. There is no proof that this setback was caused by a deteriorating economy, as was the case in Alcalá de Henares, where the chair of anatomy remained vacant for long intervals due to the insufficient and steadily declining salary of the anatomists. At the University of Salamanca, complaints from the early seventeenth century reveal that the anatomical theatre suffered neglect during the same period. In Valencia, however, there is evidence that the medical faculty continued to grow throughout the seventeenth and early eighteenth centuries. According to Amparo Filip Orts' estimates in his book *La universidad de Valencia durante el siglo XVII* (Valencia 1991), the medical

---

[63] 'No mata a muchos, principalmente en esta ciudad de Valencia se cura bien: y si despues los que estan curados buelven a tenerla, es porque ellos (como perro al vomito) vuelven a tener excesso con mugeres infectas', Ibid., p. 538.

[64] 'Que es la anotomía?: Es el arte, o ciencia, que enseña la composición del cuerpo humano, y la utilidad de sus partes no solo entre los antiguos más aun entre los doctores modernos, es en mucho tenida y estimada. Este arte o ciencia es la anatomía. Anotomía es una ciencia contemplativa, con el qual perfectamente se conocen todas las partes del cuerpo humano ... digo que la anatomía es ciencia contemplative, en cuanto a la dissección y administración della, porque hablando propiamente, la anotomía se devide en dos partes, historia y contemplación ... y la major y más cierta parte es la conteplación de la partes cuando disecamos y hazemos anotomía con nuestras propios manos o estamos mirando. Para qué el estudio y conocimiento de la anatomía? Cuatro provechos y utilidades se sacan de la anotomía ... la primera es el alabar a Dios por sus criaturas ... la segunda es conocer la parte enferma o temperamento della ... la tercera para pronosticar lo que la enfermedad puede suceder y sobrevenir al enfermo ... la cuatro para curar bien las enfermedades. Para quien? El sujeto del medico y cirujano es el cuerpo humano, luego conviene que sepamos; y aun más al cirujano que al medico porque él muchas veces cura las enfermedades del cuerpo humano, cortando, cauterizando, y si no sabe anotomía será como el carpintero ciego de su natividad, que corta el madero por donde no lo ha de cortar.' Ibid., pp. 454–55.

faculty accounted for only nine per cent of the university's expenses at the time of its inauguration in 1502, but climbed to a remarkable 40 per cent in the next two centuries.[65]

An indication of this privileged position is evidenced by the wages of the professors of anatomy during the early seventeenth century, with relatively higher salaries and fewer required dissections than their colleagues at the universities in Castile. The yearly remuneration of this chair and its obligation to perform eight annual anatomies was settled in a university reform concluded in 1611.[66] The salary of the anatomist was doubled and the number of annual dissections increased only two decades later, following a 1630 request from the newly appointed anatomist Francés Miguel Febrer.[67] After this significant rise, the payment of the professor of anatomy was more than twice the amount of 40 reales remunerated for the chair of surgery, despite the common origin and affiliation of these two university chairs since their creation in 1502. The salary lists of the medical faculty from 1630 to 1700 show the earnings of the *cátedra de anatomía* to be surpassed only by the two chairs of *hierbas y simples* and *método*.[68] Unlike the constitutions of the sixteenth century, the reforms of 1611 left behind a detailed statutory order for the chair of anatomy, with information on the everyday obligations and practice of the anatomists of the *Estudi General* meticulously recorded.[69] Interestingly, the Galenic counter-reformation in Valencia coincided with an increasing number of public anatomies performed throughout the seventeenth and early eighteenth centuries. This figure rose from eight annual dissections in 1611 to 20 in 1630

---

[65] Amparo Filip Orts, *La Universidad de Valencia durante el siglo XVII* (Valencia: University of Valencia, 1991), p. 57.

[66] 'Que la cátedra de anatomía se provea con salario de cinquenta libras solamente, y con obligación de hacer ocho anatomias al año.' Ibid., p. 40.

[67] 'Doctor Francés Miguel Febrer, catedrátich de anatomía de la Universidad obtingue gràcia de sa Magestat que se li pogués aumentar cinquanta liures de salari, més de les cinquanta que tenía la càthedra ab atendendencia de que se li havian añadit dotze anatomies més de les que tenia obligació de fer cascun any.' Ibid., p. 43.

[68] Ibid., p. 39.

[69] Ibid. p 161: 'Lo catedratic de anatomia llegirà de set a huyt, segons és costum: y sera obligat a llegir tota la historia de totes les parts del nostres cos, comencant de la material de ossibus, y après les parts que estan en las tres cavitats, après de musculis, nervis, venis, et arteriis; llegint solament la història que es pròpia de la Anatomia, y señelant los afectes de cada part, dexant la explicació y disputa de aquelles per el catedràtic de pràctica ... y sia obligat a fer moltes anatomies en lo ospital, per a mostrar als estudiants a la vista lo que se ha llegit en lo Studi. Lo temps de les anatomies serà de Tots Sants sins Cuaresma, y seran les dites anatomies la una de la cavitat natural, l'altra de la vital y l'altra y l'altra de animal; tres de venis, tres de nervis, tres de musculis et arteriis, de tal manera que cada uny faca les de les cavitats y en dos anys les de venis, nervis, et arteries. Y si falterà llisosns, les haja de refer de Sant Juan a Sant Luch, y no puga cobrar la apoca de Sant Juan, que no les haja refetes: axí com està ordenat de tots los demes catedràtics de medicina' (Constituciones de 1611, cap VIII, art. 3).

and 25 in 1707, and was matched by a tripling of the salary from 50 to 150 reales for the anatomists of the *Estudi General*. Despite these seemingly agreeable conditions for anatomical teaching, this prerequisite did not secure continuing research activity and a steady flow of publications as had been seen during the previous century. While the medical faculty eventually grew to consume almost half the university budget, it produced only five medical books during the first half of the seventeenth century, and none on anatomy until Matías García's treatise *Disputationes Medicinae Selectae* (Lyon, 1677), which was published almost a century after Calvo's 1580 surgical/anatomical treatise.

Calvo's work endeavoured to synthesise some of the disparities between Galen and the modern sceptics, and represented the last written tribute to Vesalius in the tradition of Jimeno and Collado. García's later treatise, however, lent unconditional support to the infallibility of Galenic anatomy and launched a bitter attack on William Harvey's recent theory of the general circulation of the blood. According to López Piñero, the Valencian anatomist had initially supported this discovery, but changed his mind after reconsidering the serious effects it would have on traditional medical authority: 'His ideological development towards this theme is very revealing. For many years he accepted the doctrine of Harvey as a small rectification, which did not compromise the validity and coherence of Galenism. Later, however, he became aware of the severe damage, this discovery could cause to traditional medical doctrines.'[70] The theory of 'the larger circulation', which was published for the first time in Harvey's *De motu cordis* (Frankfurt, 1628), arguably represented the most significant medical innovation of the century and was hailed by René Descartes as the greatest discovery in the entire history of medicine. García's refusal to accept this theory offered a categorical defence of Galen's physiological notion of the blood, its production in the liver and constant combustion and reproduction.

García's later attempt to refute Harvey's theory was based on his own numerous dissections and thus placed the same emphasis on direct observation over authority, which had been hailed and practised by the sixteenth-century Valencian anatomists. In spite of García's academic and, in hindsight, erroneous preconceived notion, Lain Entralgo insisted that he was 'an intelligent and experienced anatomist, as far from dry-as-dust pedantry as could well be imagined'.[71] His meticulous and practical nature is indicated in surviving accounts

---

[70]   'Su evolución ideológica frente a este tema es muy reveladora. Durante muchos años encontró acceptable la doctrina de Harvey, como una rectificación de detalle que no comprometía la validéz y coherencia del galenismo. Más tarde sin embargo, llegó al convencimiento del immense daño que tal descubrimiento podia significar par alas doctrinas médicas tradicionales.' López Piñero, 'El saber anatómico y la dissección de cadaveres humanos en la España en la primera mitad del siglo XVI', p. 106.

[71]   Pedro Lain Entralgo, *The Spanish Contributions to World Science*, Cahiers d'historie mondiale, vol. 6 (Paris: Unesco, 1961), p. 962.

of García's organisation of the anatomical teaching structure during the latter half of the seventeenth century, which has no parallel in Valencian sources from the previous century.[72] Garcia's attempts to vindicate Galen's authority through anatomical dissections came in the wake of microscopic affirmations of Harvey's theory, published in Marcello Malpighi's treatise *De pulmonibus* (1661). These were based on dissections and microscopic studies of frogs' lungs and provided the evidence of connections between the veins and the arteries, which was necessary to resolve the last uncertainties and missing links in Harvey's theory. Although García remained unwilling to concede defeat, even in the wake of such new evidence, he maintained a leading position as the most prominent Spanish anatomist until his retirement in 1687. In the scientific history of early modern Europe, that year saw the publication of both Newton's *Principia* and a demarcation line in the history of Spanish and Valencian medicine and science.[73] Not only was it the last year of García's Galenic reign over the Valencian chair of anatomy, but it was also the year of the publication of the most important call for scientific renovation by the Spanish *Novator* movement of the late seventeenth century, *Carta filosófica, médico-chímica* (Madrid, 1687), written by the Valencian physician Juan de Cabriada.[74] In the same year, the Valencian anatomist and artist Chrisóstomo Martínez departed for Paris to commence his large *Atlas anatómico* (Paris, 1689). This remarkable publication was illustrated by the author himself and was arguably among the most important anatomy

---

[72] 'Attés que inconcusement se ha observat que los estudiants de primer any de medisina donen cuatre reals, aixi per a els gavinets com per a olors, flors y preparar lo cos per a les notomies y de acó no hi haja constitució y sia rahó que se observe lo acostumat. Per co proveheixen que de huy avant los dits studiants de primer any de Medisina tinguen obligació de donar los dits gavinets, olors y demes coses desusdites, exceptats empero los que seran pobres. Y que la averiguació de que ho són toque al rector de dita Universitat, que ara és y per temps sera, essent bastant y sufficient la una acerció del estudiant al dit rector ab lo qual se li haja de despachar la ceduleta per a que no pague los dits quatre reals per ser pobre.' Orts, *La Universidad de Valencia*, p. 165.

[73] This chronological turning point is further elaborated upon in V. Peset Llorca, 'La Universidad de Valencia y la renovacion cientifica Española', *Aesclepio*, 17 (1965), p. 215: 'Con todo cabe señalar un año representativo: 1687. Es el año en que el publico español puede leer la obra del valenciano Juan de Cabriada, expression clara y definida de la nueva época; es el año en que llega Zapata, uno de los principales protagonistas del cambio, a la corte; es el año en que empiezan la academias en casa del matemático valenciano Baltasar Iñigo; y aun cabría agregar mas.'

[74] This publication can be seen as a precursor to the *polémica de la ciencia española* of the next century, and presents Spain in isolation from the intellectual currents of the rest of the continent: 'Que es lastimosa y aún vergonzosa cosa que, como si fuéramos indios, hayamos de ser los últimos en recibir las noticias y luces públicas que ya están esparcidas por Europa. Y asimismo, que hombres a quienes tocaba saber esto se ofendan con la advertencia y se enconen con el desengaño. ¡Oh, y qué cierto es que el intentar apartar el dictamen de una opinión anticuada es de lo más difícil que se pretende en los hombres!'

books of the late seventeenth century with its groundbreaking microscopic studies and representations of the structure of cancellous bone.

In 1714 a new dissection room was built within the *Hospital General*, where the number of public anatomies reached their peak following the university reforms of 1707. The most popular anatomy book of the eighteenth century, Martin Martinéz *Anatomía completa del hombre* (Madrid, 1728), nevertheless claimed that almost no anatomical activity took place outside Spain's capital and Martínez's own newly built *Amphiteatrum Matritense*. This criticism was refuted by the Valencian physician Vicent Gilabert the following year in his complimentary review of contemporary Spanish medicine, *Escrutinio physico-medico-anatomico* (Madrid, 1729), which opened with an appraisal of the Valencian chair of anatomy.[75]

The Valencian university chair of anatomy remained in existence and as a result embodied a unique case of continuity in early modern Spain, where similar chairs were either established much later or eventually abolished. The *Estudi General* of Valencia had retained its status as the Iberian centre of anatomical education since the early 1500s and continued to house a remarkably large medical faculty during the following century. Yet there was a clear disparity in its reception of recent innovations and reforms of anatomical knowledge. While the reception of Vesalius in the mid-sixteenth century showed a remarkably early interest in contemporary anatomical knowledge, the refusal to accept Harvey's theory in the latter seventeenth century contradicted the prior desire to pioneer. The two anatomical reformers shared an educational background: both had been trained at the University of Padua, which was still the leading European school of anatomy by the time of Harvey's apprenticeship under Fabrizio de Aquapendente in the late sixteenth and early seventeenth centuries. Even if archival finds dispute their claims, Pedro Jimeno and Luis Collado's alleged schooling at the University of Padua is testament to the reputation of this institution among Spanish medical students of the mid-sixteenth century. The 1559 edict of Philip II, which recalled Spanish students and professors from abroad, put an end to this association and eventually loosened the previously close ties to contemporary developments in science and medicine spearheaded by Italian universities. Even though the Universities of Naples and Bologna, as attended by Pedro Jimeno, were not included in this royal provision, there are few subsequent examples of Spanish integration and involvement in the

---

75 'Ay Professor de Anatomía, tan experto en el punto Anatómico, que no es Cathedratico de Anillo, sí muy hábil y digníssimo de la Cáthedra que desfruta ... siendo su Instituto dar y explicar materias Anatómicas, y en cada Curso hazer 25 dissecciones en el Theatro público de Anatomía, que la Ciudad tiene erigido en el Hospital General, para este fin.' V Gilabert: *Escrutinio physico-medico-anatomico* (Madrid, 1729), n.p. Cited in Martínez-Vidal and Pardo-Tomás, 'Anatomical Theatres in Early Modern Spain', pp. 266–67.

medical reforms of the Italian universities during the late sixteenth and entire seventeenth centuries.

It is ironic that the medical faculty of the *Estudi General*, which was among the earliest European institutions to defend Vesalius' corrections of Galenic anatomy, would later present itself as a radical stronghold of Galenic orthodoxy. While it was uniquely advanced in its early provisions for anatomical studies, the return to traditional medical authority was not limited to Valencia. Similar signs of medical counter-reformation can be traced in other Spanish universities with chairs of anatomy from the late sixteenth century onwards. The largest and most prominent among these institutions was the renowned University of Salamanca, which represented an analogous case of early acceptance and subsequent gradual rejection of the anatomical teaching programme introduced there in 1551 by the Valencian anatomist Cosme de Medina.

Chapter 3
Salamanca

While anatomical studies had been known in the Aragonese cities of Lleida, Valencia, Zaragoza and Barcelona since the fourteenth and fifteenth centuries, there is no evidence of similar practices at the University of Salamanca until the mid-sixteenth century. Yet despite lacking previous experience in the teaching and research of anatomy, this institution, the most prominent of the Castilian universities, launched a series of extensive reforms and initiatives in the 1550s and 1560s, which aimed to integrate anatomy and surgery into its medical faculty. The University of Salamanca was at that time perhaps the largest university in Europe, with 70 permanent chairs and almost 7,000 students during its peak in the 1580s.[1] The documentary source material from the university archive provides useful details regarding the initial phases of the establishment of chairs in anatomy, and later in surgery, and of the statutory and architectural alterations which enabled the incorporation of the new practices into the curriculum. The university's detailed administrative sources contrast with the surprisingly few publications produced by its first teachers and students of anatomy.

This scarcity of publications is peculiar given that the creation of professorships in anatomy and surgery in 1552 and 1566, the construction of the first permanent anatomical theatre on Spanish ground in 1554, and the very detailed and far-reaching alterations of university statutes in 1561 all represented radical breaks with three centuries of continuity during which Greco-Roman and particularly Arabic authorities had been left unchallenged in the medical curriculum. The medical faculty of the University of Salamanca was known as the Castilian centre of 'galenismo arabizado', despite its small size in comparison to the faculties of law and theology.[2] The Galenic medicine taught at this institution had been edited, commented upon and compiled by Arabic and Persian writers of the Middle Ages, and handed down in Latin translations largely produced in Spain from the twelfth century onwards. This late medieval intellectual framework survived well into the sixteenth century.

---

[1]   Francisco Javier Alejo Montes, *La Universidad de Salamanca bajo Felipe II* (Salamanca: Estudios de Historia/Castilla y León, 1998), p. 233.

[2]   This inferior status is dealt with in Luis Enrique Rodríguez-San Pedro Bezares' *La Universidad Salmantina del Barrocco, periodo 1598–1625* (Salamanca: Ediciones Universidad de Salamanca 1986), p. 540: 'La facultad de médica no pasaba de ser una hermana pobre, poco considerada y precaria de alumnado, frecuentada por escolares de poco pelaje y tildados de judeoconversos, que con una mentilidad practica se ocupaban de las vilezas del cuerpo.'

The primary medical university chair next to 'Prima' and 'Visperas' was that of 'Avicenna', indicating the continuing influence of the medical authority from the early eleventh century, whose *Canon* dictated the medical curriculum for first-year students in many Castilian universities. The first complete editions of his work reached Western Europe through the systematic translation of Arabic textbooks into Latin and Castilian, which was carried out in Toledo during the early twelfth century. Later professorships in anatomy and surgery were labelled 'catedrillas' or 'cátedras cursatorias', indicating that they were inferior to the permanent chairs of Avicenna, 'Prima' and 'Visperas'.[3]

A four-year bachelor's degree in 'Artes' or philosophy was required in order to study medicine. Students would then go on to master the Greek, Latin, and Arabic canons essential to contemporary medicine.[4] After four years of medical studies and two of hands-on practice, a medical student should have been familiar enough with Hippocrates, Galen and Avicenna to pass an exam before the *Tribunal del Protomedicato*. Established by the Catholic Monarchs in the late fifteenth century, this institution later restricted medical education in Castile to its three largest universities, Salamanca, Valladolid and Alcalá de Henares. The students graduated after a thorough examination before the Protomedicato, created to centralise and standardise medical practices throughout Spain. The royal mandate to create anatomical professorships at the three leading Castilian universities was a similar attempt to secure a uniform level of operational skills among future surgeons experienced in human dissection.[5] The increasing control wielded by the crown over Castilian universities during the reigns of Charles V and Philip II is evidenced by the petition to Charles V for the creation of a chair of anatomy which was submitted through the Council of Castile in 1550: 'In these Kingdoms there is a huge shortage of surgeons because no public anatomies are carried out here as in other universities where they study and teach this science. We beg your Majesty to demand that is done here, so that the surgeons can be better educated, and the sick cured more efficiently.'[6]

---

[3]   Javier Puerto, *La fuerza de fierabrás. Medicina, ciencia y terapeutica en los tiempos del Quixote* (Madrid: Editorial Just in Time, 2005), p. 96.

[4]   Pilar Valero Garcia, *La Universidad de Salamanca en la época de Carlos V* (Salamanca: Universidad de Salamanca, 1988), pp. 272–73.

[5]   Luis Enrique Rodríguez-San Pedro Bezares, *La Universidad Salamantina del Barroco* (Salamanca: Ediciones Universidad de Salamanca, 1986), p. 540.

[6]   'En estos reinos ay gran falta de cirurgía a causa de no se hazer anathomias publicas, como se hacen en otras universidades e partes donde se lee la sciencia dicha. Suplicamos a vostra Magestad mande se haga porque los cirujanos seran mas entendidos y los enfermos major curadas.' Cited in Maria Luiz López Terrada, *Medicos, Cirujanos, Boticarios y Albéitares, Historia de la Ciencia y de la Tecnica en la Corona de Castilla*, vol. III (Madrid: Junta de Castilla y León, 2002), p. 171.

The proposed renovation of medical teaching was not only an example of direct state intervention in academic affairs, but also represented a significant movement away from long-established practices which forbade medical doctors from performing invasive operations, bloodletting and other such manual interventions themselves. These measures were left to surgeons, barbers and bleeders, whose experience came from four years of hospital practice rather than a university education. The arrival of doctors with direct experience in operations and dissections by no means led to the permanent abolition of traditional medical practice. The newcomers were often seen as practitioners of an unnecessary novelty rather than the inspiration for the much-needed reform of medical education. Instead of using its own initiative to establish the new chair of anatomy, the University of Salamanca followed a provision acting on behalf of Charles V and submitted through the Council of Castile in the late summer of 1550. A 'claustro de diputados' made up of elected and appointed university associates met on 6 September to discuss the petition: 'Concerning the royal provision regarding anatomy, their Majesties ask the Council to study and discuss whether it would be worthwhile to perform anatomies in these kingdoms, as they say it is done in others.'[7] After two months, a *claustro de vicecancelario* was assembled on 16 November and an additional *claustro de diputados* was formed the following week. The *claustros* reached an encouraging conclusion: 'once the contents of the said royal proposal had been discussed and deliberated the physicians and surgeons considered that the practice of anatomies would be very advisable, beneficial and necessary, and this was the answer given by the University.'[8] The construction of Spain's first permanent anatomical theatre broke ground at the University of Salamanca a few years later and was followed by some of the most detailed statutes on anatomy from any university in the sixteenth century.[9] This wealth of documentary sources includes monthly accounts of the incorporation of anatomy into the curriculum and

---

[7] 'Lo tocante a la provision real acerca de la anatomía que sus magestades mandan que se vea e confiera si sera cosa provechosa que se aga en estos reinos, como dizen que se hace fuera dellos.' Casto Prieto Carrasco, *La enseñanza de la anatomía en la Universidad de Salamanca* (Salamanca: University of Salamanca, 1935), p. 23.

[8] 'Aviendo echo diversas vezes que se juntasen personas doctas en ciencia de medicina e de curugia e aviendo entre si conferido e platicado sobre lo contenido en la dicha provision real, a la mayor parte de los medicos e cirujanos paresció cosa muy conveniente e provechosa e necesaria que la dicha anatomia se aga y esto dio la Universidad por su respuesta.' José Àlvar Martínez-Vidal and José Pardo-Tomás, 'Anatomical Theatres in Early Modern Spain', *Medical History*, 49 (2005), p. 278.

[9] José Maria López Piñero, *La medicina, Historia de la Ciencia y de la Tecnica en la Corona de Castilla, vol. III* (Madrid: Junta de Castilla y León, Consejería de Educación y Cultura, 2002), p. 643.

also documents the appointment of Cosme de Medina as the first Salamancan *Catedrático de Anatomía*.[10]

Cosme de Medina: The First *Catedrático de Anatomía* at the University of Salamanca

It was not until a full year after the formal university approval of the royal provision in 1550 that the university rector Fernando de la Cerda informed his staff of 'bachelor Medina who has announced that if the University gives him a good salary he will teach and perform anatomy here'.[11] The news was delivered during a 'claustro de diputados' held on 5 September 1551, by which point Medina must have already left his native Valencia. He was present at the *claustro* in Salamanca on 9 September, at which the terms of his formal employment at the university were discussed and his salary was fixed at 40,000 maravedís. Medina was then subject to a probationary period from September 1551 to March 1552, when he was formally appointed to the chair of anatomy for three years. Two weeks before his formal installation as Professor of Anatomy, he exhibited his skills in anatomical teaching and dissection in *ejercicios de oposición* held before the *claustro* and an allegedly very large and curious audience. These first demonstrations were carried out on animals and included dissections of cows, pigs and dogs, which lasted for at least a week and were meticulously documented by an anonymous university notary.[12] The account of the notary indicates that the *oposición* took place in an improvised environment, with borrowed chairs and benches lining the university

---

[10]   The 1561 statutes have been published in Enrique Esparabé Arteaga, *Historia de la Universidad de Salamanca* (Salamanca: Francisco Nuñez, 1914), and the revised statutes of 1594 are published in Fransisco Javier Alejo Montes, *La reforma de la Universidad de Salamanca a finales del siglo XVI: Los estatutos de 1594* (Salamanca: University of Salamanca, 1990).

[11]   'Un licenciado Medina el qual ha dicho que si la Universidad le da un competente salario lehera e cortara e ara anatomia.' Teresa Santander, *El Doctor Cosme de Medina y su biblioteca (1551–1591)* (Salamanca: Centro de Estudios Salmantinos, 1999).

[12]   'Yo el dicho notario doy fee y testimonio verdadero en como un lunes que se contaron catorce días del mes de marzo de mill e quinientos y cincuenta e dos años dentro en el claustro de las escuelas mayores de la dicha Universidad el dicho licenciado Medina en cumplimiento de lo mandado por el dicho claustro le vi leher cerca de cosas tocantes de la anathomia, donde además de dicho claustro habia otras muchas personas estudiantes presentes. Doy fee demas de lo sobre dicho como en los corredores de las escuelas mayores su pusieron muchos bancos e asientos para donde se asentasen los señores de la Universidad para ber hacer la anathomia la cual yo el dicho notario vi hacer dos o tres días arreo donde estaban muchas personas de la Universidad asi doctores e maestros y estudiantes de manera que havia tanta gente a la ver hacer e leher que apenas podian ver muchos en que se hacia mas de que la oian leer y practicar.' Prieto Carrasco, *La enseñanza de la Anatomía en la Universidad de Salamanca*, p. 31.

corridors, which briefly worked as an operating theatre. Only some members of the large audience were able to follow the teaching properly, and subsequent initiatives to create a permanent anatomical theatre show that this was only a very temporary solution. In spite of the improvised gestalt and limited didactic value of these first public anatomies, the sheer number of spectators reveals that Medina's teaching was met with interest and enthusiasm. Another royal decree from October 1552 was submitted to facilitate the procurement of dead bodies for the anatomical teaching, and mentioned the recent appointment of an anatomist who (together with the primary teacher of medicine and the university rector) was given legal access to procure bodies of condemned criminals from the city judiciary and deceased bodies from the local hospitals 'as it is done in Italy and France. And the medical students should all pay half a real for the masses and the services held for each of the bodies they open, and they should bury them honourably.'[13] In his ambitious appointment letter produced soon thereafter, Medina charged himself with carrying out at least 30 annual dissections every year. This resolution prompted preparations for the construction of a permanent anatomical theatre to enable his high expectations to be met.[14] It was consequently necessary to establish a commission, which included Medina himself and which was led by the first chair of the medical faculty, Luis Alderete, to find a suitable home for the theatre, while the vice-chancellor and university treasurer made sure that no expense would be spared in its development.[15]

Located next to the church of San Nicolás, the theatre was completed in 1554 at a total cost of 50,000 maravedís, slightly more than Medina'a annual salary. Unfortunately, there are no written or visual sources detailing the shape and size of the building, despite its longevity. (The theatre was used throughout the seventeenth and eighteenth centuries before its demolition in 1801.) Luis Alderete embraced the extension of the faculty, declaring that the new teaching programme was in complete accordance with traditional Galenic teachings

---

[13] 'Suplicandonos vos mandasemos die sedes al catredatico de la dicha anotomia de la dicha universydad o a la persona o personas que la dicha universydad disputare pagar hazer la dicha anatomía todos los cuerpos que se condenaren a muerte después de executada la sentencia en ellos e los que murieren en los dichos ospitales de esa dicha çiudad los que al catredatico de prima e bisperas e de anotomia con el rrector de esa dicha Universidad paresçiere que conviene o a la mayor parte dellos pues ansy se hacen los estudios en universydades de ytalia y françia. y los estudiantes medicos daran cada uno medio rreal para misas e sacrificios de cada uno dellos que ansy se abrieren e le enterraran honrradamente' Archivio Histórico Nacional, *Universidades*. L 1097, f. 24 v.

[14] Vincente Beltrán de Herédia, *Cartulario de la Universidad de Salamanca* (Salamanca: Secretariado de Publicaciones de la Universidad 1972), p. 445.

[15] 'Que agan acer el edificio en el lugar donde se acer la dicha anotomía junto a San Nicolas, e que para el dicho edeficio todos los maravedis que fueron necessarios los puedan librar e libren en el acedor del dicho estudio.' Cited in Martínez-Vidal and Pardo-Tomás, 'Anatomical Theatres', p. 257.

and doctrines.[16] Alderete's statement conflicted with the soon-to-be-published *Liber de ossibus* (1555), written by Cosme de Medina's former Valencian mentor, Luis Collado, who declared recent approaches to the study of anatomy to be inconsistent with Galen's doctrine. In his defence of Vesalian osteology, Collado had praised Medina as the co-discoverer of anatomical features which questioned the validity of Galen's personal experience with human dissection.[17] Medina was thus among the new anatomical sceptics who based their theories and methods on the writings of Andreas Vesalius, and who were actively engaged in revisions of Galenic medicine. The statutes of 1561 valiantly attempted to ignore and belittle many of the contradictions which lay between *modernos* and *antiguos*, and allied contemporary scholars like Vesalius and Copernicus with Galen and Ptolemy in the curriculum of anatomy and astronomy. These decrees replaced the previous statutes from 1538 which had offered no guidelines for anatomy, the practice not having been established until several years later. Even though the 1561 regulations provide a very detailed glimpse into the routine dealings of the first anatomists at the University of Salamanca, the evidence is not consolidated by a long list of related publications authored by these newly appointed *catedráticos de anatomía*. Unlike the neighbouring University of Valladolid, where Rodriguez de Guevera's short 20-month period of anatomy teaching produced both his own and Bernardino de Montaña's treatises on their subject, no such publications emerged from the successive appointments of Cosme de Medina, Augustin Vazquez, Rodrigo de Soria and Diego Ruiz de Ochoa.

Documentary sources reveal that Salamanca's new medical chair required a great deal of hard work and dedication. Medina's appointment letter from March 1552 demanded, during his initial three-year employment, 30 dissections 'of convicted and poor human bodies', to be carried out throughout the entire

---

[16] 'Como Galeno y otros escriven ser cosa muy necessaria ver la anatomía por vista de ojos para saber conocer las enfermedades y curarlas; e por cuanto la anatomía que esta escrito en los libros es como figura e pintura de la anatomía realque se hace en los cuerpos muertos ansy es cierto que muy major se conoce viendo la propia cosa realmente que no viendola scripta ni pintada ... Por quanto en los libros que estan scriptos de la anatomía estan escriptas diversas opiniones y herrores los quales si se viese la anatomía realmente por los ojos se aviguará la que es verdad e lo que es herror, ytem por quanto las señales e curas de las enfermedades por vista de ojos o se exercitan en verlos curar ansy mismo no basta la anatomía escripta si no se bee por la vista de ojos; por los quales rezones dijo que en Dios y en sy conciencia le parece ser muy gran provecho a toda la Republica d'Espanha ansy en conocer como en curar todas las enfermedades de fisica e de cirujia, esto dijo que dara e dio por su parecer.' Luis S. Granjel, *Medicina española renascentista* (Salamanca: Ediciones Universidad de Salamanca, 1980), p. 52.

[17] Luis Collado, *Cl. Galeni Pergameni liber de ossibus ad tyrones, interprete Ferdinando Balamio, Enarratore Ludovico Collado medico* (Valencia: Typis Ioannis Mey Flandri, 1555), f. 30v.

year and not only during the cold winter months.[18] In spite of the royal provisions backing university anatomy in Salamanca (even threatening anyone who obstructed the new medical practice with fines of 10.000 maravedís) the chair of anatomy often had problems obtaining bodies of convicted criminals. A complaint from the university referenced in another royal provision from 1 November 1554 lamented that although the university 'had built a theatre at great cost' – and even though recently 'two famous thieves were sentenced to death' the local judiciary and hospitals were still preventing the Professor of Anatomy from procuring fresh corpses for his demonstrations.[19] These strenuous teaching and working conditions might excuse Medina's lack of publications. The 1561 statutes reduced the number of required dissections from 30 to the less ambitious 12 annual presentations of 'anatomia particular' (the anatomy of specific sections and individual organs of the body) and six dissections of 'anatomia universal' (the *systematic* anatomy of the skeleton, muscles, nerves, veins, etc.). Rewards were paid to those who achieved the expected number of autopsies, but reduction or full withdrawal of payment could be demanded from those who failed to satisfy the objective: 'For each systematic anatomy that is carried out, two thousand maravedís: And for each sectional anatomy one thousand maravedís. And they should only be paid for those dissections, which have been carried out perfectly.'[20] The practice of 'anatomía universal' was judged to be the most difficult and time-consuming, and performing anatomists

---

[18]   Cited in Martínez-Vidal and Pardo-Tomás, 'Anatomical Theatres', p. 256: 'Que sea obligado a hacer su anathomia en todos los meses del año oportunos conforme al miembro, para ello de todos los cuerpos humanos de justiciados e pobres de espitales en que pudiese fazer se faga e que faltando cuerpos umanos lo aga en otros cuerpos e sojectos que se le dieren e que la Universidad pague la costa si alguna ubiere e que por lo menos haga en cada un año anathomía en treinta cuerpos humanos, o en otros en su lugar, e que si más cuerpos le dieren que en más lo haga, y en cada uno destos cuerpos. En la anathomia destos cuerpos treinta se detenga tantas liciones assi en el leher como en el cortar como fueren menester.'

[19]   'En nonbre de la Universidad desa dicha çiudad de Salamanca nos hizo relaçion diziendo que por nos esta mandado que la dicha universydad su partede salario a un anotomista para que haga anotomia en los cuerpos abiertos e que para el dicho hefecto a fecho un teatro de mucha costa e por las dichas nuestras cartas susoyncorporadas vos mandamos quediesedes al catredatico de prima de mediana o al que tuviere disputado para el dicho hefecto la dicha Universidad los cuerpos de los que se justiçiaree condenaren a muerte en los quatro meses de ynbierno que son nobiembre diziembre henero y febrero e ansy mismo de los pobres de los ospitales e porque en los dichos quatro meses acahesce no aver nyngun justiçidado ny cuerpo en que se poder hazer la dicha anotomya y el salario y costa que la dicha universydad haze de en cada un ano syn hefecto e aunque por el mes de setienbre pasado ubo dos justiçiados de muerte ladrones famosos aunque fuistes requerido con la dicha provision no los quisystes dar.' Archivio Histórico Nacional, *Universidades*. L 1097, f. 24 r. and v.

[20]   'Por cada anatomía universal que hiziere, dos mil maravedís: y por cada disseccion particular mil maravedis. Y solamente se le paguen los que contare aver hecho perfecta y cumplidamente.' Esparabé Arteaga, *Historia de la Universidad de Salamanca*, p. 260.

were therefore rewarded with twice the amount paid for a successful 'anatomia particular'. Even though 'anatomia universal' was judged worthy of a higher reward, the anatomist faced severe time constraints due to the difficulty of preserving decaying bodies, even in the cold winter months, as indicated in the 1561 statutes: 'Because of the smell of the systematic anatomies they should not exceed more than two or three days, and only deal with the use and name of the parts, stating precisely the opinions of Galen and Vesalius and others, and declare whose is most accurate.'[21] The teacher of anatomy was to spend the rest of the year lecturing on the works of both Galen and Vesalius, and explaining how their theories and practices overlapped or differed from one another. The heavily illustrated *Fabrica* was a useful visual aid for students and it was in fact one of the prescribed teaching tools under the new regime: 'The professor is obliged to procure human bodies for the dissections, and if these cannot be obtained, he should show what he is lecturing using the prints of Vesalius.'[22]

Completed in 1554, Salamanca's anatomical theatre was among the earliest permanent university theatres. The Universities of Bologna, Padua, Pavia, Pisa and Montpellier had already assembled provisional wooden theatres for the anatomical demonstrations, but at Salamanca, a permanent building of brick and mortar was created for the demonstrations of human dissection. The presentations of 'anatomia universal' were apparently of greater educational value than the more specific scritiny of 'anatomia particular'.[23] These 12 stipulated 'particular' anatomies 'two of the head, two of the eyes, two of the kidneys, two of the heart, two of the muscles and veins of the arms, and two of the muscles and veins of the legs' were not open to the public, but were reserved for smaller and more intimate studies within the university hospital: 'The six general anatomies should be carried out in the anatomical theatre built for this purpose, and the twelve regional ones within the hospital or the medical faculty.'[24] Among the spectators of 'anatomia universal' was the renowned goldsmith and sculptor Juan de Arfe y Villafañe, who later wrote an account of his experience, leaving one of the only descriptions of the teaching of anatomy in this first Spanish anatomical theatre: 'After having become skilled in the study and representation of human bones, it seemed a good idea to attend an anatomical dissection of

---

[21]   'Por causa de olor en las anatomias universales no excederá de dos o tres días en ellos, solo tratando el usu y el nombre, y alegando precisamente donde lo trata Galeno y Vesalio y los demás que quisiren, declarando lo más llegado a razon.' Ibid., p. 261.

[22]   'Sea obligado el dicho catedratico a poner diligencia para auer cuerpos humanos do se hagan las dichas dissectiones, y no pudiendo auerse, lo que fuere leyendo en su lection y cathedra lo vaya mostrando en las stampas y figures de Besalio.' Ibid.

[23]   'Dos de cabeza, dos de ojos, dos de riñones, dos de corazón, dos de musculos y venas del brazo, y dos de musculos y venas de la pierna'. Ibid., p. 262.

[24]   'Las seys (anatomias) generales se han de hazer en la casa de Anatomía edificada a este fin y las doze particulares en el hospital del estudio, o en el de general de medicina.' Ibid.

human bodies. Therefore we went to Salamanca where it was practised by a professor named Cosme de Medina, and we saw the flaying of bodies of both men and women, condemned and poor. Besides being a horrific and cruel thing, we saw that it did not serve the end we were pursuing.'[25]

Arfe de Villafañe's *Varia Commensuración para la Escultura y Architectura* was published three decades after he professed to have attended Medina's lessons. It was printed in 1585 in Seville as the second volume in a compilation of two books on geometry and anatomy, followed up by treatises on zoology and architecture in a later 1587 edition.[26] In accordance with Arfe's purported training in 'general Anathomia' or 'anatomía universal', his treatise contained three long chapters on bones and muscles, and the body as a whole, and also some proof of his insights into 'anatomia particular' with chapters on isolated sections of the human body – head, torso, arms and hands, legs and feet.[27] Arfe de Villafañe's alleged eyewitness report of Medina's anatomical demonstrations was enhanced by a series of fine sketches of human anatomy and dissected body parts. This earliest fully illustrated and printed work on 'artistic anatomy' included some of the only written and visual evidence of the first decade of anatomy studies at the University of Salamanca. Arfe's accounts of anatomy as a tool for the artist showed how an interest in the mapping and research of human anatomy might extend beyond the university and its medical faculty. The goal for this 'Spanish Cellini' – an established sculptor and goldsmith, with exquisite works left behind in the Cathedrals and cities of Burgos, Seville, Avila, Lerma and Valladolid – was to gain insight into certain features of human anatomy, specifically the placements and attachments of muscles, joints and bones.[28] Parts of Arfe's account nonetheless

---

[25] 'Despues que para la demonstración de los huesos huvimos hecho toda la dilligencia dicha, nos pareció era razonable cosa ver hazer anothomia en algunos cuerpos: y asi nos fuimos a Salamanca donde a la sazon se hazia por un catedratico de aquella Universidad que llamavan el Dr. Cosme de Medina y vimos desollar por las partes del cuerpo algunos hombres y mugeres, justificados y pobres, y demas de ser cosa horenda y cruel, vimos no ser muy decente para el fin que pretendiamos, porque los muscolus del rostro y barriga, nunca se siguen en la Escultura, sino por unos bultos redondos, que diremos adelante, y los de los brazos y piernas en el natural, se ven en los vivos casi determinada y distintamente, y asi los mostraremos, con los terminus altos y baxos que el natural muestra sobre el pellejo.' Juan de Arfe y Villafañe, *Varia Commensuración para la Escultura y Architectura* (Seville, Andrea Pescioni y Juan de León, 1585), p. 141.

[26] *Libro segundo, que trata de la proporción y medida particular de los miembros del cuerpo humano, con sus huesos y morzillos.*

[27] 'Por ver como en cabeza, cuerpo y mano, en pierna, y pie la carne se ponía, atentamente en mas de un cuerpo humano vi hazer general Anathomia: quanto escribo me fué patente y llano, y mucho mas que aqui decir podría; pero solo dire lo conveniente para formar un cuerpo solamente.' Ibid., p. 141.

[28] 'Para demonstracion de esta parte hemos gastado mucho tiempo, y puesto toda diligencia, haciendo anatomia de muchos cuerpos, y aprovechandonos de tener los huesos

revealed his anatomical experience to be somewhat superficial, as his interest was focused on only the anatomical traits which had practical applications in his work as a sculptor of the human body. He was not only uninterested in the inner and hidden structures and organs, but also expressed unease with the spectacle of dissected bodies at Cosme de Medina's demonstrations. Despite the numerous problems surrounding Arfe's status as an anatomist, both Luis Lopez Alberti's *La anatomia y los anatomistas españoles del renacimiento* (1948) and *Diccionario histórico de la ciencia moderna en España* (1983) included him among the leading Spanish anatomists of the late Renaissance.

While Arfe's alleged mentor, Cosme de Medina, is not to be found among the authors who published writings on sixteenth-century anatomy, his name does appear repeatedly in documentary and administrative sources from the University of Salamanca. This source material shows how Cosme de Medina forged an impressive career within the faculty walls both during his 10 years as *catedrático de anatomia* and after the appointment had come to an end. His success was certainly not the result of an eminent educational background; the reference to *licenciado Medina* during his presentation to the *claustro de diputados* in late 1551 was erroneous, as he did not acquire his bachelor's degree in medicine until 1553. Oddly enough, the degree was obtained from the University of Valladolid and not from Salamanca.[29] The subsequent *Capitulos de la cathedra de anathomia*, which stipulated the requirements for Medina's second three-year appointment from 1555 to 1558, had been significantly revised since the first *Capitulos* in 1552. The original demands for 30 annual dissections were reduced to 24 and the contract also placed a strong emphasis on fines to be paid for unfulfilled academic obligations, referencing Medina's prior difficulties in complying with statutory demands: 'The said anatomist should be obliged to perform 12 systematic anatomies, and should be fined four ducats for every dissection he is not able to carry out, and he should perform regional dissections of 12 other corpses during all the months of the year under pain of two ducats, if he fails to carry them out.'[30]

More explicit complaints about Medina appeared after his additional appointments to the medical chairs of Articela and Visperas in 1557 and 1560, respectively, for which he received a supplementary salary. These additional

---

siempre delante. Fué con discursos largos inquirida por mi la certidumbre de esta sciencia, en que gaste gran parte de mi vida, poniendo en esto extraña diligencia: que de mi propia estancia en abscondida parte, mire gran tiempo la presencia de un cuerpo embalsamado, de los gruesos, largos y formas vi de todos huesos.' Ibid., p. 119.

[29]   Santander, *El Doctor Cosme de Medina y su biblioteca*, p. 15.

[30]   'Que el dicho anathomista sea obligado de disectar en cada un año doze cuerpos en entera anathomia universal sopena de cuatro ducados por cada uno que faltare de disectar y a de disectar otros doze cuerpos en particulas en los meses del dicho año sopena de otros dos ducados por el que faltare.' Ibid., p. 16.

responsibilities on top of his seemingly already neglected duties at Salamanca proved unsatisfactory for both the medical staff and students, as evidenced by the assembly of a *claustro de diputados* in February 1561 to discuss the student petition 'contra el licenciado Medina anotomista'.[31] Medina did retain the support of the majority of his university colleagues and his fellow *claustrales*, including the ill-fated theologian Fray Luis de Leon,[32] who voted in favour of Medina's multiple employments.

Later *Catedráticos* and the First Chair of Surgery

In 1562, Cosme de Medina's apprentice Augustin Vazquez was appointed to the chair of anatomy. Medina's own prior efforts were rewarded the following year with a promotion to 'Principal de Prima', the highest post in the medical faculty, which he held for the next three decades until his death in 1591. The fact that a practitioner of a new medical discipline was able to rise to such a prominent position in the university hierarchy is testament to the swift acceptance of anatomists by other physicians at the University of Salamanca. The 1577 abolition of the Salamancan professorship of *Avicenna*,[33] one of the cornerstones of the tripartite university structure of previous centuries, points in the same direction as evidence of significant reform for the benefit of anatomically trained doctors. A later Salamancan Professor of Anatomy, Casto Prieto Carrasco, reached this conclusion in his short 1935 history of his early modern predecessors. He noted that Diego Ruiz de Ochoa, another Professor of Anatomy, also achieved the highest post in the medical faculty after Medina's retirement in 1591.[34] However, conflicting evidence from the late 1500s indicates that successful reactionary initiatives were taken by the university authorities to maintain control and the status quo within the medical faculty. Possibly tempted by the promise of prestige, higher salaries and prominent positions, some of the anatomists were gradually integrated into the old system of traditional medicine that they had originally been appointed to reform. Indeed, Medina, in his late function as *Principal de Prima*, fiercely opposed the creation of a 'cátedra de cirugia' within the medical faculty, his

---

[31] Ibid., p. 24.

[32] After four years of imprisonment by the Inquisition between 1572 and 1576, Fray Luis de Leon allegedly opened his first university lecture in Salamanca with the famous lines 'As we were saying yesterday.'

[33] Luis García Ballester, *Historia social de la medicina* (Madrid: Akal, 1971), p. 89.

[34] 'Esta frecuencia con que los catedráticos de Anatomía pasaban a la de Prima puede estimirse como prueba de que la cátedra de Anatomía era entonces ocupada por hombres selectos; ya que el catedrático de Prima se nombraba, como todas, por oposición y votación a la que solían concurrir todos los doctores del ramo, y desde luego los mejores.' Prieto Carrasco, *La enseñanza de la anatomía en la Universidad de Salamanca*, p. 33.

resistance surprising given his own background as the first 'cátedra de anatomía'. This call for a university chair of surgery was the result of a proposal by the city council, supported by the crown, and in many ways a repetition of the same needs and alliances that had led to Medina's own professorship 16 years earlier.

Like the earlier *cátedra de anatomia*, the *cátedra de cirugia* sought to improve the education of surgeons working within the Spanish kingdom, particularly those in Salamanca and its neighbouring territory.[35] The creation of the professorship was a continuation of previous crown-instigated attempts to reform the medical education at the foremost universities of Castile, where surgery was granted university status in the late sixteenth century.[36] Despite the closeness of the two disciplines, the proposed foundation of a surgery professorship at the University of Salamanca met with resistance from Medina, who argued that the chair of anatomy was already sufficient for the improvement of surgical skills. He also claimed that even the endowment of surgery with formal university status would not alter the poor reputation and low wages of practising surgeons.[37] Medina's rejection of surgery as a university discipline consolidated his own position within the traditional medical system. When the chair was later established in spite of his protestations, Medina wrote the teaching guidelines, which were incorporated into the revised statutory orders of 1594. In these statutes the prescribed education of the surgeons adhered completely to late medieval theories and practices, with full authority given to the fourteenth-century *Chirurgia magna* and its author Guy de Chauliac, as well as the additional Galenic treatises *De apostematibus*, *De vulneribus* and *De ulceribus*.[38] An early surgical training tool has survived in the wooden anatomical mannequin made in 1571 by the local sculptor Mateo de Vangorla

---

[35]   'De algunos años a esta parte avia avido y avia en esa dicha ciudad gran falta de zurujanos y dello se avia seguido averse muerto muchas personas de heridas muy pequeñas y tales que dellas a lo que se tenia entendido no murieran si fueran bien curados y aviendo como avia en esa dicha ciudad y Universidad tantas personas doctas y suficientes no era justo faltase la curujia siendo como era cosa tan necesaria antes convendria para el bien de aquella republica y de todo el Reyno que en esa dicha Universidad ubiese catedra de zurugia pues la avia de todas las demas ciencias.' Teresa Santander Rodriguez, 'La creación de la cátedra de cirugía en la Universidad de Salamanca', *Cuadernos de historia de la medicina Española*, 43 (1965), p. 192.

[36]   Pedro García Barreno, *La medicina en El Quijote y en su enterno* in *La Ciencia y El Quijote* (Barcelona: Crítica, 2005), p. 162.

[37]   'Y si destos tales curujanos no ay muchos en esta ciudad, es porque como los estudiantes que practican la dicha arte de curujia no son naturales desta ciudad, vanse a sus tierras a ganar de comer, pareciendoles que en Salamanca se paga mal la curujia. Y los que son naturales deste pueblo, quieren mas hazer profession de medicos que de curujanos y a esta causa ay en esta ciudad pocos curujanos doctos. Lo qual no se puede remediar con nueua institucion de cathedra.' Ibid., p. 197.

[38]   Montes, *La reforma de la universidad de Salamanca*, p. 94.

for the first appointed teacher of surgery, Andrés Alcázar. This statue of questionable didactic value was probably used for the teaching of bandaging or *algebra* (bone-setting) and is today exhibited at the university museum in Salamanca.[39]

Despite the establishment of specific anatomical training for the Salamancan surgeons, the university teaching programme began to disintegrate after only

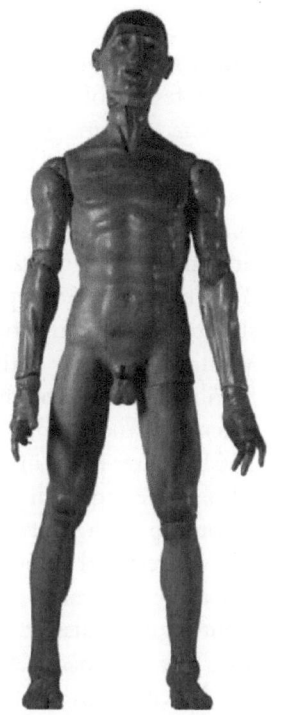

Figure 3.1 The surviving wooden anatomical mannequin made in 1571 for the University of Salamanca's first teacher of surgery, Andrés Alcázar, was carved by the local Salamancan sculptor, Mateo de Vangorla. This curious statue of questionable didactic value was arguably used for the teaching of bandage- and bone-setting (*Algebra*), and was still in use during the seventeenth century. It is today exhibited at the University Museum of Salamanca and is documented in the university archive in a reference to a sum of 60 ducats paid to the sculptor after its completion. Courtesy of the 'Museo de la Universidad de Salamanca'

---

[39] J. Ramon-Nieto, *Mateo de Vangorla: Manequi anatómico para practica de vendajes, Felipe II, un monarca y su epoca: Las tierras y los hombres del Rey* (Madrid: Sociedad Estatal para la Conmemoración de los Centenarios de Felipe II y Carlos V, 1998), p. 296. The university archive contains a contemporary reference to the payment of 60 ducats due to the sculptor following its completion: 'Tasacion y paga al escultor Vangorla de la estatua que hizo para que los estudiantes de cirugia puedan ser enseñados por ella de las ligaduras del cuerpo humano.' Salamanca, 6 de febrero 1571: 'En esta claustro el senor doctor Antonio Gallego, diputado, dio relacion diciendo en como por parte de la universidad le habia sido cometido viese una estatua que Vangorla escultor habia hecho a pedimiento del ldo. Alcázar, catedratico de cirugia para que los estudiantes por ella pudiesen ver e ser enseñados de las ligaduras e los demas requisitos tocantes a la cirugia. E asi que la hiciese tasar e ver lo que por ella la dicha Universidad deria bien pagase el dicho oficial quye la hizo 60 ducados.' Vicente Beltrán de Heredia, *Cartulario de la universidad de Salamanca (1218–1600)*, vol. V (Salamanca: Secretariado de Publicaciones de la Universidad, 1972), pp. 272–73. A contemporary inventory from Salamanca, dated 2 January 1572, lists a number of 'anatomies' among the possessions of the local artist, Odiosdado de Olivares; among these were 'a medium-sized wax figure of an anatomized man in a wooden box' and 'un rostro de metal de anatomia' and 'Otra caja con un caballo de huesso y un cuerpezico de anatomia muy pequeño' – perhaps the works of Vangorla or Olivares himself? Cited in Kelley Helmstutler di Dio and Rosario Coppel, *Sculpture Collections in Early Modern Spain* (Farnham: Ashgate, 2013), Appendix, p. 341-342

a few decades. This deterioration is certainly traceable in the careers of Medina and Ruiz de Ochoa, as well as that of Augustin Vazquez, Medina's successor as Professor of Anatomy. Vazquez's only publication, *Questiones practicae, medicae, et chirurgicae* (Salamanca, 1589), was a compendium of comments on the doctrines of Galen and Avicenna; its third and final treatise on surgery was an introduction and comment on Guy de Chauliac's *Chirugia magna* and was entirely embedded in a late medieval medical tradition. One of the first inspections of Vazquez's teachings in a 'visita de cátedras', carried out on 27 February 1563, reprimanded the newly appointed teacher for defying the statutes by teaching from Juan Valverde's Spanish textbook instead of using Vesalius' *Fabrica*.[40] This brief reference is not only an indication of Vazquez's familiarity with contemporary anatomical textbooks, but also forms a significant defence of the emphasis placed on Vesalius' texts and images by the statutes of 1561. A later *visita* in 1575 praised the pedagogical skills of Augustin Vazquez, who impressed both with his abilities to carry out anatomical demonstrations according to the statutes, and his communicative skills, which were recognised and appreciated by both students and inspectors: 'He is known to have carried out three systematic anatomies of two men and one woman, and more than eight regional dissections of a head, two arms, a leg, a heart, and a kidney, and an artery and an eye. All these anatomies have conformed well with the statutes, and have been presented very well and for the benefit of the audience.'[41]

Considering that Vazquez was penalised for his use of Valverde's rather than Vesalius' anatomy, it is curious that neither anatomist seemed to merit a mention in his medical/surgical textbook published more than 20 years later. The lack of references to Vesalius in the revised statutes of 1594 showed the same eagerness to consolidate 'pre-Vesalian' anatomy. These statutes were written at a time when the ageing Vazquez was once again in possession of the anatomical professorship for a short period. In the previous year Vazquez had been able to employ an assistant with an annual salary of 15,000 maravedís. This small expansion of the medical staff was approved by a *claustro* and later by a royal provision on 11 January 1593. Other petitions made by Vazquez and authorised by the *claustro de diputados* included a change in the anatomical teaching schedule. Vazquez lobbied for teaching to take place in the early morning instead of in the afternoon, judging the later part of the day to be hazardous to the health of the students both in warm and cold periods of the

---

[40]　Santander, *El Doctor Cosme de Medina y su biblioteca*, p. 32.

[41]　'Sabe que a fecho tres anatomyas universales de dos hombres e de una muger e mas que ocho particulares que son una cabeza dos brazos una pierna y un corazon e un riñon e una aspera arteria e un hojo las quales dichas anatomias este testigo se las a visto hacer e tener en el dicho tiempo conforme al estatuto que en este caso habla muy bien e a provecho de sus oyentes.' AUS. Visita de cátedras, 1575–77, f. 37.

year.[42] The second *visita de cátedras* of Rodrigo de Soria's chair on 20 August 1575 reported that the number of dissections carried out at Salamanca had been drastically reduced to only a few per annum.[43] De Soria's regional and systematic dissections of two humans and a dog from October 1574 to March 1575 were only a fraction of the 30 and then 24 annual dissections carried out by Medina in the early and late 1550s.[44]

Decline and Abandonment of Vesalian Anatomy at the University of Salamanca

By the turn of the century, the 1594 university statutes indicated a total abolition of the 'modern' theories and practices formerly incorporated into anatomy studies and the abandonment of reforms carried through only a few decades earlier. Following a short-lived challenge to traditional authority within the university,

---

[42] 'Hora muy rezia y muy peligrosa para la salud y aun para el aprovechamento de los estudiantes que professan la dicha facultad porque por ser hora tan ymportuna de una a dos en ynbierno y de dos a tres en verano faltan muchos a ella lo qual no harian si la hora fuesse mas acomodada.' AUS Libro de claustros, 1583–84, f. 142.

[43] This was seen in the answer given by the Professor of Anatomy, Rodrigo de Soria: 'Preguntando acerca del ditar dixo que no dita al presente pero que solia dar una teorica de un cuarto de hora y aun no llegaba a tanto preguntado por las anatomias que a echo desde el dia de San Lucas asta el dia de oy dixo que a echo anatomias de dos ojos e de un corazon e de un cuerpo humano e de todas las partes del cuerpo y que lo haze muy bien y a provecho de sus oyentes.' AUS Visita de cátedras, 1574–75, f. 25 De Soria had previously comlained about the shortage of bodies for his dissections and blamed the city hospitals for deliberately opposing the royal provision in their refusal to advise the medical faculty about recently deceased patients: 'ansí por aver los subietos, porque como los ospitales huyen de darlos, aunque les está notificada la provisión real, no quieren dar aviso quando muere alguno, paraciéndoles que pierden mucho los ospitales en darlos, e ansí es menester tener contentos e aun bien pagados los espitaleros; e ni mas ni menos que llevar los cuerpos a la casa de la Anatomía y enterrarlos, e fuera desto en carbón y olores y instrumentos e paños para limpieza dellos'. Teresa Santander Rodriguez, 'La iglesia de San Nicolas y el antiguo teatro anatómico de la Universidad de Salamanca' in Luis Enrique Rodríguez-San Pedro Bezares (ed.), *Historia de la Universidad de Salamanca. Volumen III: Saberes y confluencias* (Salamanca: University of Salamanca, 2002), p. 358.

[44] 'Preguntando diga e declare que anatomias a fecho el doctor Rodrigo de Soria en todo este año ansi universales como particulares dixo las siguientes. Dixo que en la casa de anatomia a fecho dos cuerpos de hombres e demas desto una universal de venas y arterias de un perro. E otra universal de huesos leyendo por el esqueleto. Yten particulares de dos cabezas de hombres arriba declarados e se tardaria en acer las de las piernas e brazos e cabezas e los cuerpos por mas de quince dias. Yten en las escuelas en dos hojos, dos corazones, dos rinones, desde el dia de San Lucas proximo pasado asta San Juan deste año.' Ibid., f. 77.

the medical faculty at Salamanca reverted to pre-Vesalian customs for its anatomical teaching programme. There is also evidence of the neglect and decay of the anatomical theatre. By the end of the century, the building had begun to deteriorate rapidly and in 1627 following a flood of the Tormes river, complaints were made by the then Professor of Anatomy, Jorge Enriquez, regarding the very poor state of the theatre: 'The chapel near San Nicolás where the anatomies are carried out should be restored because it is very damaged by the flood and in the windows wired bars must be inserted to keep out the cats and other animals, who enter and vandalise the bodies used for the anatomical demonstration.'[45] Throughout the seventeenth and eighteenth centuries, several laments called for a restoration or replacement of the building. Proposals to erect a new anatomical theatre launched in the 1740s and 1770s only accomplished the renovation of the existing building before its final demolition in 1801.[46] A 1771 petition for the reform of the medical and anatomical teaching programmes proposed weekly dissections and the acquisition of new instruments and textbooks, in particular the German anatomist Lorenz Heister's *Compendium anatomicum* (translated into Spanish in 1755), as well as illustrations 'by the same author, or those by Vesalius or Eustachius, which are the best'.[47] It is interesting to note that Vesalius' name reappeared in the University of Salamanca statutes more than two centuries after his first mention in 1561. The renewed focus on anatomy at Salamanca did not, however, prevent the destruction of the anatomical theatre a few decades later, on the site of which lies only a stone tablet with the following inscription: 'This theatre was built for the anatomical investigations, which were carried out on a daily basis, and consecrated to the Pure Virgin. The amphitheatres of the ancients were built for the killing of men; our theatre so that they could learn how to live longer.'[48]

The return to ancient medical tradition did not receive a unanimously warm reception. Harsh criticism appeared from within the University of Salamanca itself in the noteworthy *Retrato del perfecto medico*, which was published in 1595

---

[45] 'Se adareze la capilla de San Nicolás adonde sa hazen las anatomias, porque está muy maltratada con las abenidas del rio y en las ventanas se pongan rexas de alambre porque por ellas entran gatos y otros animales que maltratan los cuerpos de que se ace anotomia.' Santander Rodriguez, 'La iglesia de San Nicolas y el antiguo teatro anatómico de la Universidad de Salamanca', pp. 262–63.

[46] George M. Addy, *The Enlightenment in the University of Salamanca* (Durham, NC: Duke University Press, 1966), pp. 75 and 122.

[47] 'Demostrará el Cathedrático las partes, que explicáre en sus lecciones en las Tablas del mismo Autor, ò en las de Besaleo, ó Eustaquio, que son las mejores.' Ibid., p. 262.

[48] 'Para las investigationes anatómicas que se hacen cada día levantó la Universidad este teatro, consegrado a la Virgen pura. Los antiguos anfiteatros se abrían para muerte de hombres; los nuestros, para que aprendan a vevir largo tiempo.' Alberti Lopez, *La anatomia y los anatomistas del renacimiento* (Madrid: CSIC, 1948), p. 86.

by Enrique Jorge Enriquez, a Portuguese physician and lecturer in medicine at the University of Salamanca. Enriquez's book provides interesting insights into the contemporary medical practice and authority within and beyond the University of Salamanca. Completed in 1582, the manuscript was dedicated to the recently deceased Duke of Alba, Antonio Alvarez de Toledo, to whom Enriquez had been personal physician during his last years. The printed 1595 edition was introduced by a sonnet by Lope de Vega, which emphatically condemned the ignorance, greed and corruption of 'those who are doctors only in their dress and appearance'.[49] *Retrato del perfecto medico* entreated the perfect doctor to enter the medical profession in order to serve a higher ethical course, and not for the acquisition of profit, elegant clothes and titles that attracted so many 'idiotas' to partake in such a career. The perfect medical doctor was presented as a polymath and a master of Greek, Latin, Arabic, music, astrology and, of course, anatomy.[50] In his consideration of the progression of Spanish medical practice in recent decades, the author happily acknowledged the introduction and implementation of anatomy studies throughout Spain: 'A thing in Spain, which pleases me much is the recent creation of professorships that teach the anatomy of the human body, which is so necessary for the perfection of the medical doctor. Every part is laid visible to the human eye, which was not the case in past centuries. I do not know how the ancient physicians could practice medicine without the science of anatomy.'[51]

Enriquez was relentless in his praise of the few *antiguos* and many *modernos* in the field of anatomy. Among the latter, he celebrated the Italian and Valencian followers of the new anatomy, as well as his own colleagues at the University of Salamanca: 'After Galen, who excels most in this discipline? Vesalius, Realdo Colombo of Cremona, and many others of ingenuity and learning, and in our time the wise Doctor Collado from Valencia, and the clever Doctor Medina,

---

[49]   'Los que son medicos solo en el vestido.' Enrique Jorge Enriquez, *Retrato del perfecto medico* (Salamanca: Juan y Andres Renaut, 1594), prologue, n.p.

[50]   'Que sea el medico degente honrada, que sea pacifico. Que deue ser Anatomista. Y el cirujano. Si es licito hazer anatomias en cuerpos humanos.' Ibid., p. 193.

[51]   'Una cosa veo agora en España, que me agrada mucho, y es que ay Cathedras en la quales se enseña la Anathomia del cuerpo humano tan necessaria a la perfección de nuestro Medico, y cada parte se muestra al ojo, lo qual no se hazia en los Siglos passados, no se yo como podran los Medicos antiguos se medicos faltandole la Sciencia de la Anatomia ... Assi no es muy grande desastino y locura querer alguno caminar por el cuerpo del hombre con sangrias, purges y xaraues, y otras cosas sin tener conoscimiento, ni de otro hallado, ni por sabido de la composicion del hombre, apenas nos atrauemos a entrar de noche sin lumber en la casa que no auemos estado de dia., y queremos a las ciegas sin sciencia Anatomica escudriñar los rincones y venas, y escodrijos de la fabrica humana? Por cierto que es muy gran maldad, Galeno, Aristoteles y otros muchos, assi antiguos como modernos no trabajaron mucho por esta parte de la medicina.' Ibid., p. 208.

who holds the first chair of medicine at our renowned University of Salamanca, and who was succeeded in the anatomical professorship by the learned Augustin Vazquez.'[52] The contributions of contemporary anatomists from Spain and further afield were also recognised by Bernardino Montaña in his *Libro de la anothomia del hombre*, which praised 'the excellent men, both ancient and modern'.[53] The physician and theologian Fray Luis de Granada similarly wrote of 'los nuevos anatomistas' and 'medicos antiguos'. Enriquez's respect and admiration was not limited to modern anatomical achievements. He even advocated the reintroduction of the drastic measures known from the ancient medical school of Alexandria, where vivisections of human beings in the third century BC had besmirched the reputation of anatomy for the next 1500 years. Here Enriquez stood in clear opposition to the pre-Vesalian Italian anatomist Alessandro Benedetti, whose *Historia corporis humani sive Anatomice* (Venice, 1502) had fiercely condemned this controversial practice; he instead revealed himself to be in agreement with Vesalius, who, in the prologue to *Fabrica* mourned the decline of the science of anatomy since the golden Alexandrian age, when Erasistratos and Herophilos practised vivisections of condemned criminals. Enriquez was vociferous in his support of this former practice: 'Who would consider it tyrannical to anatomise the bodies of criminals and evildoers if it could benefit the just?'[54]

The documentary sources relating to anatomical practice at the University of Salamanca show a gradual return to Galenic doctrines, but contributions from modern anatomy were nonetheless still acknowledged by Enriquez in his 1595 *Retrato*. The book appeared while Enriquez held one of the professorships of the medical faculty and stands out as counter-evidence of the gradual disappearance of 'modernos' in the contemporary university statutes. In the 1594 statutes there is no trace of any prior union of ancient and contemporary anatomy. The guidelines for future professors specified only Galen's anatomical treatise, *De usu partium*, as a prescribed text, and explicitly prohibited the use of competing textbooks: 'The professor of anatomy is obliged to read the books of 'De usu partium' for two years. In the first year the first eight books, and in the second the last nine, and the professor of anatomy should not read anything

---

[52] 'Despues de Galenus, quales se auentajaron mas en esta parte? Licen. Andreas Vesalio, Realdo Columbo Cremonense, y otros muchos de gran ingenio y doctrina, y en nuestros tiempos el sapientissimo Doctor Collado medico Valentino, y el clarissimo Doctor Medina Catedratico de Prima en la insigne Uniuersidad de Salamanca nuestro preceptor, a quien va siguiendo el doctissimo Doctor Augustin Vazquez.' Ibid., p. 212.

[53] 'Los varones excelentes, antiguos y modernos.' Bernardino Montaña de Montserrate, *Libro de la anothomia del hombre. Epistola dedicatoria*, n.p.

[54] 'Que no le tenga por tyrannia anatomizar querpos humanos de hombres delinquentes, y malhechores, para que se aprovechen los justos.' Enriquez, *Retrato del perfecto medico*, p. 211.

else, because these books contain all the material that is worth knowing.'[55] The view that other books or theories were unnecessary and, indeed, prohibited was explicit evidence of a medical counter-reformation taking place within the walls of the University of Salamanca. It seems that the reformed anatomical teaching programme collided with forces within the social context of the university and that the victors were ultimately able to preserve their traditional monopoly of medical authority. In the early 1550s, a large audience had eagerly crowded on to assembled chairs and benches to witness a new approach to anatomy that had been set forth in Vesalius' *Fabrica* less than a decade before. Two generations later, the same university reverted to structuring its anatomical teaching in complete accordance with Galenic theories and practices. In the mid-sixteenth century, this institution, the largest university in Europe, was well versed in and responsive to the new intellectual currents spearheaded from Italy and France. The royal decree of 1559, which demanded the return of all Spanish scholars and students to the Iberian Peninsula, weakened and eventually broke these links, as exemplified in Luis Granjel's study of the circulation of medical students between Iberia and Montpellier in the 1500s: 'An example of the effect of this prohibition is seen when comparing the 300 Spanish students of medicine who studied in Montpellier between 1503 and 1558 with the less than ten who entered after 1559.'[56]

Further signs of regression to the old teaching model were found elsewhere and, in spite of his critical remarks on contemporary medicine, Enrique Jorge Enriquez's own work was itself an indication of a development away from the reforms realised earlier in the sixteenth century. While contemporary anatomical 'precursors' still appeared prominently in his book, the author attempted to reduce or ignore their obvious clashes with former medical tradition. Initial appraisals of the new anatomy notwithstanding, Enriquez's work offered a long treatise on humoral medicine and its intimate relationship with interfering elements of the universe, and called for future medical doctors to acquaint themselves with the doctrines of Hippocrates, Galen and Avicenna, preferably in the original languages. His direct address to King Philip II regarding the relationship between astrology and anatomy was a further indication of the return to medieval dogma at Spain's foremost university: 'I have often heard it

---

[55]   'En la cátedra de anatomía se han de leer los libros 'De usu partium' en dos años. El primero se han de leer los ocho libros primeros, y en el segundo los otros nueve, y no se ha de leer otra cosa en cátedra de anatomía, porque en estos libros se contiene toda la material que es menester saber.' Montes, *La reforma de la universidad de Salamanca*, p. 139.

[56]   'Como ejemplo de lo que esta prohibición suposo en el mundo médico basta recordar que frente a los más de trescientos escolares que estudiaron medicina en Montpellier entre 1503 y 1558, la cifra apenas supera la decena con posteridad a 1559.' Luis S. Granjel, *Discurso sobre el pasado de la enseñanza del saber y el arte medicos en la Universidad de Salamanca* (Salamanca: University of Salamanca, 1953), pp. 13–14.

said, and I agree after having read books by some astrologers that each anatomical part of the human body has a corresponding planet, which controls it. What is your Majesty's opinion on this matter?'[57] This is a possible allusion to a recently published treatise on anatomy and astrology, *Libro primero de anotomia* (Baeza, 1590), which was written by Andrés de Leon, one of King Philip's court surgeons. The textbook both described and illustrated 'the domain, which every sign of the Zodiac and planet has over the human body parts, in order to facilitate an understanding of the proper time for blood-letting'.[58] Leon's book included two short references to Vesalius and Valverde, but relied almost exclusively on Galen and Avicenna, and on the author's own astronomical calculations, which detailed the ideal times and constellations for bloodletting and purgatives. The embrace of medical astrology could be interpreted as a reactionary Spanish return to waning doctrines during the late sixteenth century and many *criticos* have advocated similar views in the lengthy *polémica de la ciencia española*.[59] John Elliott's interpretation of the political and cultural history of early modern Spain also seems applicable to the understanding of the discontinuity of previous reforms within Salamanca's medical faculty: 'By the end of the 1560s "open" Spain of the renaissance had been transformed into the partially "closed" Spain of the Counter-Reformation. In retrospect, the victory of the traditionalists appears inevitable; but at the time when the struggle opened, their eventual triumph seemed far from assured.'[60]

Almost every recent work on the development of Spanish science and medicine during the late Renaissance has recognised a rupture in the intellectual

---

[57] 'Siempre oy dezir y me acuerdo auer leido en algunos Astrologos, que sobre qualquiera dessas partes, que se veen en la Anathomia, auia un planeta que tenia dominio sobre ella, que ha parecido a v.m. desto?' Enriquez, *Retrato del perfecto medico*, p. 215.

[58] 'El dominio que cada Signo y Planeta tiene en los miembros del Hombre, para que mas facilmente se entienda lo que conviene a las Sangrias avisos para purgar.' Andres de León, *Libro primero de anotomia* (Baeza: Baptista de Montoya, 1590), 160 r.

[59] The Spanish philosopher and historian Ortega y Gasset invented the concept of a historical *tibetización* of Spain in 1920, which coincided with the contemporary references to the *leyenda negra*, which had been introduced only a few years earlier in 1913. According to the *criticos*, the basic components enabling an inclusion into a broader European Renaissance culture were wholly or partly absent in Spain by the end of the sixteenth century. For some historians both within and outside Spain, the 'siglo de oro' was instead presented as a clear continuation with the Middle Ages, as in Viktor Klemperer's 'Gibt es eine spanische Renaissance?', *Logos*, 16 (1927), p. 2.

[60] John Elliott, *Imperial Spain, 1469–1716* (Cambridge, Cambridge University Press, 1963). The Salamancan historian of medicine, Luis S. Granjel, gradually developed a similar conviction and approached the *criticos* in his synthesis *Medicina española renascentista*. This work maintained the much-discussed notion of a Spanish Renaissance, but divided the period into two opposing chronologies: an 'open' Renaissance concluding with the reign of Charles V; and a second or 'closed' Renaissance under Philip II.

history of the Iberian Peninsula following Philip II's university decree in 1559, which was further strengthened by the first Index of Prohibited Books also published that year. Like Granjel, López Piñero has traced an alarming decrease in the number of Spanish physicians with experience gained abroad.[61] While López Piñero endeavoured to resurrect the reputation of Spanish science in the early modern period, his *Ciencia y tecnica* portrayed the medical faculty of Salamanca as being in a state of almost complete decay by the end of the seventeenth century. The ruinous state of the anatomical theatre was seemingly only a symptom of the much more widespread decline of the entire institution. Documentation from 1690 reported the near-destruction of its famous university library, which had been established as early as 1254 and was praised by Piñero as 'the best in Spain' and 'seemingly the oldest university library in Europe'.[62]

It is beyond the scope of this study to analyse the alleged disruption of an institution equalled in age and prestige only by the largest and earliest universities of Italy, France and England. By the end of the sixteenth century, the university's 7,000 students arguably surpassed the size of any other academic institution in Europe. This number gradually diminished during the 1600s and was drastically reduced to only 150 students by the mid-nineteenth century. While focusing on the isolated establishment of 'Vesalian' anatomy and its eventual 'failure' within the Salamancan medical faculty, it is necessary to consider the influence of institutional ruptures and larger political context. Luis S. Granjel emphasised the turning point of 1559 as a demarcation line between two opposing 'Spanish renaissances', of which only the first represented a full integration into the cultural and scientific history of the rest of the continent, whereas the other was seen as an era of intellectual and cultural isolation. However, it should be highlighted that important Italian university cities such as Bologna and Naples were not included in the royal decree from 1559, most probably due to their placement near or within the immediate Spanish sphere of influence on the Italian peninsula. The idea of a subsequent and complete *Tibetisation* of Spanish medical science cannot be sustained, and certainly not with regard to the study of anatomy, for which Bologna was still one of the leading European centres during the mid-sixteenth century. Nor can the Inquisitorial Index of 1559 be explained as the first step towards an isolationist path. The publication of such indexes were by no means limited to mid-sixteenth-century Spain, but rather were the product of common reactions throughout the Catholic world then engaged in consolidation and Counter-Reformation. Indeed, in Italy, Pope Paul

---

[61] López Piñero, *La medicina, Historia de la Ciencia y de la Tecnica en la Corona de Castilla*, vol. III, p. 142.

[62] 'Lo mejor de España ... al parecer, la biblioteca universitaria más antigua de Europa.' Ibid., pp. 131–32.

IV's Index of Forbidden Books was published in the same year as its Spanish equivalent. It is furthermore evident that the 1561 statutes from the University of Salamanca represented a clear continuity with earlier policies and reforms from the late years of the reign of Charles V. The new challenging authorities such as Vesalius and Copernicus were placed next to ancient authorities in the university curricula, and the incorporation of a surgical programme during the same decade represented similar aspirations to improved status for applied medical sciences. Many of the royal incentives to improve medical education in general, and the anatomical training of surgeons in particular, presented the crown as allied with many of the contemporary critics of medical authorities and practices.

The most likely cause for the deterioration of the new anatomy programme at this leading Castilian university is to be found within the university walls, and most probably in opportunism, intrigues and power struggles, which have not left many traces in the historical record. Even though only scant evidence points to this conclusion, the return to a monopoly of ancient authority was arguably the result of successful counter-offensives launched by established university hierarchies, which had never fully embraced the introduction of novel university disciplines, such as anatomy and surgery. Even though the university rector, Diego de Covarrubias, was actively involved in the inclusion of Vesalius and Copernicus in the statutes of 1561, he referred elsewhere to the perils of excessive renovation of tradition, which he described as 'a dangerous thing, because it brings with it changes in the old customs'.[63] Andrea Carlino's analysis of the unsuccessful attempts to alter existing statutes on anatomy and medicine in prominent Italian universities leaned towards the Polish philosopher of science, Ludwig Fleck, and his claim that: 'Once a structurally complete and closed system of opinions consisting of many details and relations has been formed, it offers enduring resistance to anything that resists it.'[64] The events in Salamanca throughout the late sixteenth century seem to exemplify such an 'enduring resistance' against potential cracks in the existing medical tradition and paradigm, consolidated and perfected throughout centuries. Evidently its upholders had more to lose than to gain from the appearance of medical reformers within the university walls. This development was seen in Vesalius' spectacular and brief appearance in the 1561 university statutes, from which his name disappeared only a few decades later. Cosme de Medina and several of his successors in the professorships of both anatomy and surgery seem gradually to

---

[63] 'Usualmente peligrosa, porque trae con ella los cambios a las usas antiguas', cited in Americo Castro, *The Structure of Spanish History* (Princeton: Princeton University Press, 1954), p. 595.

[64] Andrea Carlino, *Books of the Body* (Chicago: Chicago University Press, 1999), p. 213.

have forged careers in accordance with previous authority and practices, rather than in opposition to these traditional norms. As indicated by the attempt to trace the gradual inclusion of the first Salamancan anatomists into the traditional university hierarchy, these 'modern' anatomists often accepted prominent positions within the 'old' system rather than existing as reformers and innovators working in its periphery. As will be shown in subsequent chapters, the course of events in Salamanca not only followed their own unique turn, but also a more general tendency towards medical consolidation and orthodoxy found in other Spanish universities towards the end of the sixteenth century.

Chapter 4
Valladolid

The study of anatomy was introduced at the University of Valladolid in 1548/49 and the process of incorporation differed markedly from that previously described at the University of Salamanca. No relevant publications emerged from Salamanca in more than two centuries of anatomical studies, whereas Valladolid produced two textbooks on anatomy after less than two years of practice. Documentary sources indicate that Aragonese universities and hospitals carried out dissections as early as the fourteenth century, but the University of Valladolid was the first Castilian institution where anatomy was taught from human dissection. In Spanish historiography, its status as the birthplace of both Castilian anatomy and the first publication on anatomy in vernacular Spanish, Bernardino Montaña's *Libro de la anothomia del hombre* (Valladolid, 1551), has often led to the misconception that Valladolid spearheaded the anatomical revolution of sixteenth-century Spain. However, a closer look reveals that Valladolid was the only large Castilian university where no permanent chair of anatomy was ever established during the sixteenth and seventeenth centuries. In fact, its anatomical practice was limited to a 20-month course taught by the physician Alonzo Rodríguez de Guevara.[1]

Alonzo Rodríguez de Guevara: Anatomical Pioneer and Political Conspirator

Guevara incorrectly presented himself as the instigator of Spanish anatomy in his only known publication, *In pluribus ex iis quibus Galenus impugnatur ab Andrea Vesalio Bruxelensi in constructione et usu partium corporis humani, defensio* (Coimbra, 1559). The work was a defence of Galen and an attack on the

---

[1]   The few written accounts on the chronology of Rodriguez de Guevara's introductory courses in anatomy at Valladolid often conflict with one another, and both 1548 and 1549 have been suggested as the first year that he taught the course. José Maria López Piñero even claimed that 1550 was the year of introduction in his *Diccionario histórico de la ciencia moderna en España* (Barcelona: Península, 1983), p. 252. This year 1550 does not correspond with Bernardino Montaña's *Libro de la anathomía del hombre*, which appeared in 1551 and which the author explicitly claimed was based on Guevara's teachings. These chronological uncertainties do not alter the fact that anatomy was already an established practice in Valencia by the time of Guevara's arrival in Valladolid.

modern anatomy advocated by Vesalius, and was finished a decade after Guevara gave his anatomical lectures at the University of Valladolid. In the prologue, the author depicted himself as a Spanish anatomical pioneer and emphasised his early training in this field during two years of prior studies in Italy.[2] Guevara's statements have been widely accepted in later accounts of Spanish Renaissance anatomy. In 1926 the German physician Fritz Lejeune published one of the few non-Spanish articles on the topic, in which he referred to Guevara as: 'The man who can rightfully be called the father of Spanish anatomy.'[3] In a 1935 article entitled 'Los anatómicos de la epoca del renacimiento', Jose Vazquez Vicente presented a similarly erroneous account of the life and importance of the physician, who allegedly 'studied in Granada, then went on to perfect his studies in Italy where he worked for some time with Valverde, whose work on anatomy he assisted in the year 1551 ... In 1550 Charles V founded the chair of anatomy in Valladolid for him, the first in Spain and the third in Europe'.[4]

There are many obvious misrepresentations and inaccuracies in these descriptions; only extremely vague circumstantial evidence suggests any acquaintance or collaboration between Guevara and Valverde, and is based only on Guevara's alleged stay in Italy during the 1540s. The exact date of Guevara's position as lecturer in Valladolid is also uncertain, and the lectureship of

---

[2]   'Quod profecto adeo Hispaniae primatum animos commouit, ut non sine vehementi admiratione in quoddam singulare elogium eruperint, quo nihil aliud Hispanis medicis deesse testarentur quandoquidem si deesset pari diligentia conquisituros affirmarent. Sed iam unde digressa est mea se conuertat oratio, quando plus ego quam caeteri in Hispania professores huic professioni augendae astrictus viderer: primum, quia primus inter Castellanos eius fundamentum ieceram: deinde quia totius huius medicorum illustris familiae oculos ad me conuersos intuebar.' Alonzo Rodriguez de Guevara, *In pluribus ex iis quibus Galenus impugnatur ab Andrea Vesalio Bruxelensi in constructione et usu partium corporis humani, defensio* (Coimbra: Juan Baverius 1559), Ad candidum lectorem, n.p.

[3]   'Der Mann, den man seiner späteren tätigkeit wegen mit Recht den Vater der spanischen Anatomie nennen dürfte. Dies war Alonzo Rodriguez de Guevara, gewöhnlich kurzerhand Guevara genannt. Vom Jugend ab hatte er besondere Neigung für anatomische Forschungen, die ihm leider ja im Spanien kaum möglich waren. So begab er sich nach Abschluss seiner medizinischen Ausbildung nach Italien in dem Wünsche, sich in anatomischen Dingen tüchtig zu unterrichten. Eine Zeit lang soll er Prosektor in Bologna gewesen sein.' Fritz Lejeune, 'Zur spanischen Anatomie vor und um Vesal', *Janus*, 31 (1926), p. 418.

[4]   'Estudió en Granada, marchando mas tarde a perfeccionar sus estudios a Italia, donde permanició algún tiempo con Valverde, al que ayudó con su obra de Anatomía que escribió en Roma en el año 1551. A su regresa a España supo granjearse el apoyo de Maximilliano, regente del Reino durante la ausencia de Carlos V, que fundó para el en 1550, la catédra de Anatomía de Valladolid, siendo esta la primera de España y la tercera de Europa, pues solo anticiparon en antiguidad las de Montpellier y Bolonia.' Jose Vazquez Vicente, *Los anatomicos de la epoca del renacimiento* in *Trabajos de la cátedra de historia critica de medicina* (Madrid: Minuesa de los Ríos, 1935), p. 221.

anatomy in Valladolid – a temporary appointment, if any – was not the first in Spain, and certainly not the third of its kind in Europe, since the Universities of Bologna, Montpellier, Paris, Padua, Pavia, Pisa and Rome all preceded Valladolid in the teaching of anatomy. Such misconceptions have been presented repeatedly and appeared as late as 1974 in an article by Luis Fernandez Martin for the prominent journal *Cuadernos de historia de medicina española.*[5] Further uncertainty surrounds the exact nature of Guevara's practice in Valladolid. The little detail provided in Guevara's own account reported that he obtained a bachelor's degree in medicine in his native Granada and during a two-year stay in Italy, after which he returned to Spain in the late 1540s to teach anatomy at the University of Valladolid. In 1552 he completed his medical education at the University of Sigüenza and in 1556 he departed for Portugal. He then spent the next three decades in Coimbra and Lisbon as a court and university physician, engaged in both medical and political affairs. Among his many biographical uncertainties, we do not know whether he was ever appointed to a real professorship in Valladolid or if his 20 months of public dissections were merely part of a temporary university initiative.

The doubtful claims to Guevara's training in Bologna can only be based on his own passing reference to a two-year stay in Italy prior to the practice in Valladolid. Even though later accounts make unambiguous reference to his studies in Bologna, his name does not appear in Vincenzo Busacchi's thorough registration of Spanish medical students who obtained their degrees from the University of Bologna during the sixteenth century. The renowned Spanish anatomist Andrés Laguna and 176 other Spanish medical doctors are listed in this survey, but the name of Alonzo Rodriguez de Guevara is conspicuous by its absence. In addition to the uncertainty surrounding Guevara's Italian education, there is no convincing evidence to confirm the claims of his collaboration with prominent anatomists such as Juan Valverde. López Piñero did consider the idea of a link between the two Spanish anatomists in Italy in his abstract on Guevara in *Diccionario histórico de la ciencia moderna en España* (Barcelona, 1983), but also raised further questions in considering why Guevara might have decided to finish his medical education in Sigüenza instead of Valladolid following his

---

5   'De estas líneas se deduce que podemos aceptar la tesis de A. Sarría Rueda cuando afirma: "La transformación experimentada de la enseñanza de anatomía en España es promovida por el doctor Rodríguez de Guevara. Formado en Bolonia, establicado en Valladolid, a él se debe la creación de la primera cátedra de anatomía en esta ciudad y la inclusion de las enseñanzas prácticas de esta material en las restantes universidades españoles, con la implantación de los métodos docentes en vigor en las escuelas europeas". De todo lo dicho se desprende que la primera cátedra española donde de una manera sistemática se introdujo la dissección anatómica como método ordinario fue la cátedra de anatomía de la Universidad de Valladolid.' Luis Fernández Martin, 'Origines de la diseccion anatomica en la Universidad de Valladolid', *Cuadernos de historia de medicina española*, 13 (1974), pp. 359–60.

anatomical lectures there. In *Defensio*, Guevara presented himself as having been the centre of attention during these lectures, which he claimed were attended by some of the most renowned names in Spanish Renaissance medicine. Yet, after 20 months engaged in public dissections, he left this impressive network of colleagues and spectators in order to continue his medical education at the much smaller and less prestigious University of Sigüenza. López Piñero describes how: 'He graduated in medicine in 1552, not at the University of Valladolid or at any other of the important medical schools then existing in the peninsula, but at Sigüenza, the prototype of "unofficial" university centres selling university degrees somewhat unscrupulously. The great Portuguese-Jewish physician Zucato Lusitano also obtained his degree there. Did Rodriguez de Guevara perhaps have problems with "limpieza de sangre"?'[6]

The sparse studies and passing anecdotes of Guevara's life and career unfortunately do not answer the questions raised by López Piñero, nor do they clarify other uncertainties in any depth or detail. The 1956 article 'Sumula de vida interlope de Alonzo Rodriguez de Guevara' by the Portuguese historian M.B. Barbosa Sueiro is the most complete account of Guevara's life to date. In this biography, Barbosa attempts to correct ambiguities and fallacies that comprise previous accounts of Guevara's life and career. Among these amendments, Barbosa clarifies Guevara's educational background and the alleged collaboration between Guevara and Juan Valverde in Pisa or Rome, as vague and unfounded ideas simply repeated by Spanish and Portuguese historians since the mid-nineteenth century.[7] Yet Barbosa's biography was not able to put an end to these fallacies in more recent publications, and even notions of collaboration with Vesalius have been suggested in speculative and confused remarks about Guevara, whose only publication, published in Coimbra in 1559, explicitly disproved Vesalius' corrections of Galenic anatomy. A 1973 article by Vesalius' biographer Charles D. O'Malley compensated for O'Malley's previous negative remarks about Spanish Renaissance anatomy and even presented Guevara as one of Vesalius' most loyal supporters.[8] Barbosa's detailed biography of Guavara helped

────────────

[6]   'El grado de licenciado lo recibió en 1552, pero no en Valladolid ni en otra de su importancia, sino en la de Sigüenza prototipo de un centro universitario "silvestre", que concedía títulos en condiciones poco escrúpulosos. Allí se graduó tambien el grán medico judío portugués Zacuto Lusitano. Tenía Rodriguez de Guevara problemas con "limpieza de sangre?"' José Maria López Piñero, *Ciencia y tecnica en la sociedad Española de los siglos XVI y XVII* (Barcelona: Labor, 1979), p. 252.

[7]   M.B. de Barbosa Sueiro, 'Sumula de vida interlope de Alonzo Rodriguez de Guevara', *Sep. de A Medicina Contemporânea*, 72(3) (1954), p. 258.

[8]   'Otro representante de investigación anatómica es Alonzo Rodriguez de Guevara, quien luego de estudiar en Italia, y unirse allí con Valverde, fue professor de anatomía en Valladolid, y mas tarde, como ya se ha dicho en Coimbra. Su actitud se evidencia en el titulo de su defenso de Vesalio y de su anatomía.' Charles D. O'Malley, 'Los saberes morfológicos en

to explain and correct some of the obvious misunderstandings surrounding the development of the anatomist's career. However, Barbosa's study paid less attention to Guevara's anatomical work than to his political intrigues as a Spanish informant at the Portuguese court, where he was employed as Royal Physician to Queen Catherine. On June 22 1556, following a royal provision by King John III Guevara was appointed to a 'cadeira de medeçina & anotomia' at the University of Coimbra where he also established the first chair of surgery in 1557 and published his *Defensio* two years later.[9] He was additionally employed at the Hospital Real de Todos-os-Santos in Lisbon, where he organised anatomical teaching programmes for medical students following another royal decree from November 20 1556.[10] In spite of his commitments to these varied assignments, Guevara had ample time to become an accomplice in the dramatic events that led to Philip II's takeover of Portugal in 1580.[11] In 1578, Guevara accompanied the Portuguese King Sebastian I on his disastrous Moroccan crusade and was taken prisoner after the King's defeat and death at Alcácer-Quibir on 4 August that year. Guevara's Castilian origin meant that he was soon freed and he was sent back to Spain, only to reappear at the Royal Court in Lisbon in early 1579. An envoy of Philip II in Lisbon named Guevara as the King's best spy at the Portuguese court in a letter sent to the Spanish monarch, dated 19 February 1579.[12] If historians have hitherto ignored or neglected this first anatomist who was active in both Castile and Portugal, it is not because his adventures do not merit closer scrutiny;

---

el Renacimiento. La anatomía' in Pedro Lain Entralgo (ed.), *Historia universal de la medicina* (Barcelona: Salvat, 1973), p. 70.

[9]   Guevara's name appears occasionally in documentary sources from the University of Coimbra. A reference from June 22 1556 'Sobre o lente de mediçina e notomia' formalized Guevara's *appintment* at the first Portuguese university chair of anatomy, instigated by royal decree and remunerated with 50.000 reales annually: 'Eu el Rei faço saber a vos Reitor & cõselhr. os da vniversidade de Coimbra q polia boa informação q tenho do Lo alonso Roiz de gueuara e por lhe fazer merçe ei por bem & me praz q elle leia nesa vniversidade na ora & tenpo q lhe por vos for ordenado hiia cadeira de medeçina & anotomia Juntam.te a qual lera em quanto eu ouuer por bem & avera delia cinquêta mil rfs. de salairo em cada hu ano'. J. M. Teixeira de Carvalho, 'A anatomia em Coimbra no século XVI.' *Revista da Universidade de Coimbra*, Coimbra, v.4, 1915., p. 260

[10]   'E asi me praz que ele faça as notomias que parecerei necessárias e vos ordenardes dos corpos mortos dos que na dita casa falecerem e asi as que se ouuerem de fazer aos corpos dos que padecerem per justiça nesta cidade e que ele auera com o dito carreguo dose mil rs.' J. M. Teixeira de Carvalho, *A Universidade de Coimbra no século xvi*. (Coimbra: Impresa da Universidade, 1922), p. 69

[11]   These varied asignments were obviously not carried out adequately. Numerous documentary sources from the university of Coimbra include references to (and complaints about) Guavara's absence from the university, as well as fines charged to the him for not being able to fulfil his teaching obligations.

[12]   'Con esta embio relacion del mal del Rey. Mande V. M. que se me dé lo que pido para el medico, porque con este tiempo es la major espia que tenemos.' De Barbosa Sueiro, 'Sumula de vida interlope de Alonzo Rodriguez de Guevara', p. 270.

rather, it is because the source material surrounding this enigmatic character is incomplete and limited to his one publication and a few remarks written by his contemporaries – who often dwell more on his clandestine political activities than his pioneering anatomical lectures in Castile and Portugal.

In his biography of Guevara, Barbosa concluded that the anatomist is unlikely to have enjoyed any fruitful collaboration with Juan Valverde, but that he could instead have become familiar with Machiavelli's ideas and writings during his two-year stay in Italy.[13] According to Barbosa, the Granadan physician made a significant contribution to the later misfortunes of Portugal, which owed not only its first known lectureship of anatomy, but also 60 years of Spanish occupation to the vigorous efforts of Guevara.[14] Barbosa even suggested that Guevara's dismissal of Vesalian anatomy in his *Defensio* may have been a cunning attempt to appease the resolutely Galenic stance of the Portuguese court physicians.[15] This hypothesis probably exaggerates Guevara's conspiratorial skills; indeed, the introduction to his *Defensio* praised both Galen and Vesalius as outstanding anatomists, and even defended the latter during the author's initial discussion of Galen's erroneous osteology. Guevara's display of sympathy towards Vesalius indicates that the author was aware of Luis Collado's own 1555 defence of Vesalius, even though Guevara's treatise served the opposite purpose, siding with Galen instead of with Vesalius. Guevara himself acknowledged the incomplete character of his work, which was not a systematic treatise on anatomy, but rather a discussion of inconsistencies in the anatomical theories set forth by Galen and Vesalius.[16]

Guevara's textbook comprised a rather repetitive summary of Vesalius' many mistakes in the anatomy of the ear, muscles, eyes, nerves and blood vessels. Guevara divided his counter-evidence into three volumes and a total of 28 chapters, each of which dealt with a different anatomical category. Rather than writing a systematic treatise on anatomy, Guevara compared and disproved some of the innovations and

---

[13] 'Além da anatomia, em outra ciencia ou arte se teria Guevara enfronhado durante a sua estadia na Itália, ilustrando uma vocacao que lhe borbulhava no imo do pensamento – a da espionagem. A política entre os pequenos estados italianos e a política interna de cada um deles orientavam-se conforme as regras do mais esmerado maquiavelismo.' Ibid., pp. 258–59.

[14] 'Alonzo Rodriguez de Guevara promoveria, efectivamente, a institucao do ensino anatómico em Portugal, mas a obra foi de acanhadas proporcoes, teve minguado influxo na evolucao da medicina portuguesa e ficou desprovida de continuidade. Se, pois, na historia da cultura e do ensino científico de Portugal o nome de Guevara se inscreve com relevo limitado, na historia do nosso infortúnio de 1580 afirma-se ele como o de um astro de primeira grandeza, pela accao lata que o arquiatra exerceu, como persistente e consumado espiao.' Ibid., p. 287.

[15] Ibid., p. 277.

[16] 'In cuius opusculi discursu de omnibus ordinate membris non agimus, sed de illis tantum de quorum situ aut substantiae modo, or tu aut insertione, circumscroptione aut usu ambigitur: si modo ta lia membra in humana fabrica reperiantur.' Rodriguez de Guevara, *In pluribus ex iis*, n.p.

corrections presented by Vesalius in *De humani corporis fabrica*, the first edition of which had claimed to divulge 200 corrections of Galenic anatomy. While Barbosa presented Guevara's Galenic stance as a deceitful attempt to appease the Portuguese court, this support of Galen can be traced back to his teaching post in Valladolid, thus preceding his arrival in Portugal by several years. Guevara's Valladolid lectures did not follow the theories or even the terminology of Vesalius, nor of any of his followers in Italy and Valencia. The names of these anatomists and their discoveries are absent from the book inspired by Guevara's teachings, Bernardino Montaña's *Libro de la anothomía* (Valladolid, 1551). Guevara's own *Defensio* offered only limited insights into his personal experience as an anatomist. The *Defensio* was written and published during his residence in Portugal, but included no details of the anatomical research he was practising at the time; rather, it focused on the author's experiences at the University of Valladolid almost 10 years earlier.[17] Among those mentioned in the text were the famous surgeon Dionisio Daza Chacón and the royal surgeons of Valladolid's Hospital de la Corte, Torres and Herrera, who were referred to by Chacón in his own work, *Teorica y practica de cirugía* (Valladolid, 1580), as his uniquely gifted mentors.[18] The cast list in Guevara's book also included the medical doctors Cespedes and Ledesma, who held prominent professorships in the medical faculties of Salamanca and Alcalá, respectively, with the latter also performing the roles of royal physician and inquisitor. Their exclusive presence was an indication of the broad interest in Guevara's pioneering public dissections, which seemingly extended far beyond Valladolid. The name of Medina among Guevara's spectators almost certainly refers to Cosme de Medina, the first Professor of Anatomy at the University of Salamanca, who later graduated with a bachelor's degree in medicine from the University of Valladolid.[19] Medina's initial public dissections, performed in 1551–52, were well attended and were recorded

---

[17]   Ibid., n.p. To get an idea of the anatomical activity at the University of Coimbra around this time the surviving 1559 university statutes fill some of the lacunae in Guevara's textbook published the same year. Apparently the anatomical training was based on eight books of Galen's *De usu partium* as well as surgical textbooks by both Galen and Guy de Chauliac. The stipulated dissections took place over a period of maximum four days between October and February; these were organized as two annual systematic anatomies carried out on human bodies, and six sectional anatomies carried out on animals: 'Ho lente de Anothomia será obrigado, do principio de Outubro até o fim de Fevereiro, a fazer duas anothomias universais ... E, alem disto, o dito anothomista será obrigado a fazer seis anothomias particulares de membros de animais brutos que lhe parecer que conforman mais com ha compreiçam humana.' Serafim Leite (ed.): Estatutos da Universidade de Coimbra (1559). Coimbra: Por ordem da Universidade, 1963, pp. 300-301

[18]   Dionisio Daza Chacón, *Teorica y practica de cirugía* (Valladolid, 1582), *Prologo al letor*, n.p.

[19]   Teresa Santander, *El Doctor Cosme de Medina y su biblioteca (1551–1591)* (Salamanca: Centro de Estudios Salmantinos, 1999), p. 16.

by an anonymous notary, but, unlike Guevara's, unfortunately did not leave behind a detailed list of bystanders. While Medina's anatomy lectures attracted mostly local students and medical professors from Salamanca, Guevara's course drew a large audience from the periphery of a strictly medical context, such as the natural philosophers Oñate and Peñeranda, and outside visitors with no formal affiliation with the university. Among these were the attending court physicians – Cartagena, Montaña and Madera – the latter was employed in Madrid, the future Spanish capital, and the other two at the court, which at that time was primarily based in Valladolid. The presence of court officials might be interpreted as a defiance of their expected daily routines and obligations, but it seems instead to have been backed by royal support, as indicated in a formal request from the royal council produced soon after Guevara's lectures took place. The very detailed royal petition presented by Maria of Austria in the absence of her father Charles V, was made in March 1551 and called for anatomy to be taught in Valladolid on a permanent basis. This document, kept in the Archivo General de Simancas and unknown until its discovery and publication in 1974, explicitly encouraged the idea of public dissections for a broader audience than those residing within the university walls.[20] The decree was the second of three provisions between 1550 and 1552 targeted at the most prominent Castilian universities in Salamanca, Valladolid and Alcalá de Henares. There is no convincing evidence, however, that this particular petition had any short- or long-term influence, even though the medical historian Anastasio Rojo Vega has indicated the existence of a permanent chair of anatomy at the University of Valladolid in his *Medicina barroca Vallisoletana* (Valladolid, 1984). In his investigations of the university's account books, Rojo Vega traced references to expenses paid for a professorship in anatomy, but it should be noted that no additional documentation supports this claim.[21]

According to Rojo Vega, a certain Pedro de Sosa held a chair in anatomy as late as 1592, when he was promoted to the first chair of medicine to replace Luis Mercado, who had advanced to Spain's highest medical position of *Protomédico general*. However, other evidence seems to invalidate this claim of anatomical continuity at Valladolid in the late 1500s. At the time of his death in 1611, the 84-year-old Mercado seemed completely unaware of any anatomical practices having taken place within his medical faculty, and was even oblivious to the fact that it had been the first Castilian institution to carry out public dissections

---

[20] 'Que hagáis saver y publicar en algunos de las dichas universidades y en otros pueblos principales cómo se ha de hazer la dicha anatomía para algún día señelado para que las personas que quisieren vengan a la ver hazer y tengan tiempo para ello, y los unos ni los otros non fagedes ende al por alguna manera so pena de la nuestra merced e de diez mil maravedís para la nuestra camera a cada uno que lo contrario hiziese.' Fernández Martín, 'Origines de la diseccion anatomica en la Universidad de Valladolid', p. 361.

[21] Anastasio Rojo Vega, *Medicina barroca Vallisoletana* (Valladolid: University of Valladolid, Secretariado de Publicaciones, 1984), p. 22.

only two generations earlier. Mercado was *Protomédico general*, former court physician to both Philip II and Philip III, and still a powerful overseer at the medical faculty of Valladolid, but rejected a proposal from the University Senate to establish a professorship of anatomy. He gave the following reason for his decision, denying the advancements made in anatomy in mid-sixteenth-century Valladolid in what seems to be a surprising display of short-term memory loss: 'In Spain good medicine has been practiced for more than 200 years without use of that discipline, and we do not have any persons in the country who are sufficiently trained in that practice.'[22]

Guevara's remarks do not indicate any lasting legacy of anatomical studies at the University of Valladolid and it seems most likely that he did not have any successors in this field for the remainder of the century. This lack of continuity makes it easier to comprehend his move to Sigüenza following his 20 months of teaching at Valladolid, which had possibly been intended to be a finite and isolated event. The unsuccessful royal provision made in March 1551 did not refer to any prior experience in this field within the university, but presented its proposal, *Para que se haga la anotomia*, as an attempt to make up for the many errors and misconceptions relating to previous medical practices.[23]

While Guevara's teaching period appears isolated and discontinuous after close scrutiny, one spectator found time during and after these lectures to produce the first anatomy book to be written in the Spanish vernacular. Bernardino Montaña de Montserrate's *Libro de la anothomía del hombre* was inspired by Guevara's lectures and was published in Valladolid on 2 November 1551, which

---

[22]   'En España se ha hecho buena medicina durante más de doscientos años sin necessidad de tal disciplina y no existen en el país personas lo sufficiente preparadas para su practica'. Cited in Àlvar Martínez-Vidal and José Pardo-Tomás, 'Anatomical Theatres in Early Modern Spain', *Medical History*, 49 (2005), p. 254.

[23]   'A vos, los alcaldes de la nuestra audiencia y chancillería que está y reside en la villa de Valladolid y a vos, el que es fuere nuestro procurador o juez de residencia y a otros qualesquier justicias de la dicha villa y a cada uno e qualquier de vos, salud e gracia. Sepades que nos fue fecha relación que una de las principales y necesarias cosas para saber y entender la ciencia de la medicina es tener conocimiento del sitio y asiento de los miembros humanos interiores y exteriores lo qual diz que se adquiere con solas sacciones y anatomias de los cuerpos humanos y no con otra y que por lo contrario los más y mayors herrores en que caen los que executan e usan la dicha arte es quando quier que son leydos y visto en los autores sin tener noticia de las dichas anatomías lo qual se remediaría si los medicos asi teóricos como manuales viesen y supiesen todos los miembros y partes de los cuerpos humanos muchas veces yendo por esas hedades y regiones y complesiones porque mediante la frecuencia y diversidad destas anatomías diz que se adquire la pláctica que es muy necessario sobre lo qual por unas nuestras cartas embiamos a mandar a las universidades de los estudios de Salamanca y Alcalá de Henares que placticasen en ello llamando personas expertas y la resolución que tomasen y orden que en ello se debía tener lo embiasen ante nuestro consejo.' Fernández Martin, 'Origines de la diseccion anatomica en la Universidad de Valladolid', p. 360.

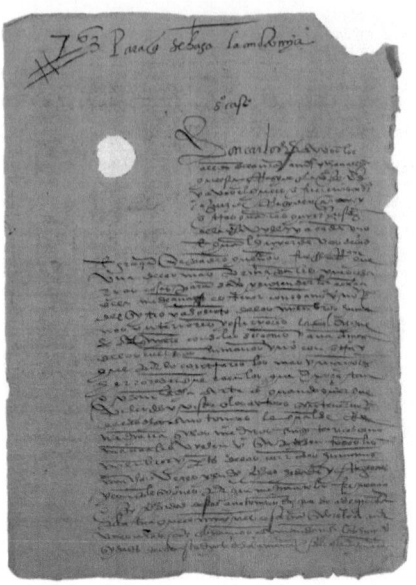

Figure 4.1   Provision *Para que se haga la anotomia*, signed in Valladolid on 16 March 1551. This royal decree by Maria of Austria – in the absence of her father Charles V – put forward a plea for the creation and consolidation of permanent anatomy studies at the University of Valladolid and was the second initiative of its kind targeted at the three leading Castilian universities in Salamanca (1550), Valladolid (1551) and Alcalá de Henares (1552). There is no evidence, however, that this petition had any short- or long-term effect, and the University of Valladolid remained without a permanent chair of anatomy throughout the late sixteenth and seventeenth centuries. Courtesy of España. Ministerio de Educación, Cultura y Deporte. Archivo General de Simancas. RGS, 155103

we would estimate to be a year after the conclusion of Guevara's anatomy course. The work demonstrates a categorical reliance on Galenic doctrines – the same attitude which appeared eight years later in Guevara's *Defensio* – and presents Valladolid as rather disconnected from the anatomical reforms carried out elsewhere in Castile and Aragon. The authors' awareness of contemporary Vesalian anatomy is only revealed in a number of crude woodcuts directly copied from Vesalius' *Fabrica*.

Bernardino Montaña de Montserrate's *Libro de la anothomía del hombre*

The full title of Bernardino Montaña de Montserrate's anatomical textbook was the rather pompous (and long-winded) Libro de la anothomía del hombre. Neuamente compuesto por el dr. Bernardino Montaña de Montserrate Medico de su Magestad. Muy util y necessario a los medicos y cirujanos que quieren ser perfectos en su arte y apazible a los hombres que huelgan de saber de los secretos de la naturaleza. En el qual libro se trata de la fabrica y compustura del hombre, y de la manera como se engendra y nasce, y de las causas porque se necessariamente muere. The book was dedicated to Montaña's patron, Luis Hurtado de Mendoza, Marquéz de Mondejar, at that time President of the Council of the Indies and later of the Royal Council of Castile, as well as a prominent member of the Royal Council of War and State. It was divided into three parts, of which the first was a treatise on anatomy with a detailed description of the development of the human body and some of the causes for its deterioration and death. This first part opened with three chapters for surgeons unfamiliar with anatomy and introduced basic anatomical terminology and definition. These chapters were followed by a further nine chapters on different sections of the body, a capite ad calcem, with independent explanations of the anatomy of the head, the facial region, the neck, the upper extremities, the vertebral column, the thorax, the abdomen, the organs of generation and the lower extremities.

The second volume was entitled Un coloqio del illustrissimo señor don Luys Hurtado de Mendoza Marquéz de Mondejar, con el doctor Bernardino Montaña medico de su Magestad acerca el dicho sueño soño su señoria de la generación nacimiento y muerte del hombre. This second part was a tribute to Montaña's learned patron and his 'dream', which consisted of a series of different questions on features and functions of the human body, written in the form of a Renaissance dialogue, like that found two years earlier in Jimeno's *De re medica* (dedicated to the Marquéz' female relative, Mencía de Mendoza).[24] In the second volume of *Libro de la anothomía*, inspired by a vivid dream, Marquez de Mondejar asked questions on humoral medicine and morphology which were readily answered by his skilful and knowledgeable physician. A third and smaller last part of Montaña's opus comprised a series of previously uncommented illustrations, most of which had been plagiarised from Vesalius' De humani corporis fabrica. Due to its use of the Spanish vernacular and the inclusion of illustrations copied directly from Fabrica, Montaña's Libro de la anothomía has been crucial to subsequent misunderstandings that the University of Valladolid was the birthplace of anatomical studies and the leading institution of Spanish anatomy in the post-Vesalian era. An isolated reading of Libro de la anothomia and its dedication to the reader might lead its reader to understand that Valladolid was the centre of anatomy studies within the Spanish realms, an institution equalled only by the Universities of Bologna and Montpellier: 'The surgeon who wishes to study it well (division or dissection) should go and practice at the universities where it is carried out regularly, such as Montpellier in France, Bologna in Italy, and Valladolid in Spain, where once again the discipline is practiced skillfully,

---

[24]   A third curious example of Mendoza patronage of Spanish scholars engaged in anatomical studies is seen in Elisa Andretta's recent study, which has found evidence of anatomical practice in the imperial Spanish ambassador Diego Hurtado de Mendoza and the Spanish humanist Juan Páez de Castro's private Aristolelian Academy in Trent during the winter 1545-1546. The event was referred to by Páez de Castro in a letter dated 25 March 1546: "este ibierno (invierno) una anotomia la mejor del mundo, y con muy pocos, y mui doctos (...) he pasado todo lo que haze a este menester diligentemente, confiriendo a Galeno, y a Vesalio." Cited in Elisa Andretta: 'The Medical Cultures of "the Spaniards of Italy": Scientific Communication, Learned Practices, and Medicine in the Correspondence of Juan Páez de Castro (1545-1552).' *Medical Cultures of the Early Modern Spanish Empire*. (Farnham: Ashgate Publishing Limited, 2014). p. 134

and with support from the council of His Majesty, by Bachelor Rodriguez, a surgeon, and an excellent man experienced in this art.'[25] In his prologue to the reader, Montaña credited Guevara, but unfortunately offered no verification of the exact background of his expertise, which has since led to much speculation and uncertainty. Interestingly, Guevara was not credited as the initiator of public dissection in Castile, but rather as the central character in the re-establishment of anatomy in Spain. Montaña thus indicated knowledge of previous anatomy studies, which may refer to the documentary evidence of such practices in Lleida, Valencia, Zaragoza and Barcelona. Fewer historical sources supported Montaña's description of Guevara's course as an initiative backed by the Royal Council, since the royal provision from 1551 did not refer to previous attempts to introduce anatomy studies in Valladolid. In addition, his indications of the permanent character of such anatomical teaching did not correspond with information found elsewhere. While Guevara's own account referred to only 20 months' lecturing, Montaña's textbook suggested that anatomical studies had continued beyond this relatively short period. Montaña even called for future Iberian surgeons to set their sights on Valladolid, which was presented as one of three European centres of contemporary anatomy studies.

While Montaña made use of the illustrative plates from *Fabrica*, he made no reference to Vesalius or to the new approach to anatomy introduced at the University of Padua. Instead, he credited the University of Bologna, by that time a far less significant institution, as the Italian centre of contemporary anatomy. Considering his strict adherence to Galenic orthodoxy, it is surprising that Montaña referred to Montpellier as the centre of French anatomy, rather than Paris, where Vesalius' former master Jacques Du Buis (Iacobus Sylvius) was still the unrivalled master and defender of Galenic doctrines, and where his fierce attack on the osteology of his ill-mannered and 'mad' pupil *Vesanus* was published in the same year as Montaña's textbook. *Libro de la anothomía* relied on writings by Henri de Mondeville, the Norman surgeon who spearheaded the teaching of anatomy at the medical school of Montpellier already in the early fourteenth century. A brief remark relating to Montaña's own journey from France to Spain in the early sixteenth century also indicated that part of the author's medical education was obtained at Montpellier's renowned medical school.[26] Montaña

---

[25]   'Y porque esta division es difficultosa de hazer como cumple, y require cirujano sabio y experimentado en ello que le haga, conviene que el cirujano que quiera bien hacella (la división o la dissección) vaya á prender este exercicio á las universidades donde se acustumbra de hacer ordinariamente, como en Francia á Mompellier, en Italia á Bolonia, en España á Valladolid, donde agora nuevamente se comienza a hacer muy artificiozamente, con auctoridad del cosejo de Su Majestad por el Bachiller Rodriguez cirujano, muy excelente hombre y experimentado en este arte.' Montaña, *Libro de la anothomia del hombre*, f. 3r.

[26]   'Ansi mismo en el año de 1513 viniendo yo de camino para España, vi en Tolosa un cauallero von un apostema duro en la ingle.' Ibid., f. 82v.

credited Guevara as the person responsible for the resurrection of anatomy in Spain, while Guevara rewarded this appreciation in his own account, which accentuated the devotion he had witnessed in the old court physician, who was among the more prominent attendees of his lectures. Guevara's 1559 account described how the ageing physician, in spite of his advanced age and painful struggles with gout, had insisted to be carried on a litter so that he might attend all of Guevara's lectures.[27]

As stated earlier, Spanish historians of medicine have widely accepted Montaña's claim that the University of Valladolid was the only medical school in Spain to practice and teach anatomy during the mid-sixteenth century. By the time of the publication of *Libro de la anothomia* in November 1551, the University of Salamanca, and perhaps the University of Alcalá as well, had already imported anatomists from Valencia to introduce the practice of dissection in Castilian institutions. Even though Montaña was of Catalan origin, the 70-year-old court physician, who was born around 1480 in Montserrate near Barcelona, was seemingly unaware of this influx of Valencian anatomists into the neighbouring Castilian universities. Montaña's textbook was devoid of any references to other contemporary authorities in the field of anatomy – with the exception of Guevara – and only revealed a dependence on Vesalius in a number of distorted woodcuts which had been clumsily plagiarised from *Fabrica*. While Montaña's work obliged the visual facet of the 'Vesalian revolution', it rejected another of its other crucial aspects and insights, namely the idea of obvious incompatibility between human and animal morphology. According to Montaña, the anatomist could reach general conclusions on the structure of the human body by considering the results of dissections of some of the larger mammals.[28] The most significant contribution of Montaña's work to the study of anatomy did not lie in its importance as an anatomical atlas of the human body – as such, it was in many ways an anachronism by the time of publication. The conspicuous novelty of this anatomical textbook should rather be seen in its deliberate and confident use of the Spanish vernacular.

In many ways Montaña's book represented one of the first steps towards a more general tendency to publish Spanish books on anatomy and surgery in the native Castilian language. This development was rooted in a deliberate national language policy that preceded other contemporary nations. On the instigation of the Catholic Monarchs, Antonio de Nebrija's national grammar

---

[27] 'Insignis doctor Montanha, qui septuaginta annos natus misere infestissima podagra vexatus, cum multas in Medica Facultate tulisset palmas, ut postremo de arrogantia triumpharet, continuó lecticula junctus lectionibus intererat.' Rodriguez de Guevara, *In pluribus ex iis*, n.p.

[28] 'Asi en el cuerpo humano como en algunos animales que son semejantes al hombre en su compestura, como los puercos en los miembros naturales y las monas en todo lo demás.' Ibid., *Dedicatoria epistolaria*, n.p.

books and dictionaries had been published in the last decades of the fifteenth century with the prophetic statement 'que siempre la lengua fue compañera del Imperio.'[29] No other European national language was promoted as early or as systematically, and only five years after Montaña's anatomy, Juan Valverde published his *Historia de la composicion del cuerpo humano* (Rome, 1556) in his mother tongue rather than in Latin, indicating the increased dominion of Spain over Italy. Valverde's introduction appealed to an audience of fellow Spaniards unfamiliar with the humanist Latin discourse, which Vesalius had taken pride in imitating, bemoaning: 'The damage, which this causes to the entire Spanish nation, partly because the surgeons (who miss the most by not understanding it) know only a little Latin, and partly because Vesalius has written in such a complicated style that can only be understood with difficulty, and only by those who have already seen an anatomised corpse before their eyes: therefore it seemed to me a wise thing to write this book in our own language.'[30] Few decades later, the Valladolid surgeon Dionisio Daza Chacón introduced his magnum opus, *Práctica y teorica de chirugía* (Valladolid, 1580), with a similarly thorough account of the reflections that had led him to abandon the universal language of European science and medicine in favour of his own. The vernacular revolution spearheaded by Montaña's work was consolidated by the time of Daza Chacón's two-volume treatise of his life and experience as an imperial surgeon, of which the first part was published in 1580 and the second in 1593. In his prologue to the reader, the author presented his ambition to break away from a long tradition of Latin treatises on surgery. Chacón explained his decision to write for a Spanish audience in their own language as necessary and worthwhile, but also difficult and laborious.[31]

---

[29]   Eugenio de Bustos Tovar, 'Nebrija, primer linguista español' in *Nebrija y la introduccion del renacimiento en España* (Salamanca: University of Salamanca, 1983), p. 206.

[30]   'Y visto el daño que desto se sigue a toda la nación Española, parte por los Cirujanos (a quien falta más haze no entenderla) saber poco latín, parte por auer escrito el Vesalio tan escuramente, que con difficultad puede ser entendido, sino de aquellos que primero algunas vezes an tenido el cuerpo delante de sus ojos, y muy bien maestro que se le declare: pareciame cosa muy conueniente, escribir esta historia en nuestra légua.' Juan Valverde, *Historia de la composicion de cuerpo humano* (Rome, 1556), Al illustr. y Reverendiss. S. Fray Joan de Toledo..., n.p.

[31]   'Antes que te dé quanta de mis trabajos y peregrinaciones (discreto lector) te quiero dezir la occasion que me movió a escrivir en nuestra lengua Española, antes que en Latin, que cierto a mi me fuera muy mas facil fin comparación, y menos trabajo hazerlo en ésta más que en aquella; y la razon es evidentissima, porque si escriviera en Latin no fuera necessario buscar la propia interpretación del vocablo que usan los cirujanos romancistas, ni traducir los textos de los antiguos y modernos que me ha sido de grandissimo trabajo.' Daza Chacón, *Teorica y practica de cirugía*, *Prologo al letor*, n.p.

Montaña, Valverde and Daza Chacón presented their publications as attempts to communicate directly with Spanish surgeons unfamiliar with new anatomical insights and discoveries published in Latin. The three authors emphasised the increased burden of work necessitated by their decision to apply a new terminology to treatises on human anatomy, which traditionally had been linked to a classical system of classification. Montaña was confident of the innovative nature of his work in the field of Spanish medicine, which according to the author had become a largely philological enterprise, based on centuries of incomprehensible doctrines: 'These days the medical doctors are so keen on Latin that they devote all their thoughts to that language ... I am pleased to write this book in vernacular, because many surgeons and other humble men unfamiliar with Latin will profit from reading it.'[32] Instead of bookish Latin doctrine, Montaña suggested a direct reading of the human body, 'the alphabet, which those who want to become doctors will have to study', and he criticised his fellow Spaniards for resisting the introduction of this new science: 'In these Spanish kingdoms no one wants to know or study it, as if it were an imprudent thing.'[33] The ageing Montaña, who had allegedly obtained his degree in medicine as early as 1508, had little respect for the majority of practitioners of medicine in his native country: 'They could more fittingly be referred to as enemies, rather than healers of nature.'[34] The only solution would be to acquaint these theorists with direct studies of anatomy, disease, diagnosis and prognosis, all of which could be obtained through a direct empirical reading of the human body. Even though Montaña remained a stern defender of Galen throughout his work, his epistemological knowledge and approach corresponded with the empirical and sceptical programme of contemporary Vesalian anatomy.[35]

---

[32] 'Los medicos estan tan aficionados al Latin, que todo su pensamiento emplean en la lengua y lo que hace al caso, que es la doctrina, no tienen más pensamiento dello que sino la leyesen ... E holgado de escrivir este libro en romance – nos dice – porque muchos cirujanos y otros hombres discretos que no saben Latin, se querrán aprender a leerlo.' Montaña, *Libro de la anathomia del hombre, Epistola dedicatoria*, n.p.

[33] 'El alphabeta por donde han de comenzar a estudiar los que quieren ser medicos ... Estos Reynos de España donde no se trata de saberla ni estudiarla más que si fuesse cosa muy impertinente.' Ibid., n.p.

[34] 'Más propiamente se pueden llamar enemigos de la naturaleza que ayudores della.' Ibid., n.p.

[35] 'Por experiencia partiendo realmente el cuerpo por sus partes y mostrando cada uno dellas por vista de ojos y por el setimiento del tacto. Por esta razon es nuestro consejo que el medico o cirujano que quisiere saber cumplidamente esta ciencia, se exercite en ver hazer anathomía real y verdadera muchas veces por incisión de manos ansí en el cuerpo humano como en algunos otros animales que son semejantes al hombre en su compostura, como los puercos en los miembros naturales, y las monas en todas las demas', Ibid., *Prohemio*, f. 2v.

In his experimental 'Castilianisation' of anatomical terminology, Montaña presented some rather absurd classifications and constructions. His chapter on osteology referred to the femur as *huesso largo*, the sternum as *huessos del pecho* and the atlas as *el primero de los dichos huessos que integran el espinazo*. Further innovative terms included the non-existing *huesso del corazon*, and a new name for the ovaries, *testiculos de la madre*.[36] Arabic terms also crept into Montaña's new terminology, such as his reference to the *mirach* in descriptions of the anterior abdominal wall, and to the *siphac* in his description of the peritoneum. A familiarity with Arabic language and medicine was not surprising from a Spanish medical doctor born around 1480 and also reveals that ongoing state initiatives to eradicate the legacies of Moorish culture had been largely unsuccessful by the mid-sixteenth century. While contemporary and later prohibitions were placed on the use of the Arabic language within the Spanish realms, Enrique Jorge Enriquez's *Retrato del pefecto medico* (Salamanca, 1595) nonetheless presented such knowledge as fundamental to the education of the ideal physician.[37]

While Montaña highlighted the novelty of his work and methods, parts of his text were in fact replicates of century-old writings on medicine and surgery. As Charles D. O'Malley has shown, Montaña's publication included direct translations of medieval texts by the French surgeon, Henri de Mondeville, similar to the English anatomist Thomas Vicary's vernacular translations of Mondeville from 1577. O'Malley saw these two contemporary translations as signs of a vernacular revolution and 'a trend, which was beginning to sweep the whole of Europe in this century. This was a factor, which must have contributed immeasurably to the improvement of standards among the apprentice surgeons, the great majority of whom could read no Latin'.[38] The idea of a general vernacular revolution having taken place by the mid-sixteenth century should not be overestimated, as groundbreaking works on science and medicine were still being written and published in Latin centuries later. In a Spanish context, however, the use of the vernacular appeared at a very early stage and was undoubtedly rooted in Spain's increasing confidence and undisputed position as the European and global superpower of the age. In Huarte de San Juan's *Examen de ingenios* (Baeza, 1575), a study on the necessary preconditions for the perfect practitioner of science and medicine, the author presented the essence of this linguistic self-assurance: 'None of the serious authors of antiquity used foreign languages to explain their theories. The Greeks wrote in Greek, The Roman in Latin, The Jews in Hebrew, and the Moors in Arabic. I therefore use Spanish,

---

[36]   Ibid., f. 61r.

[37]   'Que tenga conocimiento de la lengua Arabiga.' Enrique Jorge Enriquez, *Retrato del perfecto medico* (Salamanca: Juan y Andres Renaut, 1595), p. 158.

[38]   Charles D. O'Malley, 'Bernardino Montaña, Author of the First Anatomy in The Spanish Language', *Journal of the History of Medicine*, 1 (1946), p. 91.

which I know better than any other language.'[39] In spite of the separatist tendency in scientific and medical publications of the late sixteenth century, works by authors such as Valverde and Huarte were exceedingly popular outside of Spain. They soon reached a wide European audience who either acquainted themselves with Castilian or read their works in the numerous Latin, Italian, German, and Dutch translations produced during the sixteenth and seventeenth centuries.

Unsuccessful Attempts to Restore the University Chair of Anatomy

It may not be accidental that authors in Valladolid wrote the first vernacular books on anatomy and surgery of the post-Vesalian period. Valladolid was at the same time the birthplace of Philip II, the occasional and de facto capital of the Spanish Empire during both the sixteenth and seventeenth centuries, the seat of the Consejo Real and arguably the cradle of the purest Castilian language. It is, in this context, paradoxical that Valladolid remained the only large Castilian university town without a professorship of anatomy during the early modern period, particularly when we consider that the 1551 royal petition to the University of Valladolid vehemently insisted that a professorship should be established without further delay. Chairs of anatomy were at that time being created elsewhere in Castile, and it is surprising that Valladolid, unlike the Universities of Salamanca and Alcalá de Henares, did not act upon the royal provision. The University of Valladolid remained without a chair in anatomy, not only for the remainder of the sixteenth century but seemingly throughout the entire seventeenth century as well. A 1598 treatise by the physician Cristóbal Pérez de Herrera sought to remedy the seemingly well-known shortage of anatomists and surgeons in the Castillian medical faculties – as well as the sad state of Spain's many unemployed and unskilled poor – by proposing that every year, 8–10 poor children should be taught anatomy and surgery in Valladolid and Salamanca.[40]

During this period, at least two documented initiatives were launched at the University of Valladolid with the hope of reintroducing the short-lived anatomical lecture programme of the mid-sixteenth century. The first of these proposals was the request made by the University Senate in 1611, which,

---

[39] 'Ninguno de los graves autores antiguos fue a buscar lengua extranjera para dar a entender sus conceptos; antes los griegos escribieron en griego, los romanos en Latín, los hebreos en hebraico y los moros en arábigo; y asi ago yo en mi español, por saber mejor esta lengua que otra ninguna.' San Juan de Huarte, *Examen de ingenios* (Baeza: Juan Baptista de Montoya, 1575), Al lector, n.p.

[40] 'Y aun, en los que se hicieren en Salamanca y Valladolid, se podrían añadir otros ocho o diez en cada uno, para que se les enseñase anatomía y cirugía.' Cristóbal Pérez de Herrera, *Discursos del amparo de los legitimos pobres y reduccion de los fingidos* (Madrid: Luis Sanchez, 1598), p. 57r.

as previously described, was fiercely rejected by Luis Mercado, the officially retired yet still active overseer of the medical faculty. A similar proposition was put forward in 1626 by medical students who had found evidence of previous anatomical practices at the university, likely referring to the brief course taught by Rodriguez Guevara launched in the late 1540s: 'It would be a good idea for curing cases of disease here and within the University if there was a chair of anatomy, as there once used to be, and as there is in Salamanca.'[41]

Ironically, the plea for the establishment of a university chair such as the one in Salamanca emerged at the same time as the criticism made by Jorge Enriquez of the Salamancan professorship of anatomy, as documented in the previous chapter, regarding the poor working conditions in a ruinous and crumbling anatomy theatre.[42]

It seems that the negligence or even ignorance of the anatomical programmes introduced in the mid-1500s represented a fairly general tendency in the first decades of the seventeenth century. The unsuccessful royal provision of 1551 sought to ensure that Castile would not be left behind in the anatomical reforms launched elsewhere and offered several proposals to ensure the best imaginable working conditions during the cold winter months for future anatomists at the University of Valladolid.[43] In an account of the history of Valladolid, *Descripción e Historia Natural y Moral de Valladolid*, written in 1602 by the Portuguese traveller and chronicler Pinheiro de Veiga, the author dwelt on the fact that the city housed an unusually high number of hospitals, which could easily have

---

[41] 'Si convendría para ocasiones semejantes de la enfermedad (que) de aqui en adelante en esta Universidad (hubiera) una cátedra de anatomía pues antiguamente la solía haver... y la ay en Salamanca.'Anastasio Rojo Vega, *Medicina barroca Vallisoletana.* (Valladolid: Universidad de Valladolid, 1984) p. 21.

[42] Teresa Santander Rodriguez, 'La iglesia de San Nicolas y el antiguo teatro anatómico de la Universidad de Salamanca' in Luis Enrique Rodríguez-San Pedro Bezares (ed.), *Historia de la Universidad de Salamanca. Volumen III: Saberes y confluencias* (Salamanca: University of Salamanca, 2002),p. 263.

[43] 'En complimiento de lo qual las dichas universidades enviaron sus paresceres e vistos en el nuestro claustro y porque por ellos consta que hay mucha necesidad e conviene para la salud humana que se haga la dicha anatomía e consultado con la Serenísima Reyna de Bohemia, nuestra muy cara y muy amada hija e nieta, gobernadora de estos nuestros reynos de España por ausencia de mi el emparador y rey della, fue acordado que devíamos mandar dar esta nuestra carta para vos en la dicha razón y nos tobímoslo por bien, por lo qual damos licencia y facultad para que en los meses de noviembre, diciembre, enero y febrero de cada año se pueda haxer anatomía de un cuerpo de los que se condenasen por delitos graves a pena de muerte y se executase en ellos la dicha pena o de los que muriesen en alguno de los ospitales qual paresciere que más conviene a los medicos de la Universidad desa dicha villa y que la dicha anatomía fagan interrar el dicho cuerpo a los confrades que heran obligados a le enterrar y llevar como si no se hiziera la dicha anatomía.' Luis Fernández Martin, 'Origines de la diseccion anatomica en la Universidad de Valladolid', pp. 360–61.

hosted some of the dissections requested by the 1551 provision.[44] As well as these numerous institutions, the crown presided over its own *Hospital de la Corte* which served as an infirmary for the Royal Palace and its staff, and in which some of the most prominent Spanish surgeons of the previous century had received their education. Daza Chacón served in this hospital from 1557 onwards and later recorded his close cooperation with the two surgeons Torres and Herrera (also referenced in Guevara's list of attendees), who he described as his unequalled masters: 'In their century the world did not have better surgeons than those two.'[45] In his *Teorica y practica de Cirugía*, Chacón emphasised that the ideal surgeon must commune himself thoroughly with the new anatomy of Vesalius and Valverde, but made no reference to the anatomical studies taking place in his own city. He did not even refer to the brief course given by Rodriguez Guevara in the late 1540s, even though Guevara explicitly numbered Chacón among the audience in his own *Defensio*. The author's experience with anatomy gathered by the time of the publication of *Teorica y practica* derived from Chacón's practice as a field surgeon in Germany and Flanders, and not from his native city.

In his *Medicina española renascentista* (Salamanca, 1980), Luis Granjel represented the opposing 'open' and 'closed' tendencies of late Renaissance medicine in two of its most monumental characters, Fransisco Valles and Luis Mercado, *Protomédicos generales* and the two personal and preferred physicians of Philip II.[46] Fransisco Valles summoned the Vesalian disciple Pedro Jimeno to teach anatomy at the medical faculty of the University in Alcalá de Henares at an unknown date in the 1550s and, due to his successful treatment of King Philip's gout, was known even in his lifetime as 'el divino'. Valles was a devoted Galenist, but was also active in introducing Vesalian anatomy to the medical faculty of Alcalá de Henares. Luis Mercado presented himself as an upholder of ancient

---

[44] '20 hospitales, todos con renta, contando el de los niños de la doctrina, de los Expósitos, de los Orates, de Anton Martin, y de San Juan de Dios, y otros que tienen toda la forma de conventos.' Leopoldo Cortejoso, 'Los hospitales de Valladolid en tiempos de Felipe III', *Asclepio*, 12 (1961), n.p.

[45] 'En su siglo no tuvo el mundo mayores cirujanos que ellos.' Daza Chacón, *Practica y Teorica de Cirugia*, *Prologo al letor*, n.p.

[46] 'Las grandes figures de la segunda mitad de la centuria, como Fransisco Valles y Luis Mercado, representantes de una nueva mentalidad que ha sido calificado de escolasticismo contrareformista, sobre todo en el quehacer quirúrgico. Como ha escrito Sánchez Albornoz, en el primer Renacimiento 'maduraron en España la razón y la ciencia europea. Pero quando en pensamiento científico y filosófico modernos iban a alcanzar su eclosión ... los españoles habían entrado en el más sombrío bache de su historia. La politica Filipina – iniciada por Carlos V y desorbitada por su hijo – actuando sobre la herencia temperamental de nuestro medievo, había provocado el chispazo casi apocalíptico del cortocircuito en que se abrasó la modernidad española.' Luis S. Granjel, *Medicina española renascentista* (Salamanca: Ediciones Universidad de Salamanca, 1980), p. 14.

authority in his brusque rejection of anatomy, which he claimed was irrelevant to medical education within the Spanish realms. In his studies of this reactionary opponent of the anatomical science, whose power and influence as *Protomédico general* and royal physician extended far beyond the University of Valladolid, the Spanish historian Juan Riera argues that: 'Luis Mercado, as we will see, is the leading Spanish physician of the counter-reformation; his Galenism represents a hostile tone towards the nascent modern ideologies, rejecting every minor attack on traditional theories.'[47] Mercado obtained his medical degree from Valladolid in 1560, held the first chair of medicine in 1572 and remained the unquestioned authority of the medical faculty until his death in 1611. In his works, many of which were re-used and reprinted in the seventeenth century, Riera sees one of the most loyal Galenic defenders of the age: 'All the general characteristics that distinguish Galenic anatomy are maintained in the works of Mercado.'[48] In his study of Mercado's large opus of medical publications on plague, embryology, traumatology and surgery, Riera has come to the conclusion that Mercado did not have any hands-on surgical experience, but instead produced his large volume on this subject, *Institutiones Chirurgicae*, by compiling classical and medieval authorities. Consequently, this work was in direct opposition to *Teorica y practica de cirugía* by Daza Chacón, which was a detailed account of the everyday life and practical experience of the surgeon, and the necessity of thorough anatomical training.

In a condescending statement on 'cirugía teórica', Chacón referred to this entirely studious endeavour as 'a science .... which can be mastered by people who have never held a scissor or a needle in their hands'.[49] In contrast, he described 'cirugía práctica' as 'the real surgery, because it is learned through work and exercise with hands and instruments'.[50] With its resolute defence of empirical research and methods, Chacón's textbook conflicted entirely with the bookish approach to medicine found in Luis Mercado's writings.[51] Chacón died

---

[47] 'Luis Mercado, como podremos comprobar, es el medico español de la Contrareforma; su galenismo presenta un tono hostil frente a la naciente ideología moderna, rechazando quando pueda suponer el menor ataque a las concepciones tradicionales.' Juan Riera, *Vida y obra de Luis Mercado* (Salamanca: Cuadernos de Historia de la Médicina Española, 1968), p. 11.

[48] 'Todos los rasgos generales que singularizan la anatómica galenica perduran en la obra de Mercado, desde los esquemas antropológicos fundamentales hasta las exposiciones morfológicas.' Ibid., p. 12.

[49] 'Una sciencia que se adquire por demonstración y por el conocimiento de los principios del arte, y esta la puede saber qualquiera (y muy bien sabida), sin que en su vida aya echado mano a la tixiera, ni a la pincha'. Daza Chacón, *Teorica y practica de cirugía*, Chapter V, p. 31.

[50] 'La verdadera cirurgia, porque es saber poner por obra, y exercitar, y hazer con las manos y con los instrumentos, lo que el otro supo muy bien parlar.' Ibid., p. 31.

[51] In line with this understanding, he confidently referred to surgery as 'la más Antigua, y la más cierta parte de toda la medicina. Y un entendimiento habitual, práctico, alcancado

in 1596, aged 86, after an unusual and eventful career. Originally employed as imperial surgeon to Charles V during the military campaigns in Germany and and Flanders, he went on to work at the Royal Hospital of his native Valladolid and as Royal Physician to Don Carlos and Don Juan of Austria. He collaborated with Vesalius in the 1540s in the service of Charles V and emphasised his admiration for his younger colleague during their bloody occupation on the European battlefields, where Chacón imitated Vesalius' cure of wounds of the joints, 'which I was taught by the very learned Vesalius, together with many other things'.[52]

The two colleagues met again – at that time in the service of Philip II – in Alcalá in 1562, where they were called to save the life of the Crown Prince Don Carlos, who had fallen from a staircase and suffered a serious head injury. This dramatic incident was related in Chacón's daily account of the treatment given, a text which comprised a small, independent part of his opus, entitled *Relación verdadera de la herida de cabeza del Serenisimo Principe D. Carlos, nuestro Señor, de gloriosa memoria, la qual se acapó en fin de Julio del año 1562*. In the following decade, Chacón served as chief of the galley surgeons during the Battle of Lepanto, where the son of another Valladolid surgeon, Miguel Cervantes, received the wounds that permanently maimed his left hand. In 1580, Chacón presented his resignation to Philip II and was awarded a lifelong pension during the following years, which he dedicated to the writing of the second volume of his life and works. It is ironic that this empirical champion would eventually have to submit his work for the approval of Luis Mercado, whose authorisation introduced the second volume of *Teorica y practica* (Valladolid, 1593).

Even though Mercado claimed to be ignorant of the existence of anatomical practice in his native city, Juan Riera has traced a few examples that prove he was aware of corrections suggested by reformers of Galenic anatomy and physiology, and even rare cases of approval similar to those found in the work of Guevara.[53] Mercado nonetheless rejected most of the claims and discoveries made by these *modernos*, such as Juan Valverde's and Realdo Colombo's description of the pulmonary system or the 'smaller circulation', published in their anatomical textbooks in 1556 and 1559, but already outlined by Miguel Servet in 1553. On

---

por mucha contemplación, y mucha experiencia. La qual con instrumentos administrados, con razon y tiempo, y con la artificiosa obra de manos, rezia, y valientes, ajuntando, apartando, y consumiendo (presto, segura y jocundamente) pueda sanar las enfermedades que en el cuerpo humano pueden ocurrir'. Ibid., p. 30.

[52]   'Lo que yo he practicado y observado es, que muy mas seguras estan las heridas de junturas, que no se apuntan, que las que se apuntan: y esto aprendí del doctissimo Vesalio, y otras muchas cosas, hallandome con el en juntas de semejantes heridas.' Ibid., p. 148.

[53]   'Un par de veces apoye Mercado su autoridad en Vesalio – o Besalio como lo llama – o en Colombo, en cambio el nombre de Galeno empapa docenas de veces las *páginas de la ingente Opera* Omnia de Mercado.' Riera, *Vida y obra de Luis Mercado*, p. 28.

the death of Valles, 'el divino', in 1592, Mercado was appointed as his successor in the joint positions of personal physician to Philip II and *Protomédico general*, in which capacity he was to oversee the medical practice and education of the Spanish Empire.[54] Mercado's new appointments may have contributed significantly to the negligence of the anatomical programmes established in previous decades and arguably even helped to consolidate a resistance to similar initiatives in the future. The hostile tone traced by Juan Riera in Mercado's refusal to impose medical reforms influenced new generations of medical doctors in Spain throughout the following century. Mercado's status as the leading medical authority would remain unchallenged throughout the early 1600s due to the multiple reprints and new editions of his manuals and textbook for future physicians, surgeons and *algebristas* (bone-setters), *Institutiones chirurgicae*, *Institutiones medicae* and *Instituciones para el aprovechamiento y examen de los algebristas*, which were all printed in the 1590s. Through a 1594 royal decree, Mercado's textbooks were elevated to core curriculum throughout the Spanish kingdoms and were used as the sole works for the examination of Spanish medical students, a decision which was met with much subsequent criticism, as studied in detail by Michele L. Clouse, whose research quoted the following 1596 lament from a Salamancan faculty member: 'I have no doubt that the principal cause of the ruin of the state of medicine is this method of examination by Mercado's Institutiones.'[55]

The University of Valladolid is significant not only as the earliest setting for anatomy studies in Castile, but also for the enduring opposition towards the practice during the late sixteenth and early seventeenth centuries. The short-lived anatomy lectures begun in the late 1540s were met with enthusiasm, if one is to believe the impressive number of spectators recorded by the teacher himself almost 10 years later. Only half a century later, however, the situation had changed to such an extent that Luis Mercado could deny that anatomy studies had ever taken place there or anywhere else in Spain. This apparent ignorance of contemporary and earlier anatomical programmes was found in publications by both Montaña and Guevara, and in later secondary sources, which continuously misrepresented the anatomical practice in Valladolid as pioneering in an Iberian and even European context. Luis Mercado's own personal conviction and immense influence may have contributed to the unsuccessful incorporation of

---

[54]   There are a few indications that he was closely connected to the monarch in the preceding decades and an anonymously written (and clearly apocryphal) tradition in Valladolid later held that 'aqui (en Valladolid) dicen que el doctor Valles, se topó con el doctor Mercado y le contentó tanto que se lo llevó al rey le dijo: Señor, Valladolid tiene este medico que es mejor que yo: V. M. tome para si y honre mucho a doctor Mercado'. Cited in Javier Puerto, *La leyenda verde* (Salamanca: Junta de Castilla y León, 2003), p. 249.

[55]   Cited in Michele L. Clouse, *Medicine, Government and Public Health in Philip II's Spain. Shared Interests, Competing Authorities* (Farnham: Ashgate, 2011), p. 186.

anatomy at the University of Valladolid, yet the same pattern and drift towards medical orthodoxy was found elsewhere during the late sixteenth century, even in universities where the professorship of anatomy continued to exist. Signs of crisis and subsequent annihilation of the anatomical programmes of the mid-1500s were apparent at several other Iberian universities by the turn of the sixteenth century. Paradoxically, this setback coincided with a period that in retrospect represents a golden age in the history of European anatomy, with significant advancements all over the continent. In a Castilian context, however, this era marked the abandonment of former involvement with anatomy and dissection, as is unmistakably demonstrated in the case of Valladolid.

Chapter 5
Alcalá de Henares

The record of early anatomy studies at the University of Alcalá de Henares is even more fragmentary than sources from the Universities of Salamanca and Valladolid. In spite of the flaws in the record-keeping at Alcalá de Henares, its medical faculty – founded in 1510 – produced some of the most renowned Spanish physicians of the sixteenth century. One of these, the court physician Luis Lobera de Avila, published a medical textbook entitled *Remedio de cuerpos enfermos y la silva de experiencias* (Alcalá, 1542), which included a short treatise on anatomy. However, no 'post-Vesalian' anatomy books were published in this university town, despite the distinguished facilities of the local publisher Juan de Brocar, whose printing press manufactured a number of other significant medical books of Golden Age Spain, including the most celebrated works of Francisco Valles, who was primary professor of the medical faculty from 1557 to 1572.

Prominent former medical students from the University of Alcalá wrote and published anatomical treatises elsewhere, such as Andres Laguna's *Anatomica methodus* (Rome, 1535), Juan Fragoso's *Chirurgia universal* (Madrid, 1570), Fransisco Diaz's *Compendio de chirurgia* (Madrid, 1575), Francisco Hernandez's *La historia natural de Cayo Plinio Segundo* (unpublished manuscript) and one of the first textbooks on anatomy published in the New World, Augustin Farfán's *Tractado breve de Anathomia y Chirurgia* (Mexico City, 1579). In addition to the lack of publications from the University of Alcalá, the exact date of its introduction of anatomy studies remains elusive in the available documentary sources. Only very incomplete records offer the names of some of the professors of anatomy active at the university from the mid-sixteenth century onwards. The first evidence of the new discipline is found in a passing reference to the alleged presence in Alcalá of the Valencian anatomist Pedro Jimeno during an unspecified period in the 1550s. More solid documentation can be found in the following decade after the 1563 appointment of Pedro Marcos to a newly created professorship of anatomy. This expansion of the medical faculty was consolidated and ratified in two subsequent university reforms in 1566 and 1587. Later documentary evidence reveals the difficulties facing the new chair shortly after its creation. Evidently the *cátedra de anatomía* often remained vacant for long periods from as early as the 1570s. The unfortunate permanency of this situation was repeatedly lamented in subsequent reform initiatives during the seventeenth and eighteenth centuries.

As was the case at Salamanca and Valladolid, the introduction of this novel discipline at Alcalá was the result of a royal provision, instigated in the name of Charles V and his mother Queen Joanna. On 24 February 1552 a provision was submitted to the municipal and university authorities, which stipulated that 'the University of the city of Alcalá de Henares should practice (anatomy), and invite persons who are experts in this practice'.[1] The Crown intervened directly in the affairs of the leading Castilian universities, and particularly so at the University of Alcalá, which was the first institution of higher education founded by the Catholic Monarchs after the union of the Crowns of Castile and Aragon. Accordingly, the University of Alcalá was only half a century old by the time of its first appointment of a permanent professor of anatomy in 1563. Unlike the other two large Castilian universities in Valladolid and Salamanca, which dated back to the mid-thirteenth century, Alcalá was a young institution originally created by Cardinal Cisneros in 1509 to educate the future Spanish clergy from the Colegio Mayor de San Ildefenso, its dominant and sovereign faculty of theology.[2] In addition to the original objective, the university eventually engaged in a wide range of humanist studies, with emphasis given to the reading of Latin and Greek, as well as Hebrew, Syriac and Arabic, as highlighted in Cisneros' original constitutions. Only a short distance from the future Spanish capital, Madrid, the projected *Complutense* University received its first papal bull of approval from the Spanish Pope Alexander VI in 1499. The twin patronage of crown and curia meant that the institution was subject to repeated attempts at unsuccessful papal intervention following its foundation.[3] This situation continued until Caspar de Zuñiga's and Juan de Ovando's 1555 and 1566 reforms of Cisneros' original constitution, which downplayed papal involvement and emphasised the influence of the Spanish monarch as chief patron of the university.

---

[1]   'Una de las mas principales y necesarias cossas y saver y entender la ciencia de la medicina es tener conoscimiento e noticia de sytio e asiento de los miembros ynteriores y esteriores la qual diz que se adquiere con solas disystiones e anotomias de los cuerpos umanos e no con otro y que por lo contrario los mas e mayores herrores en que cahen los que se exerçitan e usan la dicha arte es quando quier que son leydos e vistos en los auctores syn tenner noticia de las dichas anotomias lo qual se rremediaria sy los medicos asy teoricos como manuales biesen e supiesen todos los mienbros e partes de los cuerpos humanos muchas vezes y en diversas hedades y rregiones e conplisiones para mediante la ferquençia e diversydad de la tal anotomias diz que se adquiere la pratica que es muy neçesaria sobre lo qual por nuestras cartas enbiamos a mandar a las universydades de los estudios de esa ciudad e Alcala de Henares que platicasen en ello e se mando para ello personas espertas.' Archivio Histórico Nacional, *Universidades*. L 1097, f. 23 v.

[2]   José de Rujula, *Indice de los colegiales del mayor de San Ildefenso y menores de Alcalá* (Madrid: CSIC, 1946), p. 16.

[3]   Antonio Alvar Esquerra, *La Universidad de Alcalá de Henares en a principios de siglo XVI* (Alcalá de Henares: Universidad de Alcalá de Henares, 1996), p. 13.

Cisneros' university was structured according to a Parisian model with theology as the dominant faculty, rather than law, as at the Universities of Salamanca and Valladolid, which were modelled on the University of Bologna. Theology and canonical law remained the prime focus at Alcalá, with secondary rank given to the faculties of arts, philosophy and letters. A small medical faculty was added in 1510 and was approved by another papal bull by Leo X in 1514. This modest faculty consisted initially of only two chairs, one in Avicenna and another in Hippocrates and Galen.[4] It gradually expanded during a series of successive university reforms, often initiated at the request of the Spanish monarchs themselves. Reforms carried out by Caspar de Zuñiga and Juan de Ovando between 1555 and 1566 increased the medical faculty staff from two to four professorships, and later established two further minor chairs, including one in anatomy.[5] The original number of medical chairs was thus tripled to six permanent professorships, and the second minor chair of Ovando's 1566 reform was later changed to a professorship of surgery, following another royal provision in 1594. A curious reference to the growing medical faculty of Alcalá can be seen in the *Coloquio de los perros*, written in the 1590s by Miguel de Cervantes – the most famous *alcaíno* of all time. This fictional conversation between the two dogs Berganza and Cipión touched upon the rapid growth and increased reputation of the medical school at Alcalá:

> Berganza: 'I can readily enough set down as a portentous token what I heard a student say the other day as I passed through Alcalá de Henares.'

> Cipión: 'What was that?'

> Berganza: 'That of five thousand students this year attending the university – two thousand are studying medicine.'

> Cipión: 'And what do you infer from that?'

> Berganza: 'I infer either that those two thousand doctors will have patients to treat, and that would be a woeful thing, or that they must die of hunger.'

This high number was clearly an overestimate of the size of the institution, which only exceeded 4,000 students for a very brief period during the late sixteenth century. Its faculty of medicine also remained diminutive compared to those

---

4   Ramon Gonzales Navarro, *Universidad Complutense: Constituciones originales cisnerianas* (Alcalá: Universidad de Alcalá de Henares, 1984), p. 298.

5   Stafford Poole, 'Juan de Ovando's Reform of the University of Alcala de Henares, 1564– 1566', *Sixteenth Century Journal*, 21(4) (1990), p. 600.

of law and theology, with fewer than 100 students each year in the surviving matriculation lists.[6] The Spanish sovereigns repeatedly named themselves patrons of the university, with rights to govern and oversee its internal structure and organisation. Charles V even expropriated its running costs to raise money for his own coronation as Holy Roman Emperor.[7] A 1555 reform brought about by the Bishop of Segovia, Caspar de Zuñiga, acted on behalf of the ageing monarch in order to maintain his status as chief patron, and furthermore left the medical faculty with a new structure of four chairs, one *principal* and one *segundaria* of *prima*, and two lesser chairs of *visperas*. Indeed, a few years later, Philip II referred to himself as patron of the university.[8] Under the watchful eye of the young monarch, between 1565 and 1566 the most thorough and complete reorganisation was carried out since the university's foundation by Juan de Ovando, provisor, Inquisitor and canon of the cathedral chapter of Seville.

The First Valencian Anatomists in Alcalá: Pedro Jimeno and Pedro Marcos

Among the lasting results of Ovando's reorganisation were the establishment of a permanent university chair in anatomy and a formal consolidation of the new medical professorship, which had been created in 1563 as a *cátedra menor* of medicine, and filled by the Valencian anatomist Pedro Marcos from its inception until 1570. There are indications of prior attempts to introduce anatomy before this appointment, for example, in a passing remark from a *claustro pleno* assembled in 1534, in which medical students called for the inclusion of one of Galen's anatomical texts in the medical curriculum.[9] Previous studies of the medical faculty of Alcalá, published in the late nineteenth and early twentieth centuries, took this as evidence that a programme of anatomy studies had already been established during this early period. Early biographies of Cardinal Cisneros, such as Quintanilla's *Archetypo de virtudes de Cisneros* (Madrid, 1652) and Pedro Fernández del Pulgar's *Vida de D. Fr. Francisco de Cisneros* (Madrid, 1673), even claimed that the Cardinal had created a chair in anatomy in the first university statutes.

However, there is no documentation to support the idea that anatomy studies existed in Alcalá before Pedro Jimeno's arrival in the 1550s, and even

---

6   Gonzales Navarro, *Universidad Complutense*, p. 127.

7   Ibid., p. 580.

8   'Patron que somos de la dicha Universidad ... como Rey y Señor natural.' Alonso Muñoyerro, *La facultad de medicina en la Universidad de Alcalá de Henares* (Madrid: CSIC, 1944), p. 21.

9   'Que el Dr. Reynoso en acabando los aphorismos que lea de anatomicis aggresionibus, que dicen notomía.' Ibid., p. 29.

this event is shrouded by fragmentary and contradictory evidence. López Piñero acknowledged the uncertainty surrounding events at Alcalá: 'We cannot establish with absolute certainty the exact year in which the professorship of anatomy was created at the University of Alcalá, since, as regards this point, there is an information gap from 1534 to 1559 in the available documents.'[10] Nonetheless, Piñero must have judged the existing sources to be sufficient in order to write the following hypothesis of Jimeno's influence, based entirely on a short passage in the introduction of *Claudii. Gal. Pergameni de locis patientibus libri sex, cum scholiis*, a medical textbook published in 1559 by the faculty *Principal de prima*, Francisco Valles: 'The obstacle presented by the gap in the archives may, however, be partially surmounted if we consult the works of Francisco Valles, the leading medical figure at the University of Alcalá during the sixteenth century. In his commentaries on Galen's *De locis patientibus* (1559), Valles stated that Jimeno, whom he calls "my friend", "had come from Valencia to Alcalá in order to explain the art of dissection, in which he was highly versed". Since Jimeno left Valencia in the summer of 1550, after having been professor there for three years, everything would seem to indicate that during the following academic course he became the first professor of anatomy at the University of Alcalá.'[11]

Piñero assumed that Pedro Jimeno's alleged professorship in Alcalá had long-term effects, even though the exact date and nature of his engagement still elude historians: 'The fact that Jimeno was the one to introduce the new anatomy into the university was, without doubt, the basis of a dependence on the Valencian school that lasted nearly a century.'[12] It is significant that Pedro Jimeno is definitively credited as the initiator of anatomical teaching at Alcalá in Francisco Valles' 1559 publication. As described in the previous chapter on Valencia, Jimeno claimed to have been a pupil of Vesalius at the University of Padua and went on to author an anatomical textbook, which was published in 1549. In the same year Jimeno was promoted from his *cátedra de anatómia y simples* to the higher medical position of *cátedra de practica* at the *Estudi General* in Valencia. However, his name is absent from the lists of medical professorships granted at the University of Alcalá, available in the *Registro de actos, grados y provisión de cátedras* at the Archivo Historico Nacional in Madrid. These registers document a series of Valencian anatomists who were appointed to the anatomical professorships in Alcalá throughout the latter sixteenth and early seventeenth centuries. According to the sources, the Valencian anatomist Pedro Marcos was the first Professor of Anatomy at the medical faculty of Alcalá in 1563, a decade after Jimeno supposedly arrived. Francisco Valles nonetheless

---

[10]   Jose Maria López Piñero, *Antologia de la Escuela Valenciana del siglo XVI* (Valencia: Cátedra e Instituto de Historia de la Medicina, 1962), p. 66.

[11]   Ibid.

[12]   Ibid., pp. 66–67.

referred to Jimeno as the institution's pioneering academic of anatomy. Valles also claimed to follow the guidelines put forward by Jimeno in his own later studies and teaching of anatomy, and explicitly praised his friendship with the Valencian anatomist who arrived in Alcalá at an unspecified date in the 1550s and died there not long after.[13]

López Piñero's estimation that Jimeno arrived in Alcalá during the summer of 1550 seems purely speculative and arguably too early if Francisco Valles' intervention was indeed instrumental in bringing him to the Complutense University. In 1550 Valles was still only a bachelor student of medicine in his early twenties, without the later authority of his 1557 promotion to the first chair of medicine at the medical faculty. The fact that Valles did not mention Jimeno in his *Controversiarum* from 1556, but only in his 1559 commentaries on *De locis patientibus* instead indicates that Jimeno arrived during the interval between these two publications. The *Controversiarum* was Valles' most widely read publication and appeared in several different impressions and languages throughout the next two centuries. In this work, Valles repeatedly praised Vesalius and his followers, and their recent contributions to anatomical knowledge, such as the auditory bones, which, as he emphasised, had been unknown in previous centuries.[14] While Valles occasionally disagreed with Vesalius on the exact nature and function of such discoveries, his textbook included declarations of gratitude to the famous anatomist for his groundbreaking work in osteology.[15]

Valles' textbook did not refer to Pedro Jimeno, who was awarded high praise by the same author three years later, and a chapter on the auditory bones even credited Luis Collado with the discovery of the stapes.[16] It seems unlikely that Valles would have conceded this discovery to Collado if Jimeno, 'amicissimi

---

[13]   'Nam desideras anatomen peritiam et usus omnium particularum, et morborum omnium ac simptomatum. Differentias et causas, et praeter haec, ipsa ratio dignoscendi partium internarum mala, est difficilima. Proinde ego cum superioribus annis hoc opus enarrandum suscepissem, ita me comparavi ut nullius particulae labores dicere aggrederer, quin illius formationem totam et ipse contemplarer & discipulis meis ob oculos ponerem, industria & opera cuiusdam Ximenij amicissimi mei qui nuper e Valentia Complutum, ut dissecandi artem cuius erat peritissimus profiteretur, venerat, neque multo post hic agens vita defunctus est. Ille suam operam ponebat ut ego me discipulosque meos multum exercere possem.' Fransisco Valles, *Claudii. Gal. Pergameni de locis patientibus libri sex, cum Scholiis* (Lyon: Juan de Brocar, 1559), p. 5.

[14]   'Andreas Vesalius anatomicorum peritissimus, ut sua opera testantur libro primo De humani corporis fabrica capite octavo, inquit, inuenisse se in foramina ossis lapidei, quod porus auditur est, ossicula duo, que latuerant Galenum & alios omnes dissectores ante hoc seculum.' Francisco Valles, *Controuersiarum medicarum & philosophicarum libri decem* (Alcalá de Henares: Controuersiarum medicarum & philosophicarum libri decem, 1556), f. 41r.

[15]   'Quod de numero ossium dicit: habet certe ut dicit, & est quod pro diligentia habeant omnes huic viro gratiam.' Ibid., f. 41r.

[16]   Ibid., f. 42r.

mei', as he had been described in 1559, had been present in Alcalá before the publication of *Controversiarum*. Jimeno's proposed arrival in 1550 seems ever more doubtful when considering the publication authored by another student of medicine educated in Alcalá, the renowned anatomist and urologist Fransisco Diaz. Diaz received his medical education at the University of Alcalá, where he obtained the title of *bachiller* in 1551, *licenciado* in 1555 and *doctor* in 1559. His education thus covered the entire decade that Jimeno was allegedly present at the university, but the employment of the Valencian anatomist in Alcalá received no mention in Diaz's *Compendio de chirurgia*, which began with a lengthy chapter on anatomy that concluded with a plea to extend the discipline to all urban settings: 'It is so necessary for the manual medical work, that I do not know why all towns and populous cities to not employ a great anatomist at a great salary.'[17] Diaz's *Compendio* emphasised the author's own status as *Doctor y maestro en philosophia, Por la insigne Universidad de Alcalá de Henares*. Yet, he later described how he had left his medical studies in Alcalá in order to study anatomy at the *Estudi General* in Valencia, and referred to the apprenticeship he completed there under both Pedro Jimeno and Luis Collado in *Tratado nuevemente impresso, de todas las enfermedades de los Riñones, Vexiga, y Carnosidades de la verga, y Urina, diuidido en tres libros* (Madrid, 1588). Diaz's text did not imply that Jimeno had gone to Alcalá to teach anatomy, but instead referred to his consultation with Jimeno in his native Valencia: 'Jimeno, the extremely learned Valencian, and the first who elegantly and skillfully began to dissect and carry out anatomy in the city of Valencia, where the medicine and anatomy is presently flourishing. I can boast of having spent some time in this city under the masterful guidance of doctor Collado and doctor Jimeno.'[18]

Another anatomist, Juan Fragoso, who took his bachelor's degree at Alcalá in 1552, went on to credit a number of his former teachers in his textbooks on anatomy and surgery, entitled *Erotemas Chirurgicos* (1570) and *Chirurgia Universal* (1581). He acknowledged his earlier mentors from Alcalá, such as the two former medical faculty Deans, Christobal de Vega (1545–57) and Francisco Valles (1557–72), but made no reference to Jimeno's activities at the university. It should be noted that neither Diaz nor Fragoso revealed any details of their own anatomical training as students in Alcalá. This apparent

---

[17] 'Tan necessario para la obra de manos, que no se como no se procura en todas las villas, y Ciudades populosas tener un gran Anotomico.' Francisco Diaz: *Compendio de chirurgia* (Madrid: Pedro Cosin, 1575), f. 64 r.

[18] 'Ximeno, doctissimo valenciano, y el primero que con elegancia, y gran destreza comenzó a poner la execución de cortar y a hacer anatomía en la ciudad de Valencia, donde tanto resplandece la medicina y la anatomía al presente...y no tengo yo poca jactancia de haber gastado en esta ciudad algún tiempo, y tener por maestro al peritissimo doctor Collado y al doctor Ximeno.' Fransisco Diaz, *Tratado nuevemente impresso, de todas las enfermedades de los Riñones, Vexiga, y Carnosidades de la verga, y Urina, diuidido en tres libros* (Madrid: Francisco Sanchez, 1588), f. 19r–19v.

forgetfulness might reflect the poor state of anatomy studies in Alcalá at the time of their publications in the 1570s and 1580s rather than the situation at the university two decades earlier. As indicated in the university records, the late sixteenth century encompassed a period of crisis within the medical faculty at Alcalá, where the chair of anatomy remained vacant from 1574 until 1583 with only a one-year interval in 1577–78, when Fransisco Bartolomé briefly held the chair. A *claustro pleno* assembled on 24 November 1574 presented the first in a series of complaints about this situation, remonstrations that would be repeated throughout the next centuries.[19]

While we have sufficient documentary evidence to trace a rapid and steady decline of the chair of anatomy at Alcalá, it is far more difficult to pinpoint the exact period that anatomy studies first began at the university. As previously shown, uncertainty surrounds Jimeno's supposed arrival in Alcalá, but it also clouds the nature of his appointment there, which left no trace in the documentary sources. Perhaps Jimeno's teaching was limited to a few months of introductory lectures of anatomy, as had been the case with Rodriguez de Guevara's short-lived anatomy course at Valladolid. Guavara's written defence of Galenic anatomy, which was published in Coimbra in 1559, did confirm that no anatomy studies had taken place at Alcalá or Salamanca during the time of his lectures at Valladolid 10 years earlier.[20] The royal provision 'Para que se haga anotomía', which was signed in Valladolid on 16 March 1551, ordered the establishment of anatomy at the University of Valladolid and stated that the initiative should be extended to the other two principal universities in Castile.[21] Nonetheless, a similar provision presented to the University of Alcalá in February 1552 had no immediate effect, as evidenced by the fact that a formal and permanent professorship of anatomy was not established until 11 years later. It is therefore surprising that Francisco Valles devoted the initial parts of

---

[19]   'El Dr.Carrillo dijo q. le paresce q. en esta Universidad suele aver cathedra de anatomia y a días que no se lee. Que se procure buscar persona pa. cathedra de notomia. Ansimismo dijeron q. el q. proveyeran por anatomista no le incorporen por dr. desta universidad por muchas (razones) q. pa. ella diese'. AHN. L 427, f. 211.

[20]   'Quo ad Complutensem & Salamanticensem academiam delegor, ut habitis magistrorum comitiis, quisquid re diligenter expensa seitum ac decretum esset, publicis obsignatum literis reportarem. In quibus gymnasiis in eam sententia itum est ut non solum chirurgis, verum & medicis anatonem apprimemutilem & necessariam censerent.' Alonzo Rodriguez de Guevara, *In pluribus ex iis quibus Galenus impugnatur ab Andrea Vesalio Bruxelensi in constructione et usu partium corporis humani, defensio* (Coimbra, Juan Baverius, 1559), Ad candidum lectorem, n.p.

[21]   'Destas anatomías diz que se adquire la plática que es muy necesario sobre lo qual por unas nuestras cartas embiamos a mandar a las universidades de los estudios de Salamanca e Alcalá de Henares que placticasen en ello llamando personas expertas y la resolución que tomasen y orden que en ello se debía tener lo embiasen ante nuestro consejo.' Archivo general de Simancas, RGS, 155103.

his publications from 1556 and 1559 to his own new insights into anatomical studies, and in the later publication explicitly credited Jimeno as the instigator of such studies at the University of Alcalá. In his *Controversiarum* (1556), Valles emphasised the length of his experience with anatomical studies, in which he had been engaged not once, but several times, and always in the presence of witnesses and students in order to err as little as possible in his observations.[22] It is also significant that his medical textbook, which, for the most part, dealt with humoral and Hippocratic/Galenic medicine, also included references to some of the most recent insights in anatomical research carried out in Spain and beyond. López Piñero has attempted an explanation of this apparent synthesis of medical worldviews in his *Ciencia y tecnica*: 'Valles wanted to use the data of the new anatomy in the service of traditional doctrines about disease and its manifestations with the aim of rectifying Galen's assertions.'[23]

Valles' text highlighted the errors caused by the *barbari omnes* of the medieval medical tradition, the so-called *galenismo arabizado*, and instead presented a programme of *galenismo humanista*, which had been consolidating its prominent position in Spanish medicine since the late fifteenth century. At the University of Alcalá, the philological corrections of original classical text from later additions and mistranslations went back to Antonio de Nebrija's humanist programmes of the early 1500s. Purification of the Greco-Roman tradition was also a dominant feature of the university's medical faculty. Valles fashioned himself as the renaissance physician, at once a medical doctor and a well-read master and translator of Latin and Greek. As such, he embodied the new breed of erudite physicians who initially gained renewed prestige and wealth due to their profession, but were later met with increasing critiscism by medical reformers. Among the challenges facing these medical humanists was the questioning of Galenic doctrines by groups of anatomical sceptics who had been educated at the leading universities of northern Italy. Pedro Jimeno, Luis Collado and their followers at the Universities of Salamanca and Alcalá personified the progress of anatomical reform: originating at the University of Padua, the Vesalian reforms in anatomy later entered Iberia through Valencia and eventually reached the leading medical faculties of Castile. Valles' recognition of Vesalius' and Collado's anatomical discoveries, his appraisals of the Alexandrian anatomists Herophilos and Erasistratos, and his alleged cooperation with Jimeno all reveal him to be an eager follower of the anatomical discipline that was denied

---

[22]   'Siquas controversias historia anatomes oportuit decidi, res oculis exploravi no semel, nec sine testibus, sed pluries & adscitis discipulis, munitisque, quorsum ea quere rentur, quò ita res minus posset fallere.' Valles, *Controuersiarum, Ad lectorem*, n.p.

[23]   'Valles aspiraba a utilizar los datos de la nueva anatomía al servicio de la doctrina tradicional de la localisación de la enfermedad y sus manifestaciones, con el fin de fundamentar o rectificar las afirmaciones de Galeno.' José Maria López Piñero, *Ciencia y tecnica en la sociedad Española de los siglos XVI y XVII* (Barcelona, Labor, 1979), p. 354.

any relevance by Luis Mercado, Valles' later successor as Protomedico General and personal physician to King Philip II. In a reference to recent discoveries regarding the structure of the venal system, Valles even called for the compulsory education of future medical students in such anatomical knowledge.[24] It could be argued that the distinct personalities of Valles and Mercado determined the different fates of attempts to establish anatomical practice at the medical faculties of Alcalá and Valladolid. Valles' alleged training in anatomy under the supervision of Pedro Jimeno showed how the practice was initially welcomed and subsequently maintained in Alcalá, though with evident difficulty and long intervals of vacancy. A second royal provision signed 4 April 1559 followed up on the 1552 decree and requested that anatomical teaching and procurement of bodies should be carried out according to the practice already established in Salamanca, thus indication a very limited – if any – activity in the field at Alcalá de Henares.[25]

The new chair of anatomy at the University of Alcalá established a few years after began to show signs of discontinuity, vacancy and neglect from the late 1500s onwards. In spite of these interruptions, one tradition did continue throughout the latter half of the sixteenth century and well into the seventeenth, namely the inundation of Valencian anatomists at the medical faculty of Alcalá. If Valles' account and claims are to be trusted, this influx began with Jimeno's arrival in Alcalá and untimely death there in the 1550s. In subsequent documentary evidence we see a large number of anatomists arriving throughout the late sixteenth and early seventeenth centuries. This movement was noticed by the Valencian chronicler Caspar Escolano in the early seventeenth century, as documented in his account of the deeds of his townsmen, including their crucial role in bringing the study of anatomy to Castile: 'It was unknown in Castile until Valencian doctors came to teach it in Salamanca and Alcalá.'[26] Pedro Valverde, who was appointed to the professorship in anatomy in 1591, is the only exception to this incursion of Valencian anatomists during the late sixteenth century. Valverde's name appears in an early list of Alcalá's own medical

---

[24] 'Est quod omnes medicorum sectae recipiant tude hanc anatomem.' Valles, *Controuersiarum*, f. 122v.

[25] 'E agora por parte de la Universidad de la dicha villa de Alcala de Henares nos hizo rrelaçion e diziendo que al bien comunde la dicha Universidad e de los medicos y estudiantes della conbiene que se haga anotomya de todos los cuerpos humanos que se condenaren por vos a pena de muertey se executare la sentencia e de los que murieren en los ospitales de esa dicha villa segun e de la manera que por la dicha nuestra carta susoyncorporada se mando que se hiziese en la Universidad de la dicha çiudad de Salamanca.' Archivio Histórico Nacional, *Universidades*. L 1097, f. 26 r

[26] José Maria López Piñero, 'El saber anatomico y la dissección de cadaveres humanos en la España en la primera mitad del siglo XVI', *Cuadernos de la historia de la medicina Española*, 13 (1974), p. 83.

students; this first native scholar held the chair in anatomy (and later the chair of surgery) until 1597, after which the chair remained vacant for six years until the appointment of another *Valenciano*, Dr Bardoz, in 1603.[27] The first professorships of anatomy in Alcalá were thus held by an almost uninterrupted sequence of medical scholars born and educated in Valencia. These *catedráticos de anatomía* were Pedro Marcos (1563–70), Juan Valero Tobar (1571–72), Miguel Ferri (1573–74), Jacobo de Solar (1583–85) and José Gutiérrez (1585–91). While the details of Pedro Jimeno's residence in Alcalá remain uncertain and insufficiently documented, there is more solid evidence in the case of Pedro Marcos, Alcalá's first formally appointed anatomist. Marcos held one of two minor professorships of medicine or *cátedras/partidos menores*, as recorded for the first time on 6 April 1563 in the *Registro de actos, grados y provisión de cátedras*.[28] One of these minor chairs was the first chair of anatomy, a fact which was recorded in documentation from the university compiled by its late rector, Mariano Martin Esperanza.[29]

There are no surviving statutes available from this period to provide the details and content of the anatomist's employment. The new medical chair was evidently successful, as it was ratified three years later in Juan de Ovando's thorough restructuring of teaching and general reorganisation of the university. Additional sources reveal the remuneration of the new professorship as rather modest and even insufficient. At the time of his original appointment in 1563, Marcos' annual salary was set at only 30,000 maravedís, a noticeably lower sum than the 40,000 maravedís earned by Cosme de Medina in Salamanca more than a decade earlier.[30] This example and later evidence indicate that the crisis of vacancy of the anatomy chairs in Alcalá was at least partly economic. Indeed, one of the first demands made by the newly appointed professor was for a substantial pay rise. Marcos successfully negotiated a salary increase of 40 ducats to 80 on the petition of the *claustro pleno*: 'Because he was a very skilled and good professor, and because other universities were offering him a higher salary.'[31] This is quite remarkable given the lack of prospects at the other prominent universities in Castile. No permanent professorship was ever established in Valladolid, so the alternative offer made to Marcos must have been from the University of Salamanca. This opening position would fit chronologically with Augustin Vazquez's 1567 departure from the

---

27 AHN L. 400, f. 479

28 AHN L. 525, f. 144.

29 'En 6 de abril de 1563 se nombró por primer catedrático de anatomía al licenciado Marcos y en el de crearon las dos cátedras de decretales mayores y menores.' AHN L. 1083. Mariano Martin Esperanza, *Estado de la Universidad de Alcalá*, manuscript, f. 35.

30 'Dos cátedras principales de medicina a 200 ducados, dos cátedras menores o partidos de lo mismo a 30 mil maravedíes; cátedra de anatomía 30 mil maravedíes.' Ibid., f. 26.

31 'Porque era muy doctor y buen catedrático y le llamaban a otros universidades con aumento de salario.' Muñoyerro, *La facultad de medicina de Alcalá de Henares*, p. 34.

professorship of anatomy in Salamanca following his promotion to the more prestigious medical chair of Articella.[32] The success of Pedro Marcos' pecuniary petition in 1566 kept him in Alcalá, where he held the chair of anatomy until 1570 when he made another (and this time unsuccessful) petition to raise his salary to the level known in Salamanca.[33] His 1566 request is the only efficacious one in the available documentary sources and (given the drastic reductions in the already inadequate salary of the professor of anatomy to come in the following decades and centuries), the last to successful negotiation of a pay rise at the University of Alcalá de Henares. Several ingenious proposals during the seventeenth and eighteenth centuries nonetheless endeavoured to guarantee the continuation of anatomical studies at the university. A royal provision from 1614 proposed a possible financial solution by fusing the two chairs of anatomy and surgery into one professorship.[34] Another university reform from 1665, carried through by García Medrano, further emphasised the worsening fate of the anatomists of Alcalá. While the *principal de prima* continued to receive an annual salary of 200 ducats (approximately 75,000 maravedís), and the two chairs of *vísperas*, and even the professor of surgery were each awarded 80 ducats (30,000 maravedís), the salary of the anatomist had decreased by a third to a mere 20,000 maravedís. A century after Marcos demanded a pay rise, the 1665 reform specified that an additional payment of 10,000 maravedís – increasing the salary to parity with the previous century – might be obtained as a reward for 10 annual and well-executed dissections.[35]

Decline and Vacancy of the Chair of Anatomy

The unfortunate financial situation led to many years during which the chair of anatomy was left empty in the sixteenth century, and several decades of vacancy during the seventeenth century. This dramatic decline was even noted by foreigners with only fleeting interests in Spanish affairs, such as the Italian scholar and scientist Lorenzo Magalotti, who made a journey to Iberia in the retinue of the Tuscan Duke Cosimo III. In his *Relazione del viaggio di Spagna* (1669), the distinguished traveller recorded the sad state of contemporary

[32]   Teresa Santander, *El Doctor Cosme de Medina y su biblioteca (1551–1591)* (Salamanca: Centro de Estudios Salmantinos, 1999), p. 32.

[33]   AHN Universidades, 48, Exp. 1., Doc 2.

[34]   'Item, ordenamos que el Cathedrático de la Cátedra de Cirugía que se a fundado de algunos años a esta parte y anda junta con la de Anatomía no haviendo opositor para cada una de de las dos Cathedras por si. Y sus salarios quedan señalados en el título treinta y seis de esta reformación, a de leer en los quarto primeros años materias de cirugía sin poder salir de ellas, siendo distinta material la de cada año.' AHN L. 525, f. 48.

[35]   'Mas otros 10.000 maravedies por diez disecciones anuales.' Ibid., f. 48.

Spanish medicine and used the University of Alcalá as an extreme case of neglect and backwardness: 'The whole of literature in Spain at present boils down to scholastic theology and outdated medicine as found in the works of Galen ... To prove this it is sufficient to tell you that in Alcalá – I beg you, your Excellency to mark this – in that celebrated and renowned institution, they have not taught anatomy for the last eight or ten years.'[36] Further drastic deterioration during the following century was indicated by a reform proposal from 1771, which attempted to rescue the doomed institution and to reinstate its former programme of anatomy studies.[37] The university nonetheless seems to have abandoned anatomical teaching altogether during the eighteenth century and none of its 1771 reformers could produce evidence of such practices in the preceding generations.[38] The proposed salary of 60 ducats for the chair of anatomy – 20 ducats fewer than Pedro Marcos had managed to obtain more than 200 years earlier – cannot have facilitated the attempts to reintroduce anatomical practice into the formerly distinguished medical faculty at Alcalá de Henares. According to Rector Mariano Martin Esperanza's 1802 account, *Estado de la Universidad de Alcalá*, a formal chair in anatomy was still maintained into the early nineteenth century, only a few decades before the final closure of the university itself in 1836. In Esperanza's documentation of three centuries of university appointments, the six medical chairs of Ovando's 1566 reform were formally still in existence, but the remuneration of anatomy had been reduced to only 588 reales (c. 20,000 maravedís) annually, compared with 882 for surgery and 2,205 for the two chairs in Prima.[39] Esperanza's history of the university was built on a vast supply of documentary material, and meticulously described the development and alterations of the faculties, chairs, teaching and general organisation of the university from 1500 to 1800. His attempt to secure the future existence of the university with a glorious account of its past eminence was concluded in 1805, only a few decades before the final decline and abolition of the famous institution, which would not open again until 1977. Esperanza's account exists in manuscript form in the Archivo Histórico Nacional and is based entirely on documentary sources from the university archives as they appeared in the late eighteenth and early nineteenth centuries.

---

[36] Henry Kamen, *Spain in the Later Seventeenth Century* (London: Longman, 1980), p. 313.

[37] 'Hay otras dos cátedras siempre vacas, una de cirugía latina, dotada en 100 ducados, y otra de Anatomía en 60 ducados (bien que se le han de pagar las diez dissecciones que deve hacer en cada curso). Pero como ni la dotación de las dos es insuficiente, siempre están vacantes. Com perjuicio de la Facultad médica, porque los nuevos inventos que ilustran la medicina se deven al cuchillo anatómici y aun resta que describir en este mundo menor.' Muñoyerro, *La facultad de medicina de Alcalá de Henares*, p. 36.

[38] George M. Addy, 'Alcalá before Reform – The Decadence of a Spanish University', *Hispanic American Historical Review*, 48(4) (1968), p. 582.

[39] AHN L. 1083, f. 37.

While some of these and other archival sources relating to the institution are still available at the Archivo Histórico Nacional in Madrid, others were lost during the closure of the university and in the partial removal of its archive to the capital in the mid-nineteenth century. A century later in 1939, a terrible fire destroyed the remains of the university archive of Alcalá de Henares, after the town itself had suffered two waves of destruction during the Spanish Civil War. Consequently, Esperanza's writings constitute a rare account of the institutional history of the university. One of Esperanza's descriptions refers to the employment of Pedro Marcos as the result of the second royal provision from 1559 and Juan de Ovando's subsequent reform of the university.[40] Earlier royal provisions had been presented to Salamanca (September 1550), Valladolid (March 1551) and Alcalá de Henares (February 1552), but a subsequent request was made to the University of Alcalá in April 1559. Francisco Valles' account of his anatomical training and his references to prior collaboration with Pedro Jimeno were published the same year as this later provision. Valles' references to anatomy may not have been merely coincidental, but rather an attempt to demonstrate familiarity with the practice at a time when it enjoyed support from the highest circles. It is strange, however, that a four-year interval elapsed between this 1559 provision, and the official appointment of Pedro Marcos. This situation could perhaps be compared to the two-year trial period that Cosme de Medina completed before he was formally appointed the first anatomical professor in Salamanca – and the additional two-year period before the physical framework for his teaching was put in order. There are nonetheless more discrepancies than resemblances between the respective appointments of the first two anatomists to be formally employed at the Universities of Salamanca and Alcalá. Not only did Pedro Marcos receive a significantly lower salary than the amount estimated for Cosme de Medina a decade earlier, but the framework provided for anatomical teaching was much more humble at Alcalá than at the larger University of Salmanca. Salamanca's permanent anatomical theatre represented a unique case which would not be imitated or repeated in any other Castilian university until 1700, when Madrid inaugurated its first permanent theatre of anatomy. The lecture halls playing host to Marcos' teaching were obviously plain in comparison and, as documented by Esperanza, dissections were not restricted to one permanent location, but instead were carried out in

---

[40] 'Hizo este reformador distribución de lecturas para los cinco cátedras de medicina y anatomía y mandó que ésta se emplease en la dirección de cuerpos humanos de quince en quince dias, para lo qual se mandó por probisión de 4 de abril de 1559 se diesen los cuerpos de los que condenasen a muerte o se muriesen en los hospitales de esta villa; y estando en ella el príncipe don Carlos el dia 22 de enero de 1564, el licenciado Marcos, catedrático de anatomía, pidió el al doctor Suárez, alcalde de casa y corte, el cadáber de Juan Marroquín, que en aquel día se había ajusticiado y mandó entregar y notificar a los mayordomos y oficiales de los hospitales entregasen los cuerpos muertos conforme estaba mandado por dicha real provisión.' AHN. L. 1083, f. 27.

no less than three of the town's hospitals.[41] In García Medrano's later statutes – put forward in 1665 – this tendency continued, and it is most likely that the three hospitals mentioned in Esperanza's account were the same as those referenced in these oldest surviving anatomy statutes.[42] The hospital of St Luke was the poor students' own hospice, named after the evangelist physician whose annual celebration marked the beginning of the series of public dissections to be carried out during the cold winter months. The main procedures at this university hospital were probably similar to those performed at the *hospital del estudio* of the University of Salamanca, where small-scale and more private dissections and studies of *anatomia particular* were carried out away from the public. Shortly after his formal employment in 1563, Pedro Marcos was involved in a struggle with the local authorities of Alcalá, which refused to deliver him the corpse of a recently condemned criminal. Marcos appealed to the crown, which sided with the anatomist and emphasised the obligations of the municipality to cooperate with the university in these matters as already stipulated in the 1552 provision regarding the teaching of anatomy at the University of Alcalá.[43] It is interesting to note that García Medrano's later university statutes of 1665 explicitly encouraged close cooperation between the university and the judicial and municipal authorities, thereby enabling the continued deliverances of fresh corpses for the anatomical teaching, which Pedro Marcos had struggled to obtain a century earlier. The seventeenth-century statutes from Alcalá were similar to the revised 1594 statutes from the University of Salamanca, which emphasised that only certain Galenic texts were to be used by teachers of anatomy.[44] García Medrano's statutory reforms went one step further by introducing penalties

---

[41] 'Que las ynformaciones para los graduandos de licenciados se hagan en Alcalá; que para claustro pleno se han de juntar lo menos veinte y un doctores de todas facultades; que las anatomías se hagan en los tres hospitales de esta ciudad en el tiempo del curso irremisiblemente.' Ibid., f. 31.

[42] 'Las anatomias han de hacerse por el catedrático desde el día de San Lucas hasta las vacaciones del ano siguiente en qualiquiera de los tres hospitales de Alcalá, que eran San Lucas de los estudiantes, el de nuestra señora de la Misericordia llamado de Altozana, y el de San José de la Orden de San Juan de Dios (para lo qual asi los administradores de los hospitales y los justicias de la villa no deberían poner ningun impedimento. Del aparato con que deberian anunciar esas autopsias, júzguese por el parrafo 13 de la parte copiada de Garcia Medrano.' Munuyerro, *La facultad de medicina de Alcalá de Henares*, p. 98.

[43] Alonso Muñoyerro, 'Provision de cátedras y Catedráticos de Medicina en Alcalá de Henares (1509–1641)' in *X Congreso Internacional de Historia de la Medicina. Libro de Actas. Tomo primero, fascículo II. Resúmenes y comunicaciones* (Madrid: CSIC, 1945), p. 81.

[44] 'En la cátedra de anatomía se han de leer los libros "De usu partium" en dos años. El primero se han de leer los ocho libros primeros, y en el segundo los otros nueve, y no se ha de leer otra cosa en cátedra de anatomía, porque en estos libros se contiene toda la material que es menester saber.' Cited in Fransisco Javier Alejo Montes, *La reforma de la universidad de Salamanca finales del siglo XVI* (Salamanca: University of Salamanca, 1990), p. 139.

for those who consulted supplementary textbooks outside those prescribed by the official university curriculum.[45] Even though the prestige and wages of the anatomist had decreased significantly since the sixteenth century, the 1665 statutes did emphasise the attendance of other medical professors during public lectures on anatomy. In spite of the fact that these lectures took place in very basic conditions, they were evidently still seen as important, given that the statutes included threats of penalties if other medical professors did not receive notification of the lectures at least one day in advance. The call for unification of the professorships in anatomy and surgery once again stressed the serious problems of these minor medical chairs. In spite of the flawed evidence, and both momentary and continuing signs of crisis, Alcalá de Henares was the only Castilian university beyond Salamanca where a permanent professorship of anatomy was established and maintained – though with evident difficulty – during the sixteenth and seventeenth centuries. It was also the scene of a uniquely dramatic event which united some of the most renowned physicians and surgeons of Spain, and eventually even Vesalius himself in this university town: the nearly fatal accident and subsequent cure of the Spanish Crown Prince Don Carlos.

Don Carlos was directly involved with another royal provision before his accident in April 1562. Two months earlier on 2 February 1562 a decree to the city and universities of Alcalá 'estando en ella la corte del principe Don Carlos' again requested local cooperation in the procurement of condemned and poor bodies for the anatomy studies and public dissections at Alcalá.[46] This involvement of the Crown Prince in the ratification and approval of the new

---

[45] 'El cathedratico de anatomía a de leer su curso en dos años en esta manera, el primer año se leerá "de ossibus et dissectione membrorum", "de sectione venarum", "arterianum et musculorum", y "de motu musculorum", et "de dissectione vulvae"; el Segundo año, quarto y quinto "de usurpación"; sexto, septimo, octavo, noveno "de Anotomicis administrationibus", y que si constare que ha leido otra cossa sea multado de las lecciones en que metiere otras cossas que no sea de anatomía. 13. Ytem ordenamos que el Cathedrático de Anotomía haga las dies anotomías universales y particulares que está ordenado haciéndolo saver el día antes que las a de hacer a sus discipulos y a los Vedeles para que no lo digan a otros Cathedráticos de Medicina para que asistan a berlas hacer y los bedeles para que sino las hiciere puedan multar lo que le corresponde. 14. Ytem ordenamos que el Cathedrático de la Catedra de Cirujia, que se a fundado de algunos años a esta parte y anda junta con la de Anatomía no haviendo opositor para cada una de de las dos Cathedras por si.' Muñoyerro, *La facultad de medicina de Alcalá de Henares*, pp. 95–96.

[46] 'En la villa de Alcala de Henares a dos dias del mes de febrero De mill et quinientos et sesenta y dos años estando en ella la corte del principe Don Carlos nuestro señor ante el señor liçenciado don Francisco de Castillo del consejo de su magestad/e alcalde en la su cassa e corte Martin de Cavarte sindico del colegioe universydad de esta villa de Alcala presenta? esta carta e provission rreal de su magestad sellada con un rreal sello e librada por los señores delsu consejo rreal que habla sobre la anotomia.' Archivio Histórico Nacional, *Universidades*. L 1097, f. 26 r

practice is comparable to his aunt Maria of Austria's 1551 petition for a chair of anatomy at the medical faculty of Valladolid, which was formulated on behalf of her absent father. Another account mentioning Don Carlos from 22 January 1564 (and later referred to by Esparenza) included a reference to Pedro Marcos' procurement and recent dissection of an executed Moor, Juan Marroquin, which allegedly took place not long after his official appointment and during a time when Don Carlos was present in Alcalá.[47] This first documented account of a public dissection carried out in Alcalá seems fairly reliable given that Marcos had by then been appointed *catedrático de anatomía* and knowing that the Prince made frequent and extended visits to Alcalá.[48] However, Esperanza's later reference to the university reformer Juan de Ovando in this context seems incorrect, given that Ovando did not begin his restructuring programme until 1565 and did not complete it until January 1566, two years after the execution and dissection of Juan Marroquin.[49]

Vesalius in Alcalá de Henares: The Treatment of Don Carlos' Head-Wound

An earlier visit made by Don Carlos' to Alcalá, two years before the alleged stay in 1564, was the scene of the lifesaving work carried out by Don Carlos' and King Philip's physicians and surgeons following the ill-fated prince's fall from a staircase. This unfortunate incident took place in the spring of 1562, while Don Carlos was residing in Alcalá, housed in the Palace of the Archbishop of Toledo, where he suffered an accidental fall and serious head-wound on 19 April 1562. Initially his personal physician Diego Santiago Olivares, the court physician Christobal de Vega and the surgeon Dionisio Daza Chacón oversaw his treatment. Philip II's court surgeon Hernán Lópes 'el Portugués' was later put in charge of the cure and the number of doctors involved eventually grew from three to nine as the patient grew worse in late April. Desperate attempts were then made to save his life by some of the most renowned contemporary Spanish physicians, who were called to Alcalá, where they remained until late May, when the cure had finally

---

[47]   'Estando en ella el príncipe don Carlos el dia 22 de enero de 1564, el licenciado Marcos, catedrático de anatomía, pidió el al doctor Suárez, alcalde de casa y corte, el cadáber de Juan Marroquín, que en aquel día se havía ajusticiado.' AHN, L. 1083, f. 35.

[48]   Gerardo Moreno Espinosa, *Don Carlos, El Príncipe de la leyenda negra* (Madrid: Marcial Pons Historia, 2006), p. 75.

[49]   Don Carlos and the city major allegedly gave their approval for Pedro Marcos to carry out a discreet and unadvertised dissection of the recently executed Moor: 'para que con todo secreto se haga la anatomia de su cuerpo'. Ramon Gonzales Navarro, *Felipe II y las reformas constitucionales de la Universidad de Henares* (Madrid: Sociedad Estatal para la Commemoriacion de los Centenarios de Felipe II y Carlos V, 1999), p. 100. The author unfortunately does not offer the source of this authorisation.

been pronounced successful. Among these physicians of King Philip's court was Andreas Vesalius, who arrived in Alcalá on 1 May and whose subsequent service there represents the most well-documented case of his medical activities in Spain. The day-to-day account of this dramatic event and its successful outcome was recorded by two of the nine physicians involved in the cure, and a formal report was delivered to Philip II the following year. It was authored by the Valladolid surgeon Dionisio Daza Chacón, who produced the only printed (and arguably most reliable) account entitled *Relación verdadera de la herida de cabeza del Serenisimo Principe D. Carlos, nuestro Señor, de gloriosa memoria, la qual se acapó en fin de Julio del año 1563.* An unpublished manuscript by Don Carlos' personal physician Diego Santiago Olivares was almost a direct copy of Chacón's account with only few comments and additions, and was entitled *Relación de la enfermedad del Príncipe D. Carlos en Alcalá por el Doctor Olivares medico de su camera.*

Daza Chacón's *Relación* was originally produced for the king only and was presented in Madrid on 25 July 1563, but Chacón later included the full account in his *Teorica y practica de cirugía* (1580). This extraordinary narrative portrays Vesalius and eight Spanish court physicians in a desperate day-and-night struggle to save the wounded prince. During the treatment, Don Carlos was tended by Daza Chacón and Vesalius; he was frequently visited by his weeping father (and later alleged slayer) and was constantly looked after by the mighty Duke of Alba, who is said to have sat in the same chair and worn the same clothes throughout the entire nursing period.[50] Astonishingly, Vesalius, Daza Chacón and the Duke of Alba had been united earlier in an almost similar event only three years before following the fatal head-wound suffered by the French King Henry II in a jousting accident on 30 June 1559. Vesalius had then been called to the King's deathbed as part of the royal retinue of Philip II (who was being married via proxy to the French Princess Elisabeth de Valois). Daza Chacón was present as physician to the Duke of Alba (standing in for King Philip) during the wedding celebrations in Paris. The accident had occurred during a celebratory tournament in which the lance point of the Count of Montgomery entered King Henry's open helmet. Ten days later the King died from brain injuries caused by the blow to his head, even though his skull had not been fractured. During the failed treatment, the stump of Montgomery's lance was thrust at the decapitated heads of four criminals, whose skulls were dissected to show the exact character of the King's head-wound.[51] In this instance the patient died soon after, despite

---

[50]   'El Duque de Alva que alli estuvo por mandado de su Magestad, ninguna hora ni momento en tiempo de la necessidad faltò, viendo siempre lo que se hazia, que como hombre acostumbrado à tantos trabajos de cuerpo y espirito, governando tantas vezes tantos exercitos, se le hizo facil lo que otros tuvieran por immense trabajo, porque cierto todas las noches estava velando vestido, sentado en una silla.' Dionosio Daza Chacón, *Practica y Teorica de Cirugia* (Valladolid, Ana Velez, 1582), p. 199.

[51]   Charles D. O'Malley, *Andreas Vesalius of Brussels* (Berkeley, University of California Press, 1964), p. 287.

Vesalius' eloquently presented diagnosis of his injury, likewise referenced by Chacón in his *Teorica y practica*.[52]

The previous experience shared by Vesalius, Chacón and Alba only three years earlier must have weakened their expectations of a successful outcome in the case of Don Carlos. Perhaps this prior experience was also one of the reasons for Vesalius' insistence that Don Carlos' skull should be trepanised, an idea fiercely opposed by almost all his Spanish colleagues. The Don Carlos treatment took place during an interval between Pedro Jimeno's alleged presence in Alcalá and Pedro Marcos' official employment at the university in 1563. As a consequence, this unique event failed to unite Vesalius with the first 'post-Vesalian' generation of Iberian anatomists. There is therefore no evidence of Vesalius' appreciation of the spread of his own anatomical reforms to Castile, a process independent of his own similar transfer. Moreover, there are no indications that he was acquainted with or even encountered Francisco Valles, his keen supporter and the primary professor of the medical faculty at the University of Alcalá de Henares – or any other members of the renowned medical institution. Chacón's report instead emphasised the seemingly independent professions of the physicians employed at the universities, and those at court, who constituted the entire contingent involved in the treatment of Don Carlos.[53]

Two of the nine physicians engaged in the cure, Cristóbal de Vega and Fernando de Mena, were, however, closely affiliated with the nearby medical faculty, where they had both held the two professorships of *prima* only a few years earlier. Vega was *principal de prima* from 1549 until 1557, when he was succeeded by Francisco Valles, and Mena held the second primary chair of medicine from 1553 to 1560.[54] Following his retirement from the first chair of medicine in Alcalá, Vega joined the court as *Medico de Cámera* and was awarded 150,000 maravedís annually – an exorbitant wage compared to Mena's and Valles' later salaries of 80,000 after their similar appointments in 1560 and 1572.[55] Indeed, Vega's salary of 150,000

---

[52]   'Vesalio dixo su parecer con aquel latin y con aquella facilidad que en muchas juntas (con que él tuve) vi, y cura della (que a todo esto está obligado el buen cirujano) con tanta cordura que no fue mucho quedar todos muy satisfechos y admirados.' Daza Chacón, *Practica y Teorica de Cirurgia*, p. 205.

[53]   'Los medicos y cirujanos que se hallaron en la cura del Principe, son los siguientes desde el principio hasta el fin. El Dotor Vega, el Dotor Olivares, el Licenciado Dionisio Daza : desde el Segundo dia con los dichos, el Dotor Iuan Gutierres de Santander, Medico de Camara de su Magestad, y su Protomedico general; El Dotor Portugues, y el Dotor Pedro de Torres Cirujanos de su Magestad. Despues del descubrimiento del casco, el Dotor Mena Medico de Camera de su Magestad, y el Dotor Vesalio insigne y raro varon: desde seis de Mayo, el Bachiller Torres Cirujano de Valladolid.' Ibid., p. 200.

[54]   Muñoyerro, *La facultad de medicina de Alcalá de Henares*, p. 151.

[55]   Pascual Iborra, *Historia del protomedicato en Espana (1477–1822)* (Valladolid: University of Valladolid, 1987), p. 205.

maravedís was exactly twice the amount of the agreed 200 ducats reserved for the first chair of medicine in Juan de Ovando's 1566 reforms. Their decisions to resign from university positions in favour of careers at the Spanish court were analogous with Vesalius' own change of employment. This tendency suggests that a university career was not necessarily regarded as the professional summit for medical doctors, who were often attracted instead by the prospects of increased wealth and prestige gained by working at court. The generous salary may have been poor recompense, however, for the desperate medical struggle of April and May 1562, and its uncertain outcome with potentially grave consequences for all the physicians involved. In his new position at court, de Vega was serving as Don Carlos' personal physician at the time of the accident in 1562 and was consequently the first doctor to treat the wounded prince. Neither Vega nor Mena turned to their old colleagues at the University of Alcalá in the wake of the tragic accident, and Vega instead sent an envoy to the physicians of the court.[56] As the Prince's physical condition declined significantly in late April, Mena was the one chosen to deliver the bad news to Philip II.[57]

During the initial period when only three doctors tended the wounded Prince, Daza Chacón was the first to offer a diagnosis of his head-wound: 'The wound was the size of a thumb-nail and its edges were very bruised, and we discovered that the pericranium was also injured.'[58] A schematic drawing of the wound is seen in one of three new documents (a brief avviso, the minutes draft of a letter dated 1 May, and a formal 14 May letter account from of the cure addressed to Duke Cosimo I de' Medici from his Spanish ambassador Bernardetto Minerbetti) recently unearthed for this study at the Archivio di Stato of Florence.[59] The lower margin of the avviso shows the only known illustration of the wound and its trapezoid shape, which corresponds to Chacón's reference to its thumbnail size and structure. The two other documents offer

---

[56] 'Mandó al Dotor Iuan Gutierrez su medico de camara, y su protomedico general, se partiesse luego para Alcalá, y llevase consigo a los Dotores Portugués y Pedro de Torres, Cirujanos de su Magestad, los quales llegaron a Alcalá Lunes siguiente.' Dionosio Daza Chacón, *Practica y Teorica de Cirurgia*, p. 191.

[57] 'Su Magestad visto esto, y porque el Dotor Mena Medico de su Camara le dixo, que sin duda su altexa moriria, se partió de Alcalá entre diez y once por la noche con una oscuridad y tempestad grandissima, y fuese a san Geronimo de Madrid con la pena de todos podemos entender.' Ibid., p. 195.

[58] 'Llamaronme, y descubri la herida, presentes don Garcia de Toledo su ayo, y su mayordomo mayor, y Luis Quixada Cavallerizo mayor de su Alteza, y los Dotores Vega y Olivares Medicos de Camara, y vi una herida del tamaño de de una uña del dedo pulgar, y la circunferencia bien contusa, y descubierto el pericraneo se vio que estava algo contuso.' Ibid., p. 191.

[59] ASF. MdP, vol. 5040 f. 240 v., ASF. MdP, vol. 5040, f. 235 r, ASF. MdP, f. 236 r. Thanks to Alessio Assonitis, Director of the Medici Archive Project, and Senior Research Fellow Maurizio Arfaioli for their help in recovering and transcribing this new evidence.

lenghty accounts of the Prince's illness and the attempted treatment of his injury, which grew much worse 10 days after the accident: 'On the eleventh day he was attacked by major fevers which grew so bad that His Majesty with all his entourage and almost all the principal members of his court went to Alcalá followed by Doctor Vesalius.'[60]

The 1 May minutes by Minerbetti were written on the very same day Vesalius arrived in Alcalá and did not confirm Chacón's and Olivares' later claims that only Vesalius (and not the Spanish doctors) was eager to intervene surgically on the wounded patient. Vesalius seems to have been brought to Alcalá to assist rather than to single-handedly promote the surgical procedure on the frail Prince: 'This morning (Philip II) left in great haste to see him and he brought Vesalius with him because the surgeons and physicians wanted to cut and open the wound to see if the inflammation of the neck was caused by pus from the swollen injury.'[61] The last of the physicians to appear in Alcalá were Chacón's former mentor, Bachiller Torres from Valladolid, who arrived a few days after Andreas Vesalius and Fernando de Mena had joined the medical team together with Philip II's entourage.[62] Daza Chacón's *Teorica y practica de chirugia* repeatedly praised Vesalius with whom he shared a parallel career path; as both military surgeon to Charles V, and later as physician at the Madrid court of Philip II. Chacón's *Relacion* also documented some of the disagreements between Vesalius and his Spanish colleagues, which has later led some biographers to lament the lack of appreciation for Vesalius within Spain. Among the conflicts mentioned in Chacón's account was the discord between Vesalius and some of the Spanish physicians about whether or not to perform a trepanation of the Prince's wounded skull.[63] Minerbetti's 1 May letter to Duke Cosimo

---

[60] 'Nell'undicesimo assalito da febbri maggiori in tanto peggiorò che la Maestà Sua con tutto il conseglio et quasi con tutti li principali della corte se n'andò in Alcalà et menato seco il Dottor Vesalio.' ASF. MdP, vol. 540, f. 236 r.

[61] 'Quella mattina in molta fretta è andato a vederlo et ha menato il Vessalio perché questi cerusici et phisici proponevono di voler tagliar et aprir la ferita per veder se l'enfiagion della gola nascieva da sacca che facesse la ferita.' ASF. MdP, vol. 540, f. 235 r.

[62] 'Visto estos accidentes, yo propuse en la consulta, que pues era negocio de tanta duda, que traxessen al Bachiller Torres Cirujano y maestro mio, que residia en la villa de Valladolid hombre de muchas letras, y gran experiencia ... El viernes primero de mayo partió de Madrid antes de amanecer, y llegó a Alcalá antes que curassemos a su Alteza el qual luego se curó presente su Magestad, y el Dotor Andrea Vesalio hombre Doctissimo.' Ibid., p. 193.

[63] 'El Dotor Vesalio y el Dotor Portugues visto esto, fueron de parecer, qye el daño era interior, y que no tenia otro remedio sino penetrar el casco hasta las telas; en esta opinion permanicieron tanto tiempo quando duró la calentura, tenian por burla que se tratase de otro benificio. Todos los demas fuimos de parecer, que la causa destos accidentes era una de dos, o que el huesso del casco estava purulento, y para esto era bien se legrasse por las señales dichas, y porque lunes, y martes, y todos los otros dias despues de la apercion, tornó a aperecer aquella machuela que hemos dicho

(erroneously attributed to Leonardo Nobili and without archival reference in previous accounts) concluded with the observation about the incompatibility of the famous anatomists and his unskilled Spanish colleagues: 'These Spanish doctors have until now delayed and opposed Vesalius' wishes, and those who have not seen it would not believe the poor skills of these surgeons.'[64]

As the head-wound grew worse in early May, a Moorish medical empiric from Valencia named Pintarete was called to Don Carlos' sickbed, where he arrived on 8 May. Initially high hopes were attached to the use of his homemade ointments, which, however, allegedly infected the wound and only worsened the patient's situation. Pintarete was therefore dismissed shortly thereafter, as related in Daza Chacón's account: 'When we had tried to heal with the ointment of the Moor, he went to Madrid to cure Hernando de Vega, whom he soon sent to heaven with his salves.'[65] Minerbetti's letter to Duke Cosimo from 14 May gives a different account and claims that the Morisco healer was still involved and was much more successful in the treatment than had been claimed by Chacón and Olivares, and that he was even engaged in a surgical procedure to open the wound and ease the inflammation of Don Carlos' head and body.[66] According to Chacón, the Morisco healer was soon send away, and as the Prince was thought to be almost near his end, Vesalius' drastic procedure to drill into his skull was reconsidered. The wound was cleaned and prepared for the operation, and the ensuing drilling was prevented only by Chacón's discovery that the injury was external and limited to the scalp and the pericranium, while the skull itself was unharmed. This last-minute reluctance to proceed, together with the Duke of Alba's personal intervention, prevented the operation proposed by Vesalius from being carried out. Even though Vesalius' insistence on trepanation of the skull was finally rejected, and though the patient survived without the operation, Chacón did admit that there was good reasoning behind Vesalius' recommendation of this radical invasive treatment. His account

---

en el casco: ò que la inflamacion externa se avia comunicado por las suturas a las membranas del cerebro, y en esto nos afirmamos mas, y que si avia daño dentro, q. era este, y no otro. No dexò de tener Vesalio muchos fundamentos para su opinion, los quales de lo dicho se pueden colegir : no han faltado algunos de la facultad que no se hallaron presentes, que dixeron que esto no se podia alcazar por arte, sino que acaso acertamos'. Dionosio Daza Chacón, *Practica y Teorica de Cirurgia*, p. 195.

[64]   'Questi medici spagnoli hanno indugiato tanto fino a hora a voler il Vessalio, et chi non vede non può credere la poca pratica di questi cerusici.' ASF. MdP, vol. 540, f. 235 r.

[65]   'Acordamos dar con los unguentos, y con el Morillo al traves, y el se fue a Madrid à curar Hernando de Vega, al qual con sus unguentos embiò al cielo.' Dionosio Daza Chacón, *Practica y Teorica de Cirurgia,* p. 195.

[66]   'El Moresco sfasciata la piaga et aperta con le mani vi pose dentro il naso et ve lo tenne fermo il termine di due Credi et levato suso domandò Sua Altezza se li doleva la fronte. Rispose di no. Replicò il moresco che presto con l'aiuto di Dio lo renderebbe sano in tornarono a darli brodi conservati et a cibarla dopo che cominciò a dormire quietamente.' ASF. MdP, vol. 540, f. 236 r.

also told that Vesalius had held the support of Dr Portugués, who was in charge of the Prince's treatment and who (according to Chacón, but not Olivares) continued to advocate trepanation even after the improvement of Don Carlos' condition had convinced the others of the futility of the procedure: 'When Doctor Vesalius and Doctor Portugués studied (the wound), it appeared to them that the injury was internal, and that the only proper cure would be to penetrate the skull until the membranes; while the fever lasted they never wavered in their opinion, and disparaged the idea of any other remedy.'[67]

Minerbetti's account praised the many initiatives by the royal family to appeal to divine help in Don Carlos' healing process. Allegedly Queen Elizabeth of Valois was involved in numerous prayers and religious ceremonies on his behalf, and Don Carlos' aunt Joanna of Austria defied the cold Castilian climate and walked barefoot to her Convent of Las Descalzas Reales in Madrid, while King Philip himself was affectionately tendering his son with eyes full of tears, 'a sight which could even make a stone cry'.[68] The highest praise was given to the Duke of Alba, who never wavered in his attention to the wounded Prince. According to Minerbetti, the Duke and the citizens of Alcalá were behind the initiative to have the earthly remains of a local Franciscan monk, Fray Diego de Alcalá, delivered to the patient's sickbed. In Daza Chacón's account Don Carlos later claimed that the saint's relics and the vision of this holy man addressing him had been crucial to his recovery.[69] By the time of Minerbetti's letter of 14 May, Don Carlos' health had improved significantly and by the end of June he was completely cured. In the wake of this allegedly miraculous intervention, the leader of the Franciscan order in Castile and the rector of the University of Alcalá began to campaign for the canonisation of Fray Diego. Consequently, the prominent local university finally became directly involved in the Don Carlos affair, which until then had engaged only former members of its medical faculty in their capacity as court physicians. Two of the physicians involved in the cure – Christobal de Vega and Hernán Lopéz 'el Portugués' (Vesalius' sole supporter during the lengthy period of treatment) – eventually renounced their own contributions to the treatment, instead attributing the cure to divine intervention. During interviews with the commission established to investigate the eyewitness reports concerning

---

[67]   'Sirvio esto de salir de la duda que se tenia, y assi todos, excepto de Vesalio, y el Portugues, qye nunca mudaron de parecer, entendimos que el daño era comunicado, y accidental de la calentura y de la erysipela'. Dionosio Daza Chacón, *Practica y Teorica de Cirurgia*, p. 194.

[68]   ASF. MdP, vol. 540, f. 236 r.

[69]   'Fue tanta su devocion, que segun su Alteza cuenta, el Sabado en la noche a nueve de Mayo se le aperecio el bienaventurado santo fray Diego, con sus habitos de San Fransisco, y una cruz de caña en las manos, atada con una cinta verde, pensando el Principe que era San Fransisco le dixo: Como no traeis las llagas? No se acuerda de lo que respondio, mas que le consoló, y dyxo que no moriria deste mal.' Dionisio Daza Chacón, *Practica y Teorica de Cirurgia*, p. 199.

this alleged miracle, de Vega condemned the opposing and, in his eyes, blasphemous beliefs which were held by some of his colleagues: 'He had heard it said that two or three of the physicians did not hold the cure to have been a miracle ... saying this in order to make their role in the cure seem greater.'[70]

In 1588 San Diego de Alcalá was finally recognised as the first saint of the Counter-Reformation period and was widely celebrated, even in the Americas, where many cities still bear his name. Don Carlos' cure in 1562 had been closely followed by a wide European audience, including agents and ambassadors of European courts and contemporary physicians, such as Gerolamo Cardano, who credited Vesalius with the successful treatment: 'Vesalius ... by his skill preserved Don Carlos from evident death.'[71] Vesalius' failure to convince his Spanish colleagues of his (arguably wrong) diagnosis of the prince's head-wound and the subsequent placement of Fray Diego's corpse near the sickbed of the wounded prince was commented upon by contemporaries; it also represented the culmination of Charles D. O'Malley's narrative of the battle lost by Vesalius against the Spanish orthodoxy in religious and medical matters. O'Malley referred to the treatment of the prince as 'an action characteristic, although usually not in such degree, of Spanish piety in that age of credulity and superstition.'[72] Recent studies of the royal injury and its treatment analysed from a modern medical perspective, on the other hand, affirm that Vesalius' Spanish colleagues may have been right to reject trepanation of Don Carlos' skull.[73]

In contrast to Daza Chacón's praise for the sound judgment of the Flemish anatomist, O'Malley presented the events in Alcalá as the most obvious example of the fundamental incompatibility which lay between this forerunner of modern anatomy and the backward Spanish scenery framing his late career: 'Vesalius' recognition of his own ability, bolstered by the appreciation of those at liberty to call upon him, must, when contrasted with the jealous restraint placed upon him by his lesser Spanish colleagues, have made his position extremely galling

---

[70] L.J. Andrew Villalon, 'Putting Don Carlos Together Again: Treatment of a Head Injury in Sixteenth-Century Spain', *Sixteenth Century Journal*, 26(2) (1995), p. 363. The interviews relating to the canonising process were unfortunately lost in the fire that devastated the Archivo Central de Alcalá de Henares in 1939.

[71] Cited in.O'Malley, *Andreas Vesalius of Brussels*, p. 468.

[72] Ibid., p. 300.

[73] 'This conclusion tends to vindicate generations of Spanish historians who have argued indignantly that the foreigner Vesalius, whatever his other merits, should not be credited with saving the prince. Upon his arrival at Alcalá, the great anatomist adopted the view (from which he apparently never wavered) that the injury was internal and might be cured only by drilling through to the inner membranes. Although his Spanish colleagues politely acknowledged that his views had some foundation, in the end they chose not to follow his advice. Instead, they discontinued drilling as soon as they became satisfied that the skull was healthy. In doing so, they probably saved the patient's life.' L. J. Andrew Villalon, 'Putting Don Carlos Together Again', p. 362,

and frustrating. It was a situation somewhat delineated in the Don Carlos case and its aftermath when the cure of the Prince was attributed to various saints, sanctuaries, and religious formulae. In short, the general atmosphere of Spain was hardly scientific.[74] O'Malley's critical account nonetheless disputed another anti-Spanish myth spawned shortly after Vesalius' death in October 1564 following his pilgrimage to Jerusalem. It claimed that Vesalius had fled Spain earlier that year to avoid punishment by the Inquisition after accidentally dissecting a male Spanish aristocrat who was still alive (or a woman of Philip II's court, as claimed in the French surgeon Ambroise Paré's account from 1573).[75] The tale of the Spanish nobleman was elaborated upon in Adam Melchior's *Vitae Germanorum Medicorum* (Heidelberg 1620), which included a brief biography of Vesalius, and the content of an allegedly authentic 1565 letter by the French protestant diplomat Hubert Languet, which narrated the *Vera causa peregrinationis Vesalii.*[76] In spite of O'Malley's reservations in this case, he himself produced an overly simplified explanation for Vesalius' departure from Spain in early 1564: 'It seems rather that the Galenism of Spanish medicine was one of several factors that caused Vesalius to leave the country.'[77]

An investigation of anatomy studies in Alcalá at the time of the treatment of Don Carlos instead reveals the university town as a place where Vesalian anatomy was not unknown, but had in fact been practised several years earlier by Vesalius' devoted follower, Pedro Jimeno, who arrived there at an unspecified date in the 1550s. The Valencian *Estudi General* continued this tradition by exporting several generations of newly appointed anatomists to the medical faculty at Alcalá, where only one non-Valencian professor held the *cátedra de anatomía* during the sixteenth and early seventeenth centuries. The primary chair of the medical faculty, Francisco Valles, wrote of his admiration for Vesalius, and Chacón's later publications show that the Belgian anatomist was revered by both university and court physicians. Alcalá de Henares was therefore the stage for both a written appreciation of Vesalian anatomy and a unique case of collaboration between the celebrated anatomist and his Spanish colleagues. As the cure of Don Carlos indicates, some Spanish physicians and surgeons of the court proved to be just as competent as their renowned colleague, and their rejection of his idea to open the skull of the wounded patient seemingly prevented a deadly outcome.

In spite of its initial openness towards the new anatomy, the university chair of anatomy in Alcalá nonetheless had to struggle to survive even from the

---

[74]   O'Malley, *Andreas Vesalius of Brussels*, p. 285.

[75]   Ibid., p. 468.

[76]   Adam Melchior's *Vitae Germanorum Medicorum* (Frankfurt am Main: Hered Jonae Rosae, 1620), p. 133

[77]   Ibid., p. 300.

very beginning, and it existed in an ever more rudimentary form throughout the late sixteenth· and entire seventeenth centuries. While the record of publications, statutory orders and other documentary evidence is less complete than the sources from other Castilian universities, the lists of reduced wages and deteriorating working conditions supplement much of the material found elsewhere. The 1665 revisions of the University of Alcalá statutes were similar to the 1594 revisions from Salamanca in their emphasis that only Galenic textbooks were to be used for the teaching of anatomy. Both cases demonstrate the gradual consolidation of pre-Vesalian doctrines in the only two Castilian universities with permanent chairs of anatomy. The pecuniary struggles of the university add an important facet to the enduring resistance towards new anatomical ideas and practices indicated in other sources. The financial problems inherent in supplying the newly established chair might have played an even more important part in the decline of anatomy at Alcalá. The anatomy studies initiated there in the mid-sixteenth century reached their lowest point of near-annihilation in the late eighteenth and early nineteenth centuries, when desperate and ultimately unsuccessful attempts were made to revive the former practice. This decay was only temporarily prevented during the latter half of the sixteenth and the beginning of the seventeenth centuries with the continuing influx of anatomists who had been educated in Valencia. The medical contributions from anatomists and physicians educated in Alcalá were nonetheless significant in several ways. The medical faculty left a substantial legacy in its education of physicians engaged in anatomical studies beyond the town of Alcalá (Fransisco Díaz, Fransisco Hernández and Juan Fragoso in Valencia, Guadalupe and Seville respectively) and some of the first Spanish anatomists active beyond the Iberian Peninsula (Andres Laguna in Rome, and Augustin Farfán and Juan de Barrios in Mexico City).

As late as the beginning of the eighteenth century, the medical faculty at Alcalá produced Martín Martínez, the most important anatomical contributor to the 'Novator' movement of the late seventeenth and early eighteenth centuries. Martínez was proudly described as 'hijo de aquella Universidad'[78] in an initiative to reintroduce anatomy studies at the university in 1802, and his *Anatomía completa del hombre* (1728) became the standard Spanish textbook on anatomy during the eighteenth and early nineteenth centuries. Its frontispiece depicted a beautiful and richly ornamented anatomical theatre built around 1700 in the baroque style, which indicated the renewal of Spanish activity in the study of anatomy. Martínez's former university town was not the place of his publication, however, or the location of this second Castilian university theatre of anatomy, after the structure built in Salamanca 150 years earlier. The location of this *Amphiteatrum Matritense* was the expanding Spanish capital, which

---

[78]   Muñoyerro, *La facultad de medicina de Alcalá de Henares*, p. 99.

eventually contained the University of Alcalá, most of its archive and ultimately even the town itself. Today Alcalá is little more than a suburb of Madrid, but Cisneros' renowned Renaissance University is still noticeable, particularly its beautiful facade from 1553, which has remained unaltered since the time of the arrival of the first Valencian anatomists in Alcalá de Henares

Figure 5.1   University of Alcalá de Henares. The exact year of the introduction of anatomy studies in Alcalá de Henares cannot be definitively established from the available documentary sources. The first documentation of the new discipline appears in a passing reference from 1559 to the Valencian anatomist Pedro Jimeno's practice in Alcalá few years earlier. More solid documentation can be found in the following decade after the appointment of another *valenciano*, Pedro Marcos, to a newly created professorship of anatomy in 1563. This expansion of the faculty was consolidated and ratified in two subsequent university reforms in 1566 and 1587, and coincided with the influx of anatomists from Valencia. *Photo credit*: author

Chapter 6
Barcelona

In 1559, after 150 years of preparation, Barcelona obtained its own *Estudi General*, similar to those established in Valencia in 1502 and in Zaragoza in 1583. King Martin I had made a similar attempt to found a fully developed university in the Catalan capital in 1401 and his initiative was later developed by Alfonso V in 1450. The latter attempt to supplement the existing studies of arts and medicine with faculties of grammar, rhetoric, civil and canonical law, and theology was approved by Pope Nicholas V shortly after.[1] However, well into the 1500s, only a small two-faculty institution, *Estudi General de Medicina e de les Arts liberals*, had been realised.[2] A deeply rooted university tradition existed elsewhere in the Crown of Aragon, where prominent universities had been established in both Montpellier and Lleida by the late thirteenth century. These universities were granted permission to dissect condemned criminals in 1340 and 1391, and during the fifteenth century similar privileges were awarded to schools of medicine and surgery in Barcelona, Valencia and Zaragoza in 1401, 1478 and 1488, respectively.[3] Yet, by the turn of the 1400s, none of these minor institutions had obtained university status, which would only come with the unprecedented proliferation of new Iberian universities during the sixteenth century. This development was in many ways analogous to the 'educational revolution' in contemporary Castile traced by Richard Kagan. The appearance of 26 new universities within the three Iberian kingdoms between 1475 and 1625 was unmatched elsewhere in contemporary Europe.[4]

Given its particular background as a small-scale 'medical university', the *Estudi General de Medicina e de les Arts liberals* became the nucleus of the later *Estudi General* after its inauguration in 1559. Its medical faculty was sided with the faculties of law and superseded that of theology, unlike contemporary Castilian universities, which often had only rudimentary medical faculties. The medical staff of the pre-existing school of arts and medicine was also

---

[1]   José Danon Bretos, 'Notas medicas en los libros de "Estudi General" de Barcelona (Siglos XVI–XVII)', *Cuadernos de de historia de la medicina española*, 10 (1971), p. 187.

[2]   Antonio de la Torre y del Cerro, *Provision de cátedras en la Universidad de Barcelona* (Barcelona, 1926), p. 11.

[3]   Luis Garcia Ballester, 'La cirugía en la Valencia del siglo XV, El privilegio para disecar cadaveres en 1478', *Cuadernos de historia de la medicina española*, 7 (1967), p. 167.

[4]   Between 1475 and 1625, the number of Iberian universities grew from seven to 33, with 18 in Castile, 13 in Aragon and two in Portugal.

incorporated into the new and larger institution in 1559, though it maintained a short and seemingly troublesome autonomy. An agreement was eventually made between this previous institution and the *Estudi General*, which in 1565 led to the final fusion of the two medical faculties.[5] Even though the records of matriculation have since been lost, approximately 1,700 students graduated from the *Estudi General* during the latter half of the century. This indicates a considerably smaller number of students passing through Barcelona's *Estudi General* than its counterpart in Valencia, which held up to 2,000 students at a time during the late sixteenth and early seventeenth centuries. The newly established University of Barcelona also proved less imposing in its academic output and less susceptible to medical reforms than the neighbouring University of Valencia. Yet, the staff of Barcelona's medical faculty quickly reached the level of their equivalents at the larger institution in Valencia and eventually surpassed the number of medical professors employed at the approximately 10 times larger University of Salamanca.[6]

From 1573, the medical staff of the University of Barcelona equalled Valencia with as many as eight medical professorships, twice the number appointed in Valladolid during the same period. The first *Ordinacions* of 1559 prompted the foundation of three chairs of medicine and emphasised that two dissections should be carried out within the *Estudi General* each year.[7] This modest number grew to 12 during the last two decades of the century and thus equalled or outnumbered the total number of dissections performed annually at the much larger universities of Salamanca and Valencia. The revised statutes of 1562 stated that one of the medical professorships at Barcelona should be in anatomy and surgery, along with those of *práctica* and *teórica*.[8]

Unlike the long periods of vacancy suffered by the anatomy chair at the University of Alcalá during the sixteenth and seventeenth centuries, the Barcelona professorship of anatomy was filled continuously from 1564 (with only a two-year vacancy period between 1598 and 1600) and remained in existence until the abolition of the university in 1717. Revised regulations and statutes from every decade of the late sixteenth and early seventeenth centuries emphasised the need for the consolidation and expansion of the anatomical teaching programme at

---

[5]   José Pardo-Tomás and Àlvar Martinez-Vidal, 'Anatomical Theatres in Early Modern Spain', *Medical History*, 49 (2005), p. 268.

[6]   Francisco Javier Alejo Montes, *La reforma de la universidad de Salamanca finales del siglo XVI* (Salamanca: University of Salamanca, 1990), p. 233.

[7]   'E que dos vegades l'any, o al menos una, cada hu d'ells sia obligat a fer Anathomia.' Ordinacions de 1559, f. 25.

[8]   'Item que en lo de la medicina lo que ja és estat ordenat de tres llicons annuals, co és, una de Medicina en theòrica, altra en pràctica y altra de chirurgia y anatomya.' BUB, Universidad de Cervera, libro 1, Ordenaciones de 1560 y otras, f. 25–28, cited in Antonio Fernández Luzón, 'La Universidad de Barcelona en el siglo XVI' (UAB, Doctoral Thesis, 2003), p. 409.

Barcelona, which took place in a *caseta de anatomia* from 1573 and was made compulsory for students of surgery. The establishment of the chair of *chirurgia y anatomya* in 1562 preceded a similar initiative at the University of Salamanca in 1566 to elevate surgery to university status. Throughout the late 1500s the professorships of anatomy and surgery were united in one university chair. The field of surgery consistently maintained its position as an integrated discipline within the *Estudi General* of Barcelona from its introduction in 1562, but the inclusion of this discipline was not imitated at the Universities of Valladolid and Alcalá until more than three decades later.

The seemingly progressive academic milieu of the late 1500s at Barcelona unfortunately did not leave behind a series of anatomical publications. There is furthermore a rupture in the archival record of the young university, which came to a dramatic end when it was closed by Philip V in 1717, together with a number of smaller Catalan universities such as the University of Lleida. Following the War the Spanish Succession, the *Estudi General* in Barcelona was replaced by the larger centralised University of Cervera. The surviving documentary material from all these universities is unsurprisingly scarce, given the upheaval caused by foreign invasions and the internal struggles of Catalonia.[9]

The surviving archival record of the city government, 'The Council of Hundred' or *Consell de Cent*, has fortunately left a wealth of documentary material relating to the appointment of the professors of anatomy, their salaries and teaching structure. This source material includes matriculation lists, schedules and curricula, and possibly the most intact records of statutes and regulations from any Iberian university in the sixteenth century. These statutes include frequent, if not always significant, alterations from 1559, 1562, 1567, 1575, 1576, 1588, 1590, 1596 and 1598, and have all survived in manuscript, and in one printed version of the 1596 *Ordinacions*. Most of this documentary material is kept in the *Arxiu Històric de la Ciutat* (AHCB), which houses the detailed records – surprisingly complete given the tumultuous history of Barcelona – written and stored by the *Consell de Cent*. Supplementary material and secondary sources are available at the *Biblioteca de Catalunya* in Barcelona, which is located in the former *Hospital de Santa Creu*, the stage for the public dissections of the *Estudi General* during the late sixteenth and entire seventeenth centuries. A stone from 1660 inscribed with 'Theatrum Anathomicum' still commemorates this former practice within the hospital.

---

[9] Danon Bretos, 'Notas medicas en los libros de "Estudi General" de Barcelona (Siglos XVI–XVII)', p. 188.

Francesc Micó and the Professorship of *Anatomía y simples y coses de apothecaris y cirurgia*

Despite the fact that royal privileges had enabled dissection to be performed at the *Estudi di medicina* from the early fifteenth century, very little is known about these studies and their practitioners. This is also the case with the first two professors of anatomy and surgery at the *Estudi General* during the early 1560s, Caspar Ferrer and Narcís Vidal de Vallria, who held the chair of anatomy between 1564 and 1567. Not until the employment of Francesc Micó is it possible to get a glimpse of the first generation of post-Vesalian anatomists active in Barcelona. Out of respect for Micó's reputation as a distinguished botanist as well as an anatomist, the chair of *chirurgia y anatomya* was changed to *anatomía y simples y coses de apothecaris y cirurgia* following his appointment in 1567. This chair had a precursor in Valencia in the previous two decades, where a similar professorship had attracted the young Francisco Diaz to leave Alcalá for Valencia, 'where medicine and anatomy is presently flourishing, not less than the study of plants and other curiosities', as later recorded in his *Tratado nuevemente impresso, de todas las enfermedades de los Riñones...* (Madrid, 1588).[10]

The composite university chair of anatomy, surgery and botany was the second Iberian professorship of medical botany to exist outside Valencia. Like the chair of surgery, it remained integrated as a university chair within the *Estudi General* throughout the sixteenth century, either autonomously between 1576 and 1588 or fused with anatomy, as was the case during the remainder of the sixteenth century.[11] Following Micó's departure from this joint chair of anatomy, surgery and medical botany, the professorship was often split into two and occasionally even three different chairs, thus raising the number of medical professorships at the *Estudi General* from five to eight between 1572 and 1587.[12] Glimpses of Micó's previous career were granted by Francisco Hernandez's unpublished comments on Pliny in *Historia natural de Cayo Plinio Segundo*, in which the author referred to 'Dr. Micón, who was at that time practising medicine with me.'[13] Hernandez's account was based on his own medical practice in the hospital for visiting pilgrims at the Monastery

---

[10]   Fransisco Diaz, *Tratado nuevemente impresso, de todas las enfermedades de los Riñones, Vexiga, y Carnosidades de la verga, y Urina, diuidido en tres libros* (Madrid, Francisco Sanchez, 1588), f. 19r.

[11]   Antonio Fernández Luzón, *La Universidad de Barcelona en el siglo XVI* (Barcelona: Edicions Universitat Barcelona, 2003), Appendix V, 'Nombramientos de medicina (1559–1606)', pp. 647–51.

[12]   Ibid., p. 647.

[13]   'El Dr. Micón que a la sazon practicaba conmigo medicina.' Germán Somolinos d'Ardois, Fransisco Hernández, *Obras completas* (Mexico City: Universidad Nacional de Mexico, 1958–84), p. 130.

of Guadalupe. The only anatomist to be mentioned by name in Hernandez's account was Francesc Micó, who, together with a number of unrecorded colleagues, allegedly strove to bring anatomy 'up to date'.[14] Micó's educational background was at the medical faculty of Salamanca, where he studied under Cosme de Medina and Lorenzo Alderete between 1553 and 1555 before his arrival at the Monastery of Guadalupe. During the late 1550s, Micó and Hernández worked as physicians at the nearby hospital for the pilgrims like other young doctors who came to Guadalupe for the two-year practicum, which concluded their 10 years of medical education.

The two young scholars had a shared interest in botany and together discovered a number of new botanical species during their formative years in Extremadura. On the direct orders of Philip II, Micó made a series of botanical excursions to the mountains of Catalonia and was later credited by the French botanist Jaques Dalechamps in his *Historia generalis plantarum* (1587) – and was still recognised by Linné two centuries later – as the discoverer of *Verbascum Micónii*. Micó obtained seeds and plant specimens from Iberia and the Americas, which were delivered to Dalechamp in France and to Philip II's botanical gardens. Indeed, the Spanish physician still gives his name to an entire species of plants, classified as *Micónia*. The collection and preparation of medicine from plants and herbs, commonly referred to as *simples*, underwent an institutionalisation, professionalisation and specialisation as a university discipline during the late sixteenth century, in some ways analogous to the formalisation of anatomy studies. In an Iberian context, this joint development was exemplified most clearly with the appointment of Micó to the composite chair of anatomy and botany between 1567 and 1572. Unfortunately, this first anatomist/botanist of the *Estudi General* left behind only one published work. This publication did not deal with either anatomy or botany, but instead with the importance of cooling drinks with snow, as presented in the title of this work on food conservation, diet and hygiene: *Alivio de los sedientos, en el qual se trata de la necessidad que tenemos de beber frio, y refrescado con nieve, y las condiciones que para esto son menester, y quales cuerpos lo pueden libramente supportar* (Barcelona, 1576).

This somewhat arcane subject filled an evidently important and profitable niche in the contemporary market for medical publications. It appeared in the wake of similar works by Nicolas Monardes and later Francisco Franco, who, a few years earlier, had published a pendant to Micó's work, entitled *Libro que trata de la nieve, y de sus propiedades y del modo que se ha de tener, en el bever enfriado con ella y de otros modos que ay de anfriar con otras curiosodades, que daran contento,*

---

[14] 'Exercicio que en cortar por mano ajena hombres tuve en Guadalupe, donde dexamos puesta por la bondad de dios el anatomía de su punto como hasta allí no se hubiesen cortado todo sino los miembros interiores solo.' Ibid., p. 132.

por las cosas antiguas, y dignas de saber, que cerca de esta material, en el se veran (Seville, 1571).[15] It may be an indication of the limitations of the medical milieu in Barcelona that Micó decided to produce a brief manual on the dietary and hygienic virtues of snow-cooled drinks rather than an anatomical publication for use by his fellow Catalans. In fact, there were remarkably few medical books produced in Barcelona during the late sixteenth and early seventeenth centuries, and those that were published mostly dealt with contemporary plague epidemics of the city.[16]

While little exact knowledge survives of Micó's anatomical practice at the medical faculty, he is found in other university contexts as a prominent opponent of the abolition of the *Estudi di medicina* as an independent body, and its inclusion into the *Estudi General*, united under one rector and chancellor. The discord ended with Micó's temporary exclusion from the university staff and his reappointment at the *cátedra de teórica* shortly after, as indicated in a *concordia* signed by the university and the college of doctors in 1565. Between 1564 and 1574, Micó's name appeared in a list of university teachers who were appointed as Inquisitors and overseers of the strict regulations on *limpieza de sangre* among the staff and students of the university. He was succeeded in this position by

---

[15]   Similar treatises on the hygienic and therapeutic virtues of cooled drinks and meals appeared repeatedly during the following century. In fact, these books on diet and medicine inspired an entire genre, which prospered and developed in the early 1600s. A short summary of works dealing particularly with this subject can be found in Mar Rey Bueno, 'Concordias medicinales de entrambos mundos. El proyecto sobre material médica peruana de Matías de Porres', *Revista de Indias*, LXVI(237) (2006), pp. 357–59. Her list of publications on this topic gives a good indication of the popularity of the genre, and its by no means insignificant status in the medical book market of early modern Spain: Pedro de Parraga Palomino, *La carta...en que se trata del arte y orden para conservar la salud y buen uso de bever frio con nieve* (Granada, 1612) Alonso Gonzáles, *Las utilidades de la nieve, deducidas de buena medicina* (Seville, 1622); Juan de Carnaval, *Las utilidades del agua i de la nieve, del bever frio y caliente* (Madrid, 1637); Fernando Cardaso, *Methodo curativo y uso de la nieve* (Córdoba, 1640); Miguel Fernández Pena: *Breve apologia y Nuevo discurso del Methodo que se debe observar, reprovando el agua de nieve en día de purga* (Jaén, 1641). Coinciding with these publications, a profitable Catalan transport of snow and ice from the mountainous Montseny area to Barcelona increased during the 1600s and peaked in the eighteenth century, as described by Gemma Font et al. in *El tranport de la neu del Montseny i del glac del Valles a Barcelona al segle XVIII, VI Trobada d'Estudiosos del Montseny* (Barcelona: Diputació de Barcelona, 2005), p. 198.

[16]   Works on the plague epidemics of sixteenth- and seventeenth-century Barcelona include: Onofre Bruguera, *Novae, ac infeste destillationis, quae civitati Barcinonensi ac finitimis circiter hyemale solstitium anni a Christo nato 1562 accidit, brevis enarratio* (Barcelona, 1563); Joan Rafael Moix, *Libre de la peste dividit en tre tractats. En doctrina universal preservatio, y curatio della* (Barcelona, 1587); Bernat Mas, *Orde breu y regiment molt util per a curar y preservar de peste* (Barcelona, 1625); and Joan Francesc Rossell, *El verdadero conocimiento de la peste* (Barcelona, 1632).

another professor of *anatomia y simples*, the surgeon Caspar Massaguer, who retained the rank of Inquisitor for the remainder of the century.[17]

Micó held the chair of anatomy at the *Estudi General* between 1567 and 1572, with a short interval between 1570 and 1571 when the physician Francesc Castello took over the chair, which at that time was remunerated with a modest salary of 35 libres annually.[18] The unification of the chair of *anatomia y simples* from 1567 seems to have been rooted in Micó's particular skills as both anatomist and botanist; the close relationship between these two branches of medical learning nonetheless continued after his departure from the university chair and throughout the century, and even into the next. In the documentary records, the consolidation and union of the professorship in anatomy and botany was emphasised more clearly at Barcelona than any other Iberian university. A series of revised statutes throughout the late sixteenth and early seventeenth centuries fused the two chairs as one and divided the teaching between anatomical dissections from October to February, and botanical studies and excursions during the spring and summer months. This organisation of two distinct teaching programmes was meticulously recorded in the *Ordinacions* for Micó's newly founded chair of *anatomía y simples y coses de apothecaris y cirurgia* in 1567.[19] Later the *Ordinacions* published in 1596 maintained the interrelation between *anatomia y simples*, specifying that human dissection was to be carried out within the university precincts, and botanical excursions and studies outside the city.[20] The statutes from 1567 coincided with the employment of Micó as Professor of Anatomy and included clear instructions regarding the prescription of Galenic textbooks as teaching tools, but were surprisingly vague in their guidelines for the anatomical curriculum, which was to include 'some work on

---

[17]   Caspar Massaguer held the position between 1574 and 1598. See Fernández Luzón, *La Universidad de Barcelona en el siglo XVI*, p. 543.

[18]   'Los dits magnifichs consellers, atesa la renunciació feta per mestre Francesc Micó, doctor en Medicina, segons referi lo rector del studi de la cathedra que tenia en lo dit Studi, provehiren mestre Francesc Castello, doctor en medicina, de la dita cathedra de Cirurgia y Anotomia que com dit és legia mestre Micó per temps de tres anys ab lo salari acostomat de XXXV lliures.' AHCB, Registre de deliberacions, II-72, f. 71.

[19]   AHCB, Estudi General, serie XVIII, vol 9, doc. 29, 'La qual lija un doctor expert en dites coses; co és, en lo hivern lija coses de anatomia y faca aquelles tenint ... e en la primavera lija coses de simples e cànones de Mesue; en lo restant del any coses de cirurgia.' Cited in Fernández Luzón, *La Universidad de Barcelona en el siglo XVI*, p. 438.

[20]   'E que lo lector que llegira la anatomia, sia obligat de fer cada mes dos voltes anatomies, desde lo mes de octobre fins per tot febrer, y en los altros mesos del dit any ser exides, y anar fora ciutat pera les herbes als studiants practicants sobre ellas.' *Ordinations e nov redrec fet per instauratio, reformatio, e reparatio, de la universitat del Studi general de la Ciutat de Barcelona, en lo any Mil sinc cents novanta y sis* (1596), Cap. 14, p. 32.

anatomy.[21] The 1596 *Ordinacions* used the same imprecise reference verbatim in its instructions for the teaching of first-year medical students.[22]

Antonio Fernández Luzón's *La Universidad de Barcelona en el siglo XVI* emphasised the susceptibility of the medical faculty of the *Estudi General* to contemporary reforms in European medicine: 'From humanist medicine to Vesalian anatomy and from practical surgery to new medical botany.'[23] The ambiguities in the available evidence mean that it is difficult to support the claims that Vesalian anatomy was practised in sixteenth-century Barcelona – in spite of Francisco Hernandez's prior claims that he and Micó had attempted to bring Spanish anatomy up to date. Hernandez indicated that Micó was involved in an anatomical restoration programme at the Guadalupe hospital, similar to that introduced in Salamanca and Valencia by Cosme de Medina and his forerunners Pedro Jimeno and Luis Collado. The appearance of works by Vesalius, Realdo Colombo and Gabrielle Falloppio in the vast 290-book inventory made by Micó's successor Esteve Guardiet could be seen as a further sign of a reformed anatomical teaching and research programme in Barcelona. Yet this evidence remains speculative, as these contemporary publications in the inventory appeared alongside traditional works by antique and medieval authorities. The scarcity of modern textbooks in the medical curriculum of the *Estudi General* rather suggests that the medical faculty of Barcelona was largely rooted in traditional Galenism. While the medical curriculum of Barcelona's *Estudi General* consisted almost entirely of Galenic textbooks, it was almost completely devoid of Arabic authorities and the *galenismo arabizado* of the older Castilian Universities of Salamanca and Valladolid. The only exception was the ninth-century Arabic physician Mesue, who was included in the statutes for Micó's composite chair, which stipulated a short discourse on 'medical plants, pharmacology, and the Canons of Mesue'.[24] Mesue's *Canon* also appeared in the inventory made by Esteve Guardiet, who held the professorship of both *anatomia y simples* between 1572 and 1576, and then filled the autonomous chair of anatomy until 1582. As one of the least prestigious chairs within the medical faculty, the professor of *anatomia y simples* received only 35 pounds per year, the same salary given to the two other minor chairs of *ayudante de*

---

[21]　'*Per los quals los de medicina han de ser admes en bachiller aye de auer ohit tres anys lo es en lo primer any De elementis, De temperamentis, et De naturalibus facultatibus galinii y alguna obra de anathomia.*'AHCB, Estudi General, serie XVIII, vol. 9 doc. 15.

[22]　'Com es lo any primer haya hoyt los de elementiis & temperamentis Gal. y alguna obra de anatomia.' AHCB, *Ordinations e nov redrec...*, Cap. 14, p. 107.

[23]　'Los principales renovadores de la medicina europea. Desde el humanismo médico y la anatomía vesaliana hasta la cirugía práctica y la nueva botánica médica.' Fernández Luzón, *La Universidad de Barcelona en el siglo XVI*, p. 409.

[24]　'Coses de simples, apotecaris y els Canons de Mesue.' AHCB, Estudi General, XVIII, vol. 9, doc. 29.

Galeno and *principios de Galeno*. Even though this wage was later increased to 60 pounds in the *Ordinacions* of 1588 and 1596, it remained inferior to the salary of 100 pounds paid to the *cátedras mayores* of *Hipócrates*, *Galeno* and *práctica*.[25] It therefore seems remarkable that Guardiet was able to establish a library of such considerable size, but may be explained by an investigation of his employment elsewhere in the city administration. Both Guardiet and his successor Jeroni Magaroli earned 300 pounds per annum in the prestigious post of *obrer*, taking responsibility for some of the public works of Barcelona. A significant number of the medical university staff were periodically employed elsewhere, as was the case with the two brief professorships of anatomy held by Arcàngel Queralt (1589–90) and Caspar Massaguer (1595–96), who were both recorded as members of the *Vuitena del Morbo* (overseers of public health in times of epidemics) in 1584–85 and 1590–91 respectively. Massaguer allegedly earned 80 pounds in only 20 days as a member of the *Vuitena* – almost double the modest salary he received from the university for an entire year's work.[26]

Esteve Guardiet's Professorship of Anatomy and the First 'House of Anatomy' at the Hospital General de Santa Creu

Esteve Guardiet took over the professorship of anatomy in 1572 for an annual salary of 35 pounds, a modest sum compared to the 300 pounds he allegedly earned later as *obrer* for the city administration: 'To Master Guardiet, who holds the chair of anatomy and simples 35 pound per annum. And for every anatomical dissection he carries out, he should be paid one ducat, both for those he has carried out until now, and for those he will perform later on.'[27] The limited economical attraction of the chair of anatomy was only partially improved by the promise of an added bonus paid for every dissection performed, a caveat repeated in several statutes during the late sixteenth and early seventeenth centuries. Signs of discontent and disagreement appeared later in an increase of Guardiet's salary, which also compensated the low salary of his previous years of employment. The departure of Francesc Micó from the chair of *anatomia y simples* seemingly left a

---

[25]   Ibid., p. 410.

[26]   'Las elevadas retribuciones que percibían los profesionales sanitarios encargados de envestigar las epidemias son explicables por el riesgo y el perdido de cuarentena que debían complir al terminar su misión. El Consejo de Ciento acostumbrada a pagar, a principios del siglo XVII, 4 libras y 4 sous a los medicos, y 4 libras a los cirujanos por cada jornada.' Fernández Luzón, *La Universida de Barcelona en el siglo XVI*, p. 269.

[27]   'A mestre Guardiet, la cadira de anathomia y simples ab salari de XXXV lliures l'any mes, per rahó de les anathomies li sie pagat a rahó de un ducat per quiscuna anatomia, axi de les que fins assí ha fetas com de les altres que aprés farà.' AHCB, Registre de deliberaciones, II-82, 29-XI-1573, f. 173.

transitional period of rather dubious working conditions for his successor, who was only belatedly compensated with 25 ducats by the city council in 1575 for 21 anatomical dissections carried out in the preceding four years.[28]

The annual dissections were thereafter reduced to four from the previous average of five or six as indicated by Guardiet's belated remuneration.[29] The modest number of only four public anatomies per annum could indicate that the practice was gradually falling into decline. A series of subsequent statutes prove that this was not the case, however, and contemporary preparations for a new *caseta de anatomia* and the anatomical training of students of surgery show that the practice was increasingly well established in the Catalan capital. Before Guardiet was appointed to the chair of *anatomia y simples*, he was among the doctors employed in the grand old Hospital of Santa Creu, as recorded in Josep Danon's *Visió històrica de l'Hospital General de Santa Creu de Barcelona*.[30] Francesc Micó's particular background and combined skills as both dissector and botanist paved the way for the creation of the composite chair of *anatomía y simples y coses de apothecaris y cirurgia*. Guardiet's previous employment at the *Hospital General de Santa Creu* offers a similarly logical link with a 1573 request to perform future dissections away from the poorly suited university in the more appropriate facilities provided by the hospital precincts of Santa Creu.[31] Little is known about the earliest phase of this 'house of anatomy' within the hospital, but the initiative was evidently met with approval, and the *aula de les anatomies* was conveniently placed near the *corralet* (hospital cemetery) of the Hospital of Santa Creu.[32]

According to studies of the anatomical theatre in Barcelona carried out by José Pardo-Tomás and Àlvar Martinez-Vidal, there were some early precedents for the establishment of a permanent place for the practice of dissection; documentary evidence refers to a room used for anatomical dissections as

---

[28] 'Y page deb. A Esteve Guardiet, doctor en medicina venys mes sua el quatre soue en de que el p. vent y una anatomia ha fezat per tems de quatre anys ... inclusive a de un ducat anatomia y aquelles si son pagados per lo racionae de dit studi y del que resta del salary de algunos cadire co de muster del lectoral com et acostomat.' AHCB, Registre de deliberacions, II-86, f. 151–52.

[29] 'Y al doctor fa les anatomies que quiscuna vegada farà tota la anatomia li sie donat y pagat a raó un ducat per quiscuna anatomia complida de tot lo cos y cap, les quals hajen de seer quiscun quatre solament.' AHCB, Registre de crides i ordinacions, 1575–83, IV-20, f. 77.

[30] 'Esteve Gordiet nascut al Pla de Sant Tirs. Doctor en medicina el 21 febrer de 1567 I catedràtic de l'estudi entre el 1566 I el 1580. Es metge de l'Hospital el 1572, esmentat el 'Llibre de l'inventari.' Josep Danon, *Visió històrica de l'Hospital General de Santa Creu de Barcelona* (Barcelona: Fundació Salvador Vives Casajuana, 1978), p. 131.

[31] 'E quant al que es e stat opposat comincia molt mudar e se non fer e constituir sins lo hospital general algun mort a hyt se pu guessen fer les anathomies que los metegs per millor intellegentia dels comts en medicina anos sumes de fer.' AHCB, Reg. de deliberacions. II-82, f. 132.

[32] Pardo-Tomás and Martinez-Vidal, 'Anatomical Theatres', p. 266.

Figure 6.1   The modern *Biblioteca de Catalunya* in Barcelona, housed in the former *Hospital de Santa Creu*, where anatomists from the recently established University of Barcelona (1559) carried out public dissections during the late sixteenth and entire seventeenth centuries. A stone inscription from 1660, *Theatrum Anathomicum*, still commemorates this former practice within the hospital. In 1573 a (now lost) *caseta de anathomia* was built within the precincts of the hospital for use by the joint chair of *anatomia y simples*. Dissections were performed in the cold period between October and February; the remainder of the academic year was assigned to botanical excursions outside the city. The composite chair of anatomy and botany demonstrated the curious and complex relationship between the two branches of medical learning, which were often closely linked in the early modern period – in an Iberian context, particularly so at the Universities of Barcelona, Zaragoza and Valencia. Courtesy of the 'Biblioteca Nacional de Catalunya'

early as 1481.[33] Yet there is no documentation of the size or shape of this early 'theatre', or of the later structure placed within the *Estudi General*, or even of the hospital room for dissection established in the wake of the 1573 petition. The teaching conditions for anatomy within the *Estudi General* prior to that year were only vaguely referred to as 'inappropriate' in a later petition to the *Consell de Cent*.[34] The *Consell* duly budgeted 1,000 ducats for the hospital

---

[33]   José Pardo-Tomás and Àlvar Martinez-Vidal, 'El primitivo teatro anatómico de Barcelona', *Medicina e Historia*, 65 (1996), p. 12.

[34]   'Lloch o instáncia cómmoda per a fer lo exercissi de les anathomies, les quals vuy se fan dins lo Studi general de la present ciutat, ab molta desaccomoditat aixi del dit exercici de la

Caseta from a fund established for unexpected expenses.[35] The incorporation of anatomical university teaching within an already-existing hospital followed a similar pattern to the other Iberian Universities of Valencia, Alcalá, Barcelona and Zaragoza. The University of Salamanca, which established an autonomous university theatre for teaching purposes, was a rare exception to this general rule, as was the contemporary Monastery of Guadalupe, where a hospital with no formal university affiliation carried out its own regular dissections.

Within the Crown of Aragon, the *Estudis Generals* of Valencia, Barcelona and Zaragoza all followed the same path towards consolidation of anatomical teaching within hospital grounds shortly after their respective foundations in 1501, 1559 and 1583. There are references to anatomical studies taking place in a yard within the *Hospital General* in Valencia in the 1520s, and to even earlier practices in the *Hospital General de Nuestra Señora de Gracia* of Zaragoza, where a royal provision of 1488 supported human dissection. It was not until 1524, 1573 and 1586, respectively, that the first documentary evidence referred to architectural modifications and additions made for anatomical teaching within the hospitals of Valencia, Barcelona and Zaragoza. Unfortunately, there is no description of the structure assembled within Barcelona's Hospital of Santa Creu following the 1573 request, and there is only a brief reference from 1524 to a recently dug irrigation ditch and a yard made for the dissections within the *Hospital General* in Valencia. In contrast, the builder Andrés de Capraneda produced a detailed description of the dimensions of Zaragoza's anatomical theatre, which will be dealt with in Chapter 7.

The relocation of anatomical teaching in Barcelona to the *Hospital de Santa Creu* led to the construction of the city's first purpose-built dissection room. In all likelihood, nothing resembling an anatomical theatre was established until the middle of the following century, when an investment of 140 pounds paid by the *Consell de Cent* between 1658 and 1660 secured the Catalan capital's first *Theatrum Anathomicum*, which is still commemorated in the stone inscription in the Hospital de Santa Creu.[36] Construction of the baroque amphitheatre with three wooden rows of rectangular benches, and a rotating stone dissection table began in 1673 and was completed two years later, exactly a century after the establishment of the first dissection room within the Hospital of Santa Creu. Barcelona's anatomical theatre is described in detail in a contract between its builder Sebastià Català and the city council, but the grand structure did not survive the abolition of the *Estudi General* in 1717. The theatre was built during

---

anathomia com de dit Studi General.' AHCB, Reg. de deliberacions. II-82, f. 146.

[35]   'Y les despeses a dita causa fahedores sien pagados dels mil ducats reservats per a coses extraordinàries en lo nou redrec.' Ibid., f. 155.

[36]   A. Cardoner, 'La construcción de un anfiteatro anatómico en Barcelona en el siglo XVII', *Medicina clínica*, XXXVIII(5) (1962), p. 389.

the final phase of the late seventeenth century, coinciding with long periods of war, which nonetheless coincided with a golden age of anatomy studies in Barcelona. Significant contributions from some of Barcelona's professors of anatomy preceded the *Novator* movement of the late 1680s. The most prominent among these anatomists was Joan D'Alos, whose *De corde hominis disquisitio physiologica-anatomica* (Barcelona, 1694) was based on the anatomy he had studied during the three preceding decades and presented a systematic defence of Willam Harvey's theory of the circulation of the blood. This challenge to traditional Galenism remained under constant attack in the neighbouring and formerly far more progressive University of Valencia. Francesc Feu, another anatomist from Barcelona, began his career within the renovated facilities of the Hospital de Santa Creu before moving to Madrid, where he was appointed as the first anatomist in the capital and became a professor at the *Hospital General* in 1689. There are no similar examples of significant precursors or contributors to anatomical knowledge among the anatomists of the previous century, who instead followed local career paths in the city administration or in the medical faculty of the *Estudi General*. The professorship of anatomy nonetheless began to receive increased attention during the mid-1570s. An employment letter from 1575 specified that Esteve Guardiet was to receive a raise of five pounds – creating a total annual salary of 40 pounds – and underlined his obligations to train students of surgery in anatomy.[37] In 1575 the integration of anatomy and surgery was further emphasised, and a familiarity with anatomy and dissection was confirmed as the very foundation of the education of future surgeons.[38] The increased prestige and expectations of the chair of anatomy were matched by the rise in the number of annual dissections from four to 12 during the period between 1575 and 1588, outnumbering the eight annual dissections carried out at that time at the University of Valencia.

---

[37] 'La de anathomia annual a mestre Steve Guardiet, ab que hage de legir mès avant algun libre de chirurgia adaptant-se ab la lectura y llengua a la capacitat dels studiants, ab salari ordinary de 40 lliures.' AHCB, Registre de deliberacions, II-91, f. 124.

[38] 'Lo qui legirá anathomia lija chirurgia. E que lo cursant lija a son temps los simples. E que se continue la lico de Ypóchretes. Item, per quant la anathomia és fonament de la chirurgia y sie cosa molt convenient per aquella part de medicina, statuhïren, per có, y ordenaren los dits magnifichs consellers y prohómens que lo qui legirà la cathedra de anathomia en lo yvern y al temps cómmodo per a fer anatomia, que lo mateix lector en lo situ hage de legir algun libre de chirurgia adaptant-se ab la lectura a la capacitat des oïnts.' AHCB, Registre de crides I ordinacions, IV-20, 1575–83.

An Insider's Account: Jeroni Magarola's *Republica original sacada del cuerpo humano*

On 14 September 1581 Esteve Guardiet left the chair of anatomy, apparently with some reluctance and ambivalence, since he was succeeded by Jeroni Magarola on the condition that he could return to his former chair at any time.[39] Two and a half months later, on 29 November, Guardiet *did* return to the professorship of anatomy, but died the following year. He was replaced by Arcángel Queralt, who held the chair for a year before the employment of Jaume Ortenada, who retained the professorship with only a few intervals during the late sixteenth and early seventeenth centuries. As a consequence of Esteve Guardiet's swift return to his former chair in 1581, Jeroni Magarola stands out as the Iberian anatomist with the shortest period of employment of the late sixteenth century. Ironically, he was also the only one from Barcelona to leave behind a published work vaguely relating to anatomy: *Republica original sacada del cuerpo humano* (Barcelona, 1587). By the time of its publication, Magarola had held a number of prominent positions within the *Estudi General*, as professor of *práctica* and later as rector during the early 1580s. He had begun his studies at the *Estudi General* more than two decades earlier – allegedly after some years of study under Rondelet, 'Guillermo Rhondoleto, medico excellentissimo y maestro mío', at the University of Montpellier – and he was among the first medical students to graduate from the young University of Barcelona in 1565.[40] His *Republica original sacada del cuerpo humano* emphasised the close ties between anatomy, surgery, and botany, which were formalised by the unification of the three medical chairs under his predecessors Micó and Guardiet.[41]

If Magarola's *Republica* is in any way indicative of the medical approaches and conceptions at the *Estudi General*, it is difficult to ascribe a general willingness to accept contemporary anatomical reforms to late-sixteenth-century Barcelona. Magarola's treatise included hardly any references to empirical studies of human anatomy based on dissection and direct observation, and the author even maintained the Galenic notion of the *rete mirabile* under the brain as a bridge between the body and the soul.[42] He ignored or was perhaps unaware of the fact

---

[39] 'Per quant al temps que's donà a mestre Magarola, metge, la cadira de Anathomia, se oferi y prometé que sempre y quant mestre Guardiet, doctor en medicina, volgués llegir dita cadira la qual ja abans llegia, la tornaria dit mestre Magarola e renuntiara dita cadira al dit mestre Guardiet.' AHCB, Registre de deliberaciones, II-90, f. 192.

[40] Fernández Luzón, *La Universidad de Barcelona en el siglo XVI*, p. 460.

[41] 'El medico tiene por familia al boticario, curujano, adroguero, herbolario, lapidario, al agricultor, edificador, y al merceder: y finalmente todos aquello que hazen y procuran cosas, por los quales la salud del hombre se conserua y se procura.' Jeroni Magarola, *Republica original sacada del cuerpo humano* (Barcelona: Pedro Malo, 1587), p. 45.

[42] 'Embia parte del at celebro por la arterias llamadas carotides: y deste mas y mas perfecionado se haze el espiritu animal (en una como red del dicho celebro dicha por los

that this 'wonderful network' had been doubted as a feature of human anatomy in Berengario de Carpi's *Isagogae breves* (1522), and was later disregarded by Vesalius and other renowned anatomists throughout the mid-sixteenth century. Magarola's book focused on a series of Platonic microcosmos/macrocosmos analogies and correspondences between the healthy body and the well-run state, which were repeated only a few years later by Marco Antonio de Camos in his *Microcosmia y Govierno universal del hombre christiano para todos los estados* (Barcelona, 1592). Magarola made a rare reference to contemporary anatomical developments by crediting Bernardino Montaña de Montserrate's dialogue *El sueño del Marques de Mondejar*, which comprised the second part of Montaña's anatomical textbook *Libro de la anothomia del hombre*.[43] Despite his, albeit brief, employment as a professor of anatomy, Magarola's anatomical descriptions were not rooted in direct observation, but were instead based on idealised interrelations in nature, between the heavenly bodies and the bodily organs, as exemplified in his depiction of the solar influence on the nervous system.[44]

In his descriptions of the imitation of art in nature and nature in art, Magarola referred to the auditory ossicles, *martillo* (the malleus) and *ayunque* (the incus), described by Vesalius in 1543, but was seemingly unaware of the third auricular *estribo* (the stapes), which had been described and depicted by Pedro Jimeno and Juan Valverde in 1549 and 1556.[45] This recently discovered 'stirrup bone' would actually have supported Magarola's complicated theory of imitation between *naturalia* and *artificalia*, which he underlined with his reading of Bernardino Montaña's interpretation of Marqués de Mondéjar's dream. Yet the author of *Republica original sacada del cuerpo humano* was plainly unaware of this and other current developments within contemporary studies of anatomy. His appointment to the professorship of anatomy does not indicate that the university was particularly receptive to the renovations of Galenic medicine carried out elsewhere in Spain and Europe. Resistance to these reforms was noticeable in Magarola's publication and also in the statutes from the *Estudi General*. The reaction from Barcelona was less vigorous than in the 1594 and 1665 statutes of the Universities of Salamanca and Alcalá de Henares, which explicitly forbade the consultation of anatomical textbooks other than Galen's

---

anatomicos Rete Mirabile). El qual segun Galeno es el inmediato y principal instrumento (del espiritu hablo) del anima racional: mediante el qual govierna todo el cuerpo, y la facultad animal embia estos espiritus, como rayos de claridad y fuerca por todo el.' Ibid., p. 102.

[43] Ibid., p. 160 This dialogue was included in Montaña's work and consisted of an imagined discourse between the anatomist and his benefactor and patron, Luis Hurtado de Mendoza, Marqués de Mondéjar, and like Magarola's later work was entirely embedded in traditional Galenic medicine.

[44] 'El sol que esta en este cielo, es la facultad animal, porque influye por irradiacion en los nieruos, almenos enlos que no son huecos.' Ibid., p. 118.

[45] Ibid., p. 184.

De usu partium, but repeated dogmas of Counter-Reformation medicine were nonetheless found in Magarola's work and were consolidated in the subsequent statutes of 1596, 1629 and 1638. The reinforcement of Galenism would not be questioned openly within the *Estudi General* until Joan D'Alos's defence of William Harvey at the turn of the seventeenth century.

The Chair of Anatomy During the Late Sixteenth and Seventeenth Centuries

The *Ordinations* of 1596 continued the reward scheme, first introduced in 1572, of one ducat for every anatomical dissection.[46] While rewards for anatomists continued at the same rate, the 1596 statutes reduced the number of dissections from 12 each year, as stipulated in the 1588 statutes, to two monthly dissections to be performed between October and February, thus making an annual total of 10. The remaining seven months of the year were reserved for botanical studies and excursions outside the city: 'The lecturer of anatomy is obliged to carry out two anatomies each month, from the month of October until the end of February, and in the other months of the year he should bring his students outside the city in order to gather and lecture on medical plants.'[47]

The richly detailed statutes of 1596 provide an insight into the teaching schedule for students of medicine, mentioned only briefly in the prior regulations from 1587. In these earlier statutes anatomy lessons took place between 2 pm and 3 pm, and were taught by Felip Pinyol, who held the chair of anatomy for a brief period between 1587 and 1588.[48] The teaching of *Práctica* between 4 pm and 5 pm adhered to a lecture programme 'based on the practice of Jacobus Sylvius.'[49] The appearance of this fiercely anti-Vesalian Parisian professor in the teaching programme offers yet another indication of the Galenic orthodoxy prevalent in the *Estudi General*.[50] Jacobus Sylvius was the only contemporary

---

[46]   'Que al doctor qui fara les anatomies, sia donat a raho de un ducat per anathomia (...) Cum conste, y sia notorilo profit y utilitat gran ques reb del ferse anatomies, y de la cognitio de tot lo cos, y deles parts dell, que ab ditas anatomies se fa ys reb : per tant statuyren y ordenaren (com tambe estaua ordenat en lany MDLXXXI) que al catedratich qui fara les anatomies, per quiscuna vegada que fara anathomia, li sia donat, y pagat un ducat, a que sia dita anatomia sia complida, y de tot lo cos, y cap.' *Ordinations e nov redrec fet per instauratio, reformatio, e reparatio, de la universitat del Studi general de la Ciutat de Barcelona, en lo any Mil sinc cents novanta y sis* (Barcelona, 1596), Cap. 14, p. 71.

[47]   'E que lo lector que llegira la anatomia, sia obligat de fer cada mes dos voltes anatomies, desde lo mes de octobre fins per tot febrer, y en los altros mesos del dit any ser exides, y anar fora ciutat pera les herbes als studiants practicants sobre ellas, y que dites tres cathedratichs menors, y la de chyrurgia, tinga de salari quiscuna cada any sexanta lliures.' Ibid., p. 72.

[48]   Fernández Luzón, *La Universidad de Barcelona en el siglo XVI*, p. 410.

[49]   'De la práctica de Jaume Silvi.' Ibid., p. 412.

[50]   Ibid., p. 458.

medical scholar to be mentioned in the myriad statutes and regulations from the late sixteenth century. In its guidelines for the chair of *Práctica*, the University of Barcelona once again set itself apart from the contemporary *Estudi General* of Valencia, where Sylvius' Galenic osteology was fiercely rejected and ridiculed in Luis Collado's 1555 *Liber de ossibus*. This university chair was held by Joan Francesc Rossell, whose examining board was filled by Francesc Micó and Esteve Guardiet during his graduation in medicine in 1579. Rossell later became a renowned physician and a leading diplomat at the courts of Philip III and Philip IV.[51] The 1596 *Ordinacions* did not differ significantly from the 1587 statutes, but changed the time of the anatomical lectures from the afternoon to a one-hour morning lecture from 8 am to 9 am.[52]

Once again, the 1596 statutes were quite explicit in their specifications of the titles and number of Galenic textbooks to be used for general medical teaching, and the regulations of surgical teaching included Guy de Chauliac's *Chirurgia magna* in the curriculum. Yet no similar requirements appeared for anatomy, which was merely subject to the vague request for 'alguna obra de anatomia'[53] to be used in teaching. A further revision of these statutes appeared only two years later, paying increasing attention to the anatomical instruction of future surgeons within the *Estudi General*. These 1598 *Ordinacions* emphasised how future applicants for the professorship would henceforth prove their anatomical skill in trial dissections of three human corpses. This public examination was judged by a committee of three professors employed within the medical faculty: 'No professor can hold the chair of anatomy without having carried out three human dissections in the presence of three doctors from the medical faculty.'[54]

---

[51] Francesc Amorós i Gonell, *Correspondencia diplomatica de Joan Francesc Rossell, 1616–1617* (Barcelona: Institut d'Estudis Catalans, 1992), p. 22 and Nazario González, *Una historia abierta* (Barcelona: Edicions Universitat Barcelona, 1998), p. 147. Rossell's treatise on plague, *El verdadero conocimiento de la peste, sus causas s enales, preservación y curación* (Barcelona: Sebastian i Iaime Mathevad, 1632), was one of the most widely read works on the plague during the seventeenth century. Rossell's medical activity, however, was overshadowed by his position as *cap conseller* and ambassador of the *Consell de Cent* at the Madrid courts of Philip III and Philip IV.

[52] 'Haya de llegir lo de natura humana y los tres libros de temperamentis y laltra y ambe sia de Galeno, I haja de llegir los libres de differentiis causis, morborum & symptomatum, y la tercera sia de anatomia y simples, y ultra destes cathedres la de chyrurgia : y que los que llegiran dites cáthedras, les llijan les hores seguents, com és de set hores a les vuyt demati la cathedra de principis de medicina, y la cathedra de anatomia y simples demati de vuyt a nov hores, y de nov a dou la cathedra major de Gal. y de dou a onze la cathedra de chyrurgía y après dinar de dues a tres la cathedra seg. De gal y de 3 a 4 la cathedra de Hippocrates y de 4 a 5 la cathedra de pratica.' AHCB, *Ordinations e nov redrec...*, pp. 71–72.

[53] AHCB, *Ordinations e nov redrec...*, Cap. 14, p. 107.

[54] 'Y que ningun doctor se puga opposar a la cathedra de anatomia, que primer no sie exercitat en fer al manco tres anatomias de cossos humans, en presentia de tres doctors collegiats.' AHCB, *Registre de deliberacions*, II-197, f. 193.

These procedures were probably tied to the training of future surgeons in anatomy. Similar practices were seen at the Universities of Salamanca, Alcalá and Valladolid, where chairs of surgery were established in 1566, 1594 and 1599, respectively. In Barcelona the integration of anatomy and surgery was well established, dating back to the chair of *chirurgia y anatomya*, which had been established in the wake of the 1562 corrections of the founding statutes from 1559. Apart from two short intervals in 1573–74 and 1586–88, the chair of surgery remained united with the professorship of *anatomia y simples*, and only in 1596 was the first autonomous chair of anatomy established at the *Estudi General*. The *Ordinations* of 1629 required future surgeons to study anatomy for two years, even though the statutes themselves admitted that this stipulation could be difficult to fulfil.[55] Later statutes from 1638 consolidated this decree, but reduced the anatomical practice period for the surgeon to one year. A basic one-year training in anatomy was necessary for anyone who wished to enrol as a student of surgery at the *Estudi General*, and the statutes also encouraged active involvement in the autopsies taking place within the precincts of the Hospital General de Santa Creu.[56] According to the statutes, the anatomical training of surgeons took place within the old facilities of the hospital, continuing a practice begun by Esteve Guardiet in the early 1570s. The dissection room of the hospital nonetheless remained inadequate for this purpose and did not undergo any serious restoration until the 1638 initiative to refurbish it for the education of future anatomists and surgeons.[57] A complete renovation was not accomplished until 1673, and the original interior of the dissection room, dating from 1573, probably remained largely unaltered by the time of the 1629 and 1638 statutes. According to José Pardo-Tomás' and Àlvar Martinez-Vidal's research and article on 'El primitivo teatro anatómico de Barcelona', two attempts to renovate this dissection room in the late 1650s and early 1670s was the direct result of an increasing number of students of surgery attending the lectures on anatomy in the wake of the 1629 and 1638 statutes.[58]

---

[55]   *Ordinations 1629*, pp. 63–64. Cited in Fernandez Luzon, *La Universidad de Barcelona en el siglo XVI*, pp. 433–34.

[56]   'Que haya de hauer practicat por lo menos hu de complit en lo hospital general.' AHCB, Estudi General, Serie XVIII, Vol. 8, *Ordinacions e nou redress de la Universitat del Studi General de la ciutat de Barcelona fetes an lo any 1638*, f. 103.

[57]   AHCB, Estudi General, Serie XVIII, Vol. 8, *Ordinacions e nou redress de la Universitat del Studi General de la ciutat de Barcelona fetes an lo any 1638*, f. 103.

[58]   'Así pues, las necesidades inherentes a la formación de cirujanos se convierten en un nuevo estímulo al mantenimiento de una docencia práctica de la anatomía a lo largo del siglo XVII. Vale la pena subrayar este hecho, al menos en el caso barcelonés, ya que no suele hacerse valer este componente hasta épocas posteriores y se la hace coincidir con los presupuestos ilustrados típicos del siglo XVIII. Por el contrario, la coincidencia de que la formación de los cirujanos era

The turbulent years during the Catalan Revolt (1640–59) saw both French and Spanish invasions of Barcelona and put the renovation programme of the anatomy theatre on hold until 1657–59, when an overall rebuilding of the dissection hall was begun. The renovation took place almost 30 years after the 1629 statutes, which specified the anatomical training for students of surgery within the *Hospital General de Santa Creu*, and 20 years after the call for new dissection facilities in the statutes of 1638. According to A. Cardoner's article, 'La construccion de un anfiteatro anatómico en Barcelona en el siglo XVII', repeated attempts were made to move forward with the refurbishment of the *aula de les anatomies* during the long interval and in 1644 the Consell de Cent budgeted 80 pounds for the renovations.[59] Construction work did not begin until after the 'Peace of the Pyrenees' between France and Spain in 1659, but dissections continued to take place during the turbulent period, as proven in the research carried out by José Pardo-Tomás and Àlvar Martinez-Vidal.[60]

The 1638 statutes included not only schedules for the medical lectures, but also detailed instructions for the organisation of anatomical teaching, which had been absent in earlier statutes. These teaching requirements indicate that the didactic programme and methods at Barcelona differed markedly from the previous divisions of lectures on anatomy into *anatomia particular* and *anatomia universal* at the Universities of Valencia, Salamanca and Alcalá de Henares. For the first time, a book title appeared in the *anatomia y simples* curriculum, namely Galen's *De simplicium medicamentorum facultatibus*. Yet the statutes still did not include specific references to anatomical textbooks, such as *De usu partium*, which comprised the core curriculum at the Universities of Salamanca and Alcalá de

---

imprescindible a la 'utilidad pública y buena policía del principado era general a mediados del siglo XVII.' Pardo-Tomás and Martinez-Vidal, 'Anatomical Theatres', pp. 16–18.

[59] 'Los snrs Consellers tenian facultat de far una stancia en lo hospital gen. de san Creu de pont ciutat para far los anatomies. Y con los administrados de dit hospital per la falta tenian de aquella ne hajan feta una stancia para fer ditas anatomies y para acabar aquellas ad co perfectio se den hajan offert als senors Consellers que … la costa de anatomia se ha de fer en lo hospital an offert los administradors de aquella casa para los rations en lo proposito.' AHCB, *Deliberacions, Consell de Cent*, II-153, f. 374.

[60] Pardo-Tomás and Martinez-Vidal, 'El primitivo teatro anatómico de Barcelona', p. 18. 'Esta continuidad de la prácitica disectiva en el Corralet puede aformarse a la luz de tres pruesbas documentales. La primera, la constituyen las autopsias llevadas a cabo con motivo de la mencionada epidemia de 1651 para determinar en el cuerpo de un hombre y una mujer jóvenes y de cemplexión saludable, si la muerte había sido debida por la peste como de hecho se confermó. La segunda prueba a favor de la continuidad es que, en 1654, el entonces catedrático de anatomía, Joan Maresch, solicitó el pago de la disecciones effectuadas, aunque al parecer no constaran las doce reglamentarias. La tercera prueba documental data de mayo de 1656 y alude a la celebración de los exámenes para cualificar a los doctores de medicina en la art de Anatomia, necesarios desde 1598 para poder optar a la cátedra universitaria; en este, como en los anteriores documentos, se vuelve a mencionar el patio anejo al cementerio como lugar de las disecciones.'

Henares. Unlike the Universities of Valencia and Zaragoza, which had important research programmes in anatomy and pathology, the somewhat provincial *Estudi General* tended to ignore anatomical discoveries made elsewhere. This isolation from contemporary intellectual currents was emphasised by the vague curriculum, which consisted only of Galen's *De simplicium*. The guidelines for anatomical teaching were similarly elusive and shifted freely between certain anatomical regions and categories, concluding with vague instructions for a final study of the bones of the human skeleton. It is unclear how much time was set aside for anatomy students to practise dissection. The reference in the 1638 statutes to 'One day of anatomy and medical plants' indicates a markedly shorter allocation than for the public dissections carried out in Salamanca, which would last for up to three days.[61] The statutes once again emphasised that the teaching of *simples* was to form an additional part of the teaching of anatomy, hence the reference to Galen's *De simplicium medicamentorum facultatibus*. This instruction once again separates Barcelona's *Estudi General* from the contemporary universities of Castile, where the professorship of anatomy worked either autonomously or jointly with surgery. The prescribed order of the dissections of 11 different anatomical categories, beginning with the surface musculature and finishing with the skeleton, shows an emphasis on *anatomía universal* and indicated that only one body at a time was systematically anatomised during a relatively short period. Yet this sequence conflicts with information presented elsewhere in the statutes, which emphasised that 11 different dissections of particular anatomical traits should take place within a long period from September to February, possibly with one daily lecture on each anatomical category. The similarly vague reference to studies of diseases traced through this practice indicates some empirical knowledge of anatomical pathology, and morphological changes in the human body caused by illness, reminiscent of Juan Tomas Porcell's autopsies of plague victims in Zaragoza during the previous century.[62]

---

[61]   Enrique Esparabé Arteaga, *Historia pragmática é interna de la Universidad de Salamanca* (Salamanca: Francisco Nuñez, 1917), p. 261.

[62]   'De la cathedras menors una dia de anathomia y de medicamentos simples en lo qual se enseye de santa Creu de settembre fins al principi dela quaresma La differentia de instruments y parts del cos huma lo sito, lloch y un de aquellas y los affections o malaties ales quals almanco mes ordinariament estan subjectas. Y del principi de la quaresema fins a st. Juan de Juny llega des simples medicaments interpretant la libre quart y quint De simplicium medicamentor facultatibus de Gal. o altres tractats dedita material ben vistos obligant al doctor que llegira dita cathedra fassa des del octobre al febrer ootse dissections o anathomias para que los studiants comprouen a la vista lo que per la audicio y lectura de poundsauran entes la primera sia dela cavitat animal, la segun de la cavitat vital la tercera de la cavitat natural de les parts continents atotas les conventas, la quarta dels museles del gras, la quinta dels museles dela carra, la sexta dels demes museles del cos huma, la septima totas las venas, la octaua de totas las arterias, la nova dels vint y nov parells de nervis dela espinal medusla. La decima del sise parell de nervis del cervell, la onsena de las partes dedicadas

As shown in the 1638 statutes, the medical faculty of Barcelona was still actively engaged in anatomical teaching. Students and teachers were encouraged to attend the public dissections, as was also the case in the 1665 statutes from the University of Alcalá de Henares.[63] The twin professorship of *anatomia y simples* once again manifested itself very literally in the 1638 statutes, as in earlier contexts, through the obligation of the professorship to include an afternoon lecture on medical botany after the completion of the morning dissection.[64] The concluding remarks suggested the establishment of a small botanical garden of medical plants and thus called for further consolidation of the twin professorship, which had hitherto depended on excursions outside the city for the teaching of botanical medicine. During the refurbishment of the anatomical facilities between 1657 and 1673, such a garden was created, and a reference by the eighteenth-century botanist José Quer still mentioned a botanical garden located within the hospital precinct of Santa Creu.[65] The requirements of tickets at the entrance of the anatomy theatre, and possibly even entrance fees, gives an indication of the interest surrounding these demonstrations, which were overseen by the highest authorities within the university administration, either in person or through witnesses, as stipulated in the statutes. Their presence was furthermore an attempt to secure that the controversial practice of anatomical dissection was carried out correctly and properly.

---

a la generacio en la donna o en lo home, la dutsena del instrument dela vista y demes instruments dels sentiments y que ditas anathomias si en integras. Ultra deles quals cosas de ensenyar atots los estudiants la historia dels ossos.' AHCB, *Odinacions, 1638*, f. 55.

[63] 'Ytem ordenamos que el Cathedrático de Anotomía haga las dies anotomías universales y particulares que está ordenado haciéndolo saver el día antes que las a de hacer a sus discipulos y a los Vedeles para que no lo digan a otros Cathedráticos de Medicina para que asistan a berlas hacer y los beadles para que sino las hiciere puedan multar lo que le corresponde.' Cited in Alonso Muñoyerro, *La facultad de medicina de Alcalá de Henares* (Madrid: Consejo Superior de Investigaciones Científicas, 1944), pp. 95–96.

[64] 'Y la dia que se haura de fer anathomia haja de posar un billet ala porta de la universitat diene la hora, lo lloch y que sema de fer la anathomia y que haya de ferse de ditas anathomias lo catedratich al vicerector ab dos testimonies que la han vista y que dit vicerector les haja de absentar paraque lo refferesca al rational dela universitat a effecte ques sabia si cumple absa obligatio y les hipose en son credit y quelo vedell notifique al catedratich de chirurgía que ha la anathomia para que dit catedratico fassa avisar per dos estudiants sens totes les botigas de metier chirurgians preque los jovens si vaian y se aprofixen, y que para fer ditas anathomias se fassa una instancia en lo hospital acomodada y para que nos perden les llissons dela matinada si fassan las ditas anathomia a la tarde desde missa quaresma a fins a st. Juan de Juny deu herbolisations elegant los dies de festaper no faltar a la lectura y que para facilitar ditas herbolisations se precure que en lo hort dels caputxins que es molt capas se plantan moltas exquisitas herbas y es cert ques fara y conservara ab molt pocha costa de la ciutat y sera cosa de molt gran profit y utilitat.' AHCB, *Ordinacions e nov... MDCXXXVIII*, Cap. 55, f. 55.

[65] Fernández Luzón, *La Universidad de Barcelona en el siglo XVI*, p. 456.

The increased support of the chair of anatomy, which we can trace in the mid- to late seventeenth century, led to the renovation of the crumbling teaching facilities and circulation of both scholars and ideas emerging from the anatomical school of Barcelona. Francesc Feu, who left the *Estudi General* in 1677 in order to found the first professorship of anatomy in Madrid, was a luminous example of this dispersal of anatomists who had been educated in Barcelona to other Iberian universities. This circulation had no precedent in the previous century, in which only the Universities of Valencia and Salamanca produced similar 'founders' of chairs of anatomy throughout Aragon and Castile. The *Estudi General* of Barcelona was also among the first in Europe to produce contemporary publications defending Harvey's theory of the circulation of the blood (in an Iberian context equalled only by the University of Zaragoza).[66] Traces of decaying working conditions still appeared in some of the statutes and regulations from the late sixteenth century, recording reductions of annual dissections and overdue payments. However, Barcelona had evidently abandoned its policy of intellectual isolation from other Iberian and European medical schools in the late seventeenth century – a period which elsewhere in Spain represented the gradual decline and abolition of anatomy studies. While it is difficult to find support for the idea that the sixteenth-century *Estudi General* of Barcelona was an early practitioner of Vesalian theories and practices, this institution was evidently among the European strongholds of Harveyan anatomy and physiology during the mid- to late seventeenth century. This defence asserted by the Catalan physicians Frances Morelló and Joan d'Alos was

---

[66]   The early susceptibility of William Harvey's theory of the larger circulation is dealt with in a brief article by Jaume Pi-Sumyer entitled 'Joan d'Alos and the Doctrine of the Circulation of the Blood', *Yale Journal of Biology and Medicine*, 28 (1955–56), which describes the unique status of the *Estudi General* in openly defending the new doctrine during most of the seventeenth and eighteenth centuries: 'the Spanish medical schools shared in the attack and their most prominent leaders, headed by Matías García, author of Dispositiones anatomicae ab Harveyo suscitatoe motu cordis arteriarum et sanguinis (in folio 1648), and Andrés Piquer, the Spanish Hypocrates (1711–1772) turned Valencia, and later Madrid into very active centers opposing the circulation thesis; Salamanca, Alcalá and other universities took the same stand. The medical school of Barcelona may be considered an exception because of two remarkable men, Frances Morelló and Joan d'Alos'. A recent article by Álvar Martínez Vidal, entitled 'Harvey. Dal vecchio al nuovo mondo' in *Harvey e Padova. Atti del convegno celebrativo del 4o centenario della laurea di William Harvey* (Padua: Antilia, 2002), moderates the former view of the early reception of Harvey's theory in Barcelona as unique. In fact, the royal physician of Philip IV and Charles II, Gaspar Bravo de Sobremonte, published a similar defence, *De sanguinis circulatione et de arte sphygmica* (Madrid, 1662), which was followed by Juan de Cabriada's *Carta filosófica medico-chymica*, published in Madrid in 1687. Harvey's theory of the circulation of the blood received its first published defence in Spanish America in 1727 in *Evidencia de la circulación de la sangre*, printed in Lima and written by the Sicilian physician Federico Bottoni. Álvar Martínez Vidal, *Harvey. Dal vecchio al nuovo mondo* (Padua, 2002), pp. 363–82.

in sharp contrast to attitudes at the contemporary *Estudi General* of Valencia; the medical historian Pedro Lain Entralgo later described how, during the same period, this former Iberian centre of Vesalian anatomy deteriorated into 'a centre for the most uncompromising support of the theories of Galen' and 'the avowed enemy of such important novelties as the discovery of the circulation of the blood'.[67]

A different opinion was set forth in a recent history of Barcelona by Professor Jaume Sobrequés i Gallicó from the *Universitat Autònoma de Barcelona*. In line with previous Catalan views of the negative historical development of sixteenth- and seventeenth-century Barcelona, the author concluded that the Catalan capital was largely secluded from the revolutionary scientific insights of the early modern period, as opposed to neighbouring Valencia: 'The intellectual activity was diminished in the field of letters, politics, history, and scientific investigation, a fact that disconnected Barcelona from the great scientific revolutions of the modern era. The same was not the case with nearby cities, such as Valencia.'[68] This understanding of a general exclusion from contemporary intellectual currents is only partially applicable to the field of anatomy, as shown in the differing receptions of Vesalius and Harvey at the *Estudi General* of Barcelona. The opposing reaction to these two 'fathers of modern anatomy' in Barcelona and Valencia show the shifting intellectual currents within the Crown of Aragon during the sixteenth and seventeenth centuries. While Vesalius was regarded as a hero by his Valencian followers many years before he appeared in Spain in 1559, he is not to be found in any contemporary medical references from the *Estudi General* founded in Barcelona that same year. Furthermore, while other Iberian university documents offered explicit references to Vesalius and contemporary anatomical reformers, this influence on the medical practitioners in Barcelona can only be traced indirectly in the book inventory of Esteve Guardiet. The influence of contemporary anatomical sceptics and reformers did not strike a chord within the medical school of Barcelona, which left almost no traces of contemporary anatomists in its statutes, regulations and publications from the late sixteenth and early seventeenth centuries.

Opposing positions were also taken by Barcelona and Valencia towards Jacobus Sylvius, who was embraced by the former institution, but discarded by the latter. The universities also responded differently to the anatomical renovation programmes of the mid- to late sixteenth century, proposals which

---

[67]   Pedro Lain Entralgo, 'The Spanish Contributions to World Science', *Cahiers d'historie mondiale*, 6 (1961), p. 962.

[68]   'La actividad intellectual se decautó mas por las letras, la politica, la historia y la erudición que por la investigación en el terreno de las ciencias, hecho que desmarcó a Barcelona de las grandes revoluciones científicas modernas de la edad moderna. No pasó lo mismo en otras ciudades cercanas, como Valencia.' Jaume Sobrequés i Gallicó, *Historia de Barcelona* (Barcelona: Plaza & Janes Editores, 2008), p. 137.

were seemingly ignored in Barcelona, but embraced in Valencia, from where teaching in anatomy was impressed upon the most prominent universities of Castile. A century later, another divergence may be noted in the conflicting receptions of the Harveyan theory on the larger circulation, which was actively supported at the University of Barcelona, but fiercely rejected by the contemporary *Estudi General* in Valencia. In conclusion, the anatomists of the *Estudi General* in Barcelona contributed actively to the accumulation and circulation of new anatomical knowledge during the late 1600s, but remained effectively cut off from the anatomical reforms of the prior century.

Chapter 7
Zaragoza

The foundation of the University of Zaragoza in 1583 followed the pattern of Valencia and Barcelona, where minor colleges of arts and medicine eventually obtained university status in the sixteenth century. Fragmentary evidence indicates the existence of a small *Estudi* in Zaragoza as early as the thirteenth or fourteenth century. In 1476 the college was referred to as *Estudio General de artes* and *universitas magistrorum* in contemporary royal and papal sanctions. In 1542 Charles V granted this minor university the privilege to extend beyond its existing faculty of arts and called for the establishment of faculties of theology, civil and canonical law, medicine and philosophy. The royal initiative was approved by two papal bulls in the following decade. However, problems with funding and a long-held dispute with the neighbouring University of Huesca – and later Philip II's wavering support – delayed the realisation of the proposed measures by several decades. The university was not inaugurated until 1583, when its official establishment was forced through by an alliance of local oligarchs and clerics led by Archbishop Pedro Cerbuna.

Anatomy Studies in Zaragoza Before and After the Creation of the University

Like the university itself, the requirements made in the 1583 statutes of anatomy were built on a tradition dating back to the late fifteenth century. A royal privilege signed by King Ferdinand in 1488 encouraged a seemingly pre-existing practice of dissection at the *Hospital de Nuestra Señora de Gracia*. This early document enabled the physicians and surgeons of the *Cofradía de San Cosme y San Damián* to perform as many dissections as they found appropriate on available bodies within the hospital precincts.[1] The provision was unusual not

---

[1] 'Primo que toda vegada por los Meges, y Cirujanos de la dicha Cofraria, o, por los Meges, y Cirujanos que visitaran en el Spital de Sancta Maria de Gracia sera deliberado obrir, o, anatomizar algun cuerpo muerto en el dicho Spital lo pueden obrir, o, anatomizar todo, o, en parte agora sea de hombre, agora de mujer tantas cuantas vezes en cada un anyo a ellos sera visto, sin que incorrer en pena alguna. Empero que en tal obra ayan de ser clamados los Meges, y Cirujanos de la dicha Cofraria para que hi sean, los que hi querran ser, y contribuir si algunos gastos acerca de aquello se avran de facer, y que en tal anatomizacion ninguna persona de qualquire estado, o, condicion sea no presuma ni ose poner empacho alguno sus pena de mil sueldos aplicaderos la una parte...del

only in the permission it gave for an unrestricted number of dissections, but also in its forewarning of heavy fines to be paid by anyone who tried to obstruct the practice. Such vehement royal support indicates that there had been opposition to and legal ambiguity concerning the controversial practice. The intervention must have been an invaluable tool for the subsequent institutionalisation of anatomical studies, yet medical practitioners in the grand city hospital of Zaragoza still lacked a proper medical faculty. Several references indicate that Miguel Servet, the famous discoverer of the pulmonary system, received his earliest training in Zaragoza, but these claims are undocumented and rather speculative.[2]

Until the foundation of the university in 1583, the *Hospital de Nuestra Señora de Gracia* was the only physical and institutional framework of a seemingly long-established tradition of surgery and medicine. In this context, active studies of human anatomy preceded the foundation of the university by almost a century. The first statutory orders of 1583 emphasised that the practice was integral to the future educational programme, even if no reference was made to a formal chair of anatomy until a few years later. The staff of a small pre-existing medical school, the *Colegio de Cosme y San Damian*, were incorporated into the newly established university in June 1584, only a year after its foundation.[3] These doctors included Juan Valero Tabar, who has previously been mentioned as a professor of anatomy at the University of Alcalá de Henares between 1571 and 1572. Tabar was described as an eminent anatomist at the new university, and as court physician to Philip II in Lázaro de Soto's *In librum Hippocratis De dieta commentationes* (Madrid, 1594), in a reference to 'anatome primum, Doctorem Tabar Caesaragustanum primariae medicinae cathedrae moderatorem, & Regis

---

Señor Rey.' Asunción Fernández, *Documentos para la historia de las profesiones sanitarias: El colegio de medicos y cirujanos de Zaragoza (Siglos XV–XVIII)* (Zaragoza: Colegio Oficial de Médicos de Zaragoza, 1997), p. 41.

Thanks to José Pardo-Tomás and Álvar Martinez-Vidal, CSIC, Barcelona for providing the document on the *Construccion de una casa de anathomia en Zaragoza*. AHPZ (Archivo Historico de Protocolos de Zaragoza), Martin Español, año 1586, ff. 143–46. Transcripción (pp. 167–68) y comentarios pp. (7–9) de Angel Sanvicente (1981).

[2]   References to Miguel Servet, the ill-fated discoverer of the pulmonary system, and his early education in Zaragoza appear repeatedly in biographies and in several works on the history of Zaragoza. Yet nothing is known about the early education of Servet in nearby Villanueva de Sigena, as related in Charles D. O'Malley's 1959 biography: 'He may have attended some church school, and possibly the University of Zaragoza some sixty miles distant from his home.' Charles D. O'Malley, *Miguel Servetus* (Philadelphia: American Philosophical Society, 1959), p. 10. Fernando Solano's *Historia de Zaragoza* from 1976 follows long-established speculation: 'Parece ser que la mayor parte de sus biógrafos afirman que a los trece años Servet o Serveto se trasladó a Zaragoza a fin de continuar sus estudios.' Fernando Solano: *Historia de Zaragoza* (Zaragoza: Ayuntamiento de Zaragoza, 1976), p. 187.

[3]   Ibid., p. 214.

nostri Philippi secundi Medicum'. De Soto's text referred to the renown enjoyed by Tabar for his construction of lifelike anatomical models. This claim is not supported elsewhere, but is nonetheless valuable as evidence of an unfortunately lost teaching tool from the history of early Iberian anatomy, of which only Mateo de Vangorla's anatomical mannequin from the University of Salamanca still survives. The figurines were probably made in the contemporary tradition of *ecorché* representations of surface anatomy and may have imitated the movable automata-figures built for clocks and wunderkammers. Tabar's anatomical figures certainly exceeded the surviving mannequin from Salamanca in their detailed representations of every important category of systematic anatomy, such as bones, muscles, nerves and veins, as well as the natural movement of all the joints, which were even applauded by members of the royal family.[4]

A certain Dr Juan Sanz – of whom little is known – was another of the first professors appointed to the chair of anatomy. His name appeared in the 1583 statutes as one of the first three professors of medicine at the newly founded university. Several physicians who held the chair of anatomy in the last decade of the sixteenth century, namely Juan Garcés, Juan Gil Capilla and Domingo Eugenio y Dueñas, are mentioned in contemporary matriculation and employment lists. They also appear in an account of the founding years of the university, which has survived in the recently published manuscript *Lucidario de la Universidad y Estudio General de la Ciudad de Zaragoza* by Dr Diego Fraylla, rector of the university between 1595 and 1596. Fraylla's 1603 account of the first 20 years of the University of Zaragoza documented another case of the vacancy suffered by the recently established chair of anatomy, as seen periodically or permanently at contemporary Castilian universities. The author nonetheless claimed that this unfortunate state of affairs was only a brief and temporary situation at Zaragoza.[5]

---

[4]   'Hic enim maxima cum ratione volens faetorem atque horrorem (qui ex dissectione cadauerum contrahitur, & nostris sensibus sese offert, quem nos naturaliter auersamur & fugimus) vitare, magno studio plurium annorum curriculo comparato statuas efformabat, ex corio membranúlis, ossibus, & aliis rebus compositas, venis, arteriis, neruis, cartilaginibus ex serica materia factis, refertas, uniquique; propium colorem & modum substantiae (quoad fieri potest) tribuens, & demum ossibus, musculis, glandulis, carne, & cute, tam vera, quam non vera fabricatas, ut animantes videri dicas: nam & suos motus singulae partes edunt, musculis suas ipserum partes moventibus. Quod quanta cum admiratione non solum Regis nostri, qui inter suos familiares Medicos illum ipsum annumerare dignatus est) sed doctissimorum virorum curiam regiam frequentantium) quantoque; cum labore & rei familiaris, salutisque; prosperae iactura perfecerit, dictu facile (nisi videas) non erit. Atque utinam nostro hoc aevo iam curiosi huius partis Medicinae promocantur anatomici, qui a praedicto viro instructi fabricam humani corporis sine perturbatione, horrore ac faetore studiosis omnibus ostendant. Sed pro dolor: vita functus est, dum haec scribo.' Lázaro de Soto, *In librum Hippocratis De dieta commentationes* (Madrid: Luis Sanchez, 1594), f. 34.

[5]   'La cátedra de Anatomia tuvo el doctor Eugenio de Dueñas. Ahora no la tiene nadie porque los señores jurados la quitaron. Tiénela ahora, que la volvieron, el doctor Juan de Medrano.

The first generation of formally appointed anatomists at the University of Zaragoza were of local descent and were educated either in nearby Huesca or at the medical faculty of the new university.[6] While these early appointments have left almost no trace beyond their names, the contemporary university statutes have survived. Even though they are not as detailed as the contemporary statutes of Salamanca, Valencia and Alcalá de Henares, these early records from Zaragoza offer a glimpse into the organisation of anatomical teaching into systematic and regional dissection of human bodies.[7] Their explicit references to the use of images and prints in the absence of bodies for dissection were similar to the Salamanca statutes from 1561 and were almost identically phrased.

The Zaragoza statutes contain references to the support needed from the viceroy of Aragon 'to obtain the human bodies of some condemned men, licensed by the viceroy', demonstrating the university's attempts to involve not just local administrative and judicial authorities but also the highest regional authorities in the running of the newly founded institution. The objective had the further advantage of acquiring powerful allies in Zaragoza's ongoing struggle with the nearby University of Huesca, which (obviously well aware of the fading influence of the University of Lleida in the wake of the foundation of the *Estudi General* of Barcelona only a few decades earlier), was clearly opposed to the creation of a new university in its vicinity. The hope of such a university alliance with the viceroy of Aragon may have been overly ambitious, as emphasised by Fernando Solano Costa in his study of the new university and its founder Pedro Cerbuna. Apparently the creation of the university was met with categorical resistance by the same viceroy whose support was sought in the 1583 statutes.[8]

Es el salario de ella 60 libras.' Diego Fraylla, *Lucidario de la Universidad y Estudio General de la Ciudad de Zaragoza* (Zaragoza: Institucion 'Fernando el Catolico', 1983), p. 92.

[6]   M. Jimenez Catalan, *Memorias para la Historia de la Universidad Literaria de Zaragoza. Tomo II* (Zaragoza: Typ. La Académica, 1926), pp. 436–40.

[7]   The study of anatomy was incorporated into the earliest curriculum on medicine, in 1583, as can be seen in the *Fundación y creación de la Universidad y Estudio General de la ciudad de Zaragoza y los Estatutos y Ordinaciones della en el año MDLXXXIII. De los catedráticos de Medicina*: 'Que los catedráticos de Medicina hayan de ler de los libros del Arte de Medicina y Ipocrates y Galeno, repartiendo entre sí los dichos libros y lecturas que dentro de quarto años que es el curso entero de Medicina, lean los libros que tratan de la material y doctrina medicinal más práctica y necessaria y hagan alguna anatomía de algún cuerpo humano o parte del, procurándose por parte de la Universidad de haber cuerpos humanos de algunos ajusticiados con licencia del virrey o del hospital de los regidores y faltándolo muestren en las figuras y estampas que dello hay.' Ibid., p. 429.

[8]   'Entre los adversarios más decididos de la nueva Universidad figuró el conde de Sástago, en aquel entonces, 1580, virrey de Aragón, que no solamente no apoyó economicamente a la nueva Universidad, sino que en sus informes al rey siempre se mostró negativo: "si lo que hace falta a Aragon, escribía a Felipe II, es gente que labre los campos, gente que sirva a los ricos, gente que haga calzas y zapatos. Gente que sepa, para que? No se logrará sino aumentar los vagos, crear

The university's survival in the face of such opposition remained dubious for the remainder of the century, even after two visits to Zaragoza by Philip II in 1585 and 1591. Even though Charles V and Pope Julius III had called for the establishment of a fully developed university as early as 1542 and 1555, respectively, the new institution in Zaragoza was not officially approved until a royal visit in 1599 by the recently crowned Philip III. Two years after the official inauguration of the university, it was still doubtful whether it would survive with its insufficient resources and constant struggles with hostile neighbouring institutions, as recorded by the traveller Enrique Cock in 1585.[9]

Detailed information about the construction of its first anatomical theatre in March 1586 appeared during this unfortunate phase in the early history of the University of Zaragoza, only a month before the privileges granted previously by Charles V and Julius III had been declared invalid by the *tribunal de los catalanes* in a verdict that was supportive of the University of Huesca. Subsequent appeals to the Council of Aragon and the royal tribunal in November the same year achieved the withdrawal of these annulations and enabled the architectural alterations of the new university to proceed. It is uncertain whether the *Casa de anathomia* had actually been established by this time; a contract between its constructor and the city council in March 1586 had required building work to be finished in two months. Like other Iberian universities (with the University of Salamanca as a sole exception), the anatomical theatre of Zaragoza was not placed within the university itself, but instead in nearby and pre-existing hospital precincts. After the 1583 inauguration of the University of Zaragoza, the city hospital continued to be the centre of the anatomical education of university students and would remain so throughout the next three centuries until its destruction during the *Guerra de la Independencia*. A *Capitulación para hazer la casa de Anathomia* from 1586 called for the establishment of permanent facilities for dissection within the hospital precincts only three years after the foundation of the University of Zaragoza. Detailed information regarding the size and appearance of this architectural addition to the hospital complex fortunately survives to this day, unlike the hospital itself and its extensive archive, which were destroyed during the French siege and invasion of Zaragoza in 1808. The almost complete annihilation of the city hospital and most of the university left Zaragoza with devastated archives and only scattered documentary sources for later historians.[10]

---

viciosos, despoblar más los campos y extender la miseria; demasiado saben ya para que se les facilite saber más". Fernando Solano Costa, 'Pedro Cerbuna y el funcionamiento de la nueva Universidad' in *Historia de la Universidad de Zaragoza* (Madrid: Nacional, 1983), p. 101.

9   'El año de 1583 se instituyó una Academia, pero no sé si será duradera, por los pequeños salarios de cada a año paga a los maestros della, mayormente teniendo las Academias de Huesca y Lérida tan cerca.' Ibid., p. 104.

10   This unfortunate situation was emphasised by Asunción Fernández (Director of the *Unidad de Historia de la Medicina* at the medical faculty of the University of Zaragoza) in her *Documentos*

With regard to sixteenth-century anatomy, however, the University and general hospital of Zaragoza offer two sources, which are unique in both an Iberian and broader European context. One of these sources provides one of the most detailed descriptions of any anatomical theatre in the early modern period, left behind in a contract between the city council and the builder of the *casa de anathomia*, Andrés de Capraneda, dated 27 March 1586. The other source – Juan Tomás Porcell's *Información y curación de la peste de Caragoca y praeservación contra la Peste en General* (Zaragoza, 1565) – is an extraordinary and uniquely detailed account of anatomical practice at the general hospital in Zaragoza.

Anatomical Pathology within the Hospital de Nuestra Señora de Gracia: Juan Tomás Porcell's *Información y curación de la peste de Zaragoza y praeservación contra la Peste en General*

Both before and after the establishment of a full university and medical faculty in early modern Zaragoza, its chief medical institution was the *Hospital de Nuestra Señora de Gracia*. Founded in the mid-fifteenth century by King Alfonso V and enlarged during the next century, it would eventually become one of the largest Iberian hospitals of the sixteenth and seventeenth centuries. At its peak the general hospital housed more than 400 permanent beds, an infirmary, an asylum for the insane, a botanical garden and a cemetery, as well as a staff of allegedly multilingual nurses for the caretaking of foreign patients. The establishment of a cemetery and the presence of foreigners facilitated the supply of corpses for dissection, providing poor patients with no means to pay for an official funeral elsewhere or deceased foreigners with no relatives to claim their bodies. The most detailed historical study of the hospital, Aurelio Bacquero's *Bosquejo historico del Hospital Real y General de Nuestra Señora de Gracia de Zaragoza* (1952), placed the dissection facilities next to the morgue and a small chapel of the hospital cemetery.[11]

---

para la historia de las profesiones sanitarias: El colegio de medicos y cirujanos de Zaragoza (Siglos XV–XVIII): 'Ni el más mínimo rastro de esto archivo nos ha quedado. He indigado en todas las fuentes de informacion posibles, pero ha sido inútil, solo una referencia de Chiarlone y Mallaina (1865) a que compraron en Madrid un libro intitulado "Indice del archivo del colegio de medicos de Zaragoza", lo qual, aunque por un lado confirma su existencia, por otro lado es una información bastante desalentadora.' Fernández, *Documentos para la historia de las profesiones sanitarias*, pp. 21–22.

[11]　Unfortunately, the author gave no indication of the original source of his information: 'El cementerio que estaba tapiado y dividido en dos partes: la una destinada a a enterramientos, y la otra, que por tener las estaciones del Via Crucis y la estatua de la Madre de Dios, servía de calvario, encontrándose en su interior el depósito de cadaveres y el aposento donde los medicos y catedráticos de la Universidad hacían las anatomias.' Aurelio Baquero, *Bosquejo historico del Hospital Real y General de Hospital de Nuestra Señora de Gracia* (Zaragoza: Institución Fernando el Católico, 1952), p. 54.

Juan Tomás Porcell's studies of anatomical pathology at Zaragoza provided no similar details of the area and hospital facilities where dissections were performed, and there are no descriptions of a permanent dissection room prior to the contract made between the Zaragoza city council and the builder of the *casa de anathomia* in 1586. Consequently, Porcell was not officially appointed as an anatomist after his arrival in Zaragoza on an unknown date in the late 1550s or early 1560s, since the Aragonese city did not yet house a medical faculty where anatomy could be formally taught and performed. While there was no official professorship of anatomy in Zaragoza at this early stage, the royal provision from the previous century and Porcell's detailed account provide us with evidence of an active tradition in the field. The strong constitution of Porcell, a young Sardinian physician practising in Zaragoza, meant that he was promoted to a position as the leading medical authority within the city hospital during a 1564 plague epidemic, which killed three of his superiors and colleagues, and an estimated third of the 30,000 inhabitants of Zaragoza. According to Porcell, the medical reputation of Zaragoza and its general hospital was already well known at that time throughout the university towns of Iberia, where he had spent the previous stages of his medical career.[12] In his remarkably early work on anatomical pathology, Porcell recorded many intimate details of his own dissections of five victims of the 1564 plague within the *Hospital de Nuestra Señora de Gracia*. These accounts included information regarding the patients' age, sex and pathological symptoms, as well as the varying difficulties and the time invested in each dissection. The author was fully aware of his pioneering status and made no attempt to understate the importance and novelty of his anatomical enterprise, as emphasised repeatedly in the account which was addressed directly to King Philip II: 'None of those who have written about the plague until now have had the courage to open and anatomise the body of a plague victim, in order to study the bad humours, origin, and seat of the disease, and the cause of its calamities, like this loyal vassal of your Majesty, doctor Juan Tomás Porcell has done: During the plague of Zaragoza I opened and carried out anatomical studies of five bodies.'[13]

---

[12] 'Despues de hauer gastado Illustre señor, la mas y major parte de mi vida por escuelas y universidades, estudiando y leyendo en ellas, para obtener y venir en alguna cognició de Philosophia y Medicina, escogi por el major y mas famoso lugar, en donde huuiesse de hazer mi asiento y morada, la famosa y leal ciudad de Zaragoza: en la qual exercitando y continuando mi estudio, profession y lectura y lectura en medicina (aunque he tenido hartos emulos y contrarios).' Juan Tomás Porcell, *Información y curación de la peste de Caragoca y praeservación contra la Peste en General* (Zaragoza: Casa de la viuda de Bartholome de Nagera, 1565), *Al muy illustre senor don Bernardo de Bolea*, n.p.

[13] Ibid., f. 64v: 'Porque nenguno dellos, ni de quantos hasta hoy han escripto de peste, ha tenido tanta caridad que abriesse hombre alguno que de peste se huuiesse muerto, y hiziesse anathomia en el: por ver el humor peccante y malo, su origin y asiento, y la causa de los accidentes

Porcell's *Información y curación de la peste de Caragoca y praeservación contra la Peste en General* included an account of the symptoms, diagnosis and prognosis of the plague and the containment of (and potential cure against) the epidemic. His account also returned repeatedly to the practice of dissection as a vital tool for the accumulation of empirical knowledge and increased understanding of this infectious disease. Chapters 3 and 4 of the treatise's total of 15 chapters were entirely devoted to Porcell's account of the dissection of the five plague victims within the city hospital and were subtitled: 'En que trata de las cinco Anathomias que se hizieron, y delo que en ellas se hallo digno de consideracion, y notar.' As emphasised by Porcell in his address to the Spanish monarch, the practice of tracing the characteristics of plague though systematic autopsies of its victims was a task that had never been undertaken by his predecessors in the field of anatomy. To underline the unique nature of his work, the only illustration in Porcell's treatise was a representation of the author performing an autopsy, thereby emphasising the importance of this precarious practice. Like many previous and later treatises on plague, Porcell's *Información y curación* was produced speedily to satisfy the clamour of the medical book market for fresh information about the recent epidemic. This was also the case with the subsequent 1640 and 1652 plague epidemics in Zaragoza.[14] Porcell's book appeared in March 1565, only a few months after the plague had struck Zaragoza between early May and December 1564. Porcell had been put in charge of the caretaking of plague victims within the *Hospital de Nuestra Señora de Gracia*, where several of his former colleagues and superiors had been struck down by the disease.[15]

---

della como esta leal y fiel vassallo de vuestra Magestad, el doctor Joan Thomas Porcell ha hecho: que en esta peste de Zaragoza ha abierto cinco, y hecho anathomia en ellos.'

[14]   Fernando Zubiri Vidal and Ramon Zubiri de Salinas, *Las epidemias de peste y cólera morbo-asiático en Aragón* (Zaragoza: Diputación Provincial, Institución Fernando el Católico, 1980), p. 65. As mentioned in previous chapters, several larger epidemics of plague produced medical accounts of the events shortly afterwards. Zaragoza was no exception to this general rule, as in the *Tratado de la peste de Zaragoza en el año 1652, compuesto por el. Licenc. Joseph Estiche, Cirujano del insigne Colegio de Medicos y Cirujanos de la Imperial Ciudad de Zaragoza* (Pamplona: Diego Zavala, 1655). Porcell's treatise is therefore only one of several printed accounts of the plague epidemics of Zaragoza in the sixteenth and seventeenth centuries.

[15]   'Por haberse muerto los cirujanos que curaban los pobres heridos de peste en el Hospital General de la dicha ciudad de Zaragoza, y el fisico que los visitaba haberse herido y adolescido de dicho mal, desde los primeros de mayo hasta los ultimos de Julio, y entonces no hallar medico nu cirujano alguno que, o por dinero, o por caridad juntamente con dinero los quisiese visitar ni curar – tanto era el miedo que en ellos reinaba, por la muchedumbre de enfermos que al hospital acudía, y haber estado los pobres enfermos sin ser curados ni visitados tres o cuatro días … los jurados de dicha ciudad me enviaron a llamar y me encargaron y rogaron tuviese en bien de visitar dichos dolientes de peste en dicho hospital.' Porcell, *Información y curación de la peste de Caragoca*, f. 1r.

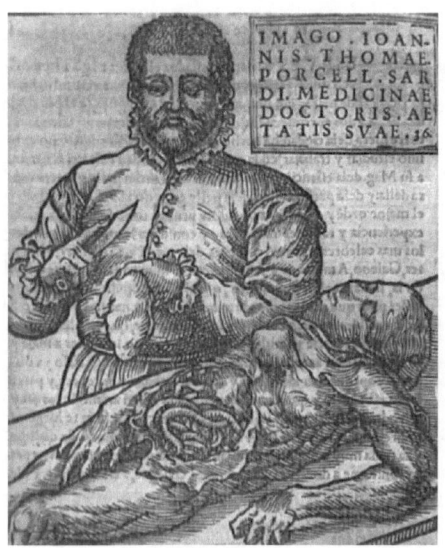

The account provides a vivid insight into Porcell's attempt to look after at least 800 and perhaps up to 2,000 patients, twice or even five times the number catered for by the 400 beds within the hospital, and to oversee each twice a day with only four available assistants.[16] Porcell allegedly decided

Figure 7.1 Juan Tomás Porcell's *Información y curacion de la Peste de Zaragoza y preservacion contra la Peste en General* (Zaragoza, 1565). While Porcell's account of his own autopsies of plague victims in the *Hospital de Nuestra Señora de Gracia* in 1564 did not include any written references to Vesalius, the text's sole illustration nonetheless leaned towards the title page of Vesalius' *Fabrica* in its representation of an anatomist single-handedly engaged in human dissection. In the image Porcell was depicted in the same posture as Vesalius and was similarly involved in the dissection of a female body with its abdomen cut open. This illustration of Porcell was in fact so similar to the original that it appears to be the same image from a different perspective. Only a supplementary text revealed the true identity of the anatomist: Juan Tomás Porcell of Sardinia, age 36. Courtesy of 'Universidad de Salamanca (España). Biblioteca General Histórica'

---

[16] 'Con el favor divino y el orden que con dichos dolientes en dicho hospital se ha tenido, no solo ochocientos, más aún, dos mil, puede visitar y curar un doctor solo con cuatro cirujanos ... los visitaba y curaba a todos dos veces al día, tres y cuatro horas por la mañana, y otras tantas por la tarde...viendo algunas orinas, tocando los pulsos y tumores, siquiera apostemas, y hallándose siempre presente al tiempo de curar y nunca consentir que curen los cirujanos son que el medico esté presente ... Segun la realidad por haber abierto y hecho anatomias en cuerpos diferentes que se han muerto de dicho mal, y haber visto al ojo y claramente conoscido el humor malo y predominante, sus asientos y origin, y a qué parte inclinaba, y las causas de los grandes y bravos accidentes que consigo traía. Según la experiencia por haber visitado desde los primeros de mayo hasta el mes de diciembre de dicho año de 1564, los pobres enfermos de peste en el Hospital General de la dicha ciudad, en donde con el grande número de enfermos que ordinariamente ha habido (porque ha llegado día de 800 con los convalescientes, que estaban fuera de la ciudad, en la torre) medianamente se ha podido practicar y ejercitar lo que los autores en semejante enfermedad mandar hacer y guardar; y de todo ello escoger lo major y más seguro.' Ibid., *A la S.C.R.M del potentissimo, y invectissimo Monarcha don Philippe. Por la diuina clemencia Rey de las Españas,&c. De la India Oriemtal,& Protector y Restaurador de la fe. &c.* n.p.

to base his own studies of the disease and autopsies of five diseased bodies on extensive previous experience with anatomical dissections: 'more than fifty of which I have carried out until now'.[17] Unfortunately, he did not support his claims with any details of the time and place of these alleged prior dissections. His account instead provided reports of pathological changes of both the surface morphology and the inner organs of the infected victims. One of the objectives of these studies was an ambitious attempt to find an effective cure once the disease had been properly described and understood. Porcell may well have exaggerated the high number of dissections he claimed to have performed, but this alleged experience could be explained by the traditional encouragement and support of anatomical practice within *Hospital de Nuestra Señora de Gracia* since the 1488 provision of King Ferdinand of Aragon, or by the statutory requirements for 24 annual dissections at the University of Salamanca, where Porcell received his medical education.

Porcell's treatise described the five anatomies of plague victims carried out on a 28–30-year-old pregnant woman, a 33-year-old man, a 12-year-old girl, a 24-year-old woman and a 25-year-old man. The account of each case showed both exemptions and affirmations of the clinical detachment of the early modern anatomist, which often appeared in contemporary accounts of human dissection. In Porcell's unique treatise, the age, sex and deathbed symptoms of each victim was registered and narrated, often with empathetic attempts to normalise what must have been a horrific ordeal for all parties. Porcell's compassion for the dead and dying was demonstrated in the first autopsy, which was performed on a pregnant woman whose similarly infected six-month-old foetus was removed from the womb of her dying mother and sprinkled with baptismal water before expiring.[18] The subsequent description of the autopsy was a matter-of-fact and purely technical account of the pathology of the inner organs of the recently deceased woman.[19]

---

[17]   'Passan de cincuenta las que hasta hoy he hecho.' Ibid., f. 7r.

[18]   'La primera, pues, anatomía que hice, fue de una mujer preñada de seis meses. La qual, estando visitando y curando los muertos murió, y como la criatura estuviese viva e le saltese dentro de la barriga, para que dicha alma tuviese agua de baptismo y se salvase, la abrí luego y le saqué la criatura que aun boqueaba, y como el vicario de los heridos de peste estuviese presente, tuvo agua de baptismo y luego murió. Esta mujer era de edad de veintiocho hasta treinta años; tenía el tumor o apostema debajo el brazo izquierdo, muy grande, ancho, y llano; tenía grandes ascos y vomitos allente de otros accidentes que padescia; murió al cuarto día de su dolencia.' Ibid., f. 4r.

[19]   'Todos los miembros nutritivos, hígado, bazo, riñones, tripas, estaban tan buenos, en color, sustancia y magnitud, quanto en un hombre sanísimo se puede desear. Porque los abrí y reconocí todos, y primero las tripas quitadas fuera del cuerpo según y como require y lo manda Galeno, y no hallé más de que, en la última y penúltima tripa ... habia unas pocas de hiesces. Segundariamente abrí el hígado, sacándole afuera, y lo deshice todo mirando y considerando en él, ansi la carne, como las venas y la sangre que en él estaba.' Ibid., f. 4v–5r.

Surprisingly, Porcell made no attempt to compare Galen's anatomical descriptions based on dissected animals with the corrections discovered in his own studies, as was encouraged by the contemporary statutes of Salamanca, from where Porcell had graduated some years prior to his employment at the general hospital of Zaragoza. With the exception of a few prominent Spanish court physicians, no references were made to contemporary medical authorities. Porcell seemed instead to have been deeply influenced by another tradition consolidated at the University of Salamanca, namely the dependence on Avicenna in the medical curriculum. This tradition was evident in a discussion on the character and origin of bile, in which the author presented his intention of amalgamating the humoral pathology of Galen and Avicenna.[20] Such reliance on Avicenna was in clear opposition to attitudes at the recently founded University of Barcelona, where his name was absent from the curriculum, and only the physician and botanist Mesue represented any continuity with a previous tradition of Arabic medicine. Porcell was deeply rooted in *galenismo arabizado*, a tradition consolidated within his former university. He nevertheless presented a challenging yet ambiguous observation on the futility of slavish reliance on ancient medical authorities: 'here we do not follow the writings of Hippocrates, because he is Hippocrates, and even less so Aristoteles because he is Aristoteles ... but only those of their claims, which conform to reason and truth'.[21]

The Sardinian physician from Cagliari referred repeatedly to his prior educational background at the University of Salamanca after the transfer to Spain, where he allegedly became a devoted apprentice to Lorenzo Alderete, who held the primary chair of medicine at the University of Salamanca. A result of this background was a constant appraisal of the former, recently deceased master, who was involved in the creation of both the anatomical professorship and anatomical theatre of Salamanca in the early 1550s.[22] Porcell's account did not refer to Cosme de Medina, the first Professor of Anatomy at his former university, who had recently been promoted to the first chair of medicine at the University of Salamanca. This apparent forgetfulness might indicate a rivalry between Alderete's two former apprentices, in contrast to Porcell's appraisals of other Spanish colleagues, such as the three physicians involved in the treatment of Don Carlos' head-wound three years earlier: Cristóbal de Vega, Fernando de

---

[20] 'Y ansi digo que si Auisce. En algo erro, fue enlo que siguio y imito a Galeno, y esto, por no tener verdadera traducion de todas sus obras, como nosotros hoy en dial as tenemos: porque lo quell trato y escriuio de si mesmo, si bien sentiende, lo trato tan doctamente y tambien, que no hay en que poder lo calumniar. Tornando pues a mi proposito ques conciliar a Auis. con Gal.' Ibid., f. 23r.

[21] 'No seguimos la sentecia de hipocrates porque es hipocrates ni menos aristoteles porque es aristoteles ... sino porque lo que dizen u escriuen es conforme a razón y verdad.' Ibid., f. 12r.

[22] 'Lorenzo de Alderete que esta en el cielo doctor y cathedratico de prima de medicina enla famosa y insigne Universidad de Salamanca maestro y preceptor mio meritissimo.' Ibid., f. 54r.

Mena and Juan Gutierrez de Santander.[23] Porcell's treatise made no reference to Andreas Vesalius, who had also been involved in Don Carlos' treatment and who left Spain the same year Porcell was engaged in his autopsies of plague victims inside the *Hospital de Nuestra Señora de Gracia*. Vesalius' absence in the account is somewhat surprising: Porcell depicted himself in the same posture as Vesalius in the only illustration in his entire treatise, which provided a visual representation of the author/anatomist perfoming a dissection on a female body. Porcell's dress, hairstyle and beard were almost indistinguishable from the depiction of Vesalius on the original 1543 title page of *Fabrica* and the author was similarly placed at the left side of a dissection table where a woman was shown with her womb cut open and abdomen visible to the reader. The image of Porcell was in fact so close to the original Vesalian title page that it appears to be the same image simply seen from a different perspective. Only a supplementary text revealed the true identity of the physician involved in the dissection of the young woman on the table: *Imago Ioannis Thomae Porcell Sardi medicinae doctoris aetatis suae 36.*

Porcell's anatomical pathology, with its systematic autopsies of morphological changes caused by illness, represented an epistemological leap forward from a mere mapping of human anatomy and towards an active study of pathological changes. Later statutes from the *Estudi General* in Barcelona indicated the existence of similar studies of diseases traced through autopsies. Yet Porcell's work remained ultimately unique in an Iberian context and even elsewhere in Europe. Treatises on anatomical pathology were few and rare after the Florentine physician Antonio Benivieni produced his pioneering work on the hidden causes of disease, *De abditis nonnullis ac mirandis morborum et sanationum causis* (Florence, 1507), a publication traditionally acknowledged as the first textbook on pathology. Lopez Piñero's short article – 'La obra de Juan Tomas Porcell y los origines de la anatomia patologica moderna' – is to this day the most systematic of only a handful of articles on Porcell and presents his treatise as the origin of modern pathology.[24] This assertion is obviously overstated, as is Piñero's reference to Fransico Diaz as 'the father of urology' elsewhere in the same article. Porcell's autopsies did, however, represent a significant step forward from observations of visible pathological variations made in dissected bodies by mere chance towards a comparative and deliberate examination of the causes and characteristics of certain illnesses witnessed through postmortem dissections of deceased patients.

---

[23] 'No se pueda dezir por v. Ma. ni por Don Carlos principe d'España por tener tan celebres y doctissimos varones por medicos, quanto para la consideracion y restauracion de la salud de la real persona de vuestra Ma. y de la alteza de nuestro principe, conviene y es necessario, a saber en doctor Gutierrez, y el doctor Mena, y el doctor Christoual de Vega.' Ibid., p. 66v.

[24] José Maria López Piñero, 'La obra de Juan Tomas Porcell y los origines de la anatomia patologica moderna', *Medicina Espanola*, 52 (1965).

As emphasised in Porcell's fifth and final autopsy of the plague victims, his objective was not merely to seek new knowledge and corrections to the Galenic corpus on humoral pathology, but also to affirm the infallibility of ancient medicine when the proper opportunity arose. The latter tendency was seen in his description of the autopsy performed on a 25-year-old surgeon apprentice or *mancebo*, in which Porcell claimed to have witnessed 'with my own eyes what I have often read in the many works of Galen'.[25] The concluding remarks made by Porcell regarding his five autopsies did not indicate that his primary objective was to present a correction of ancient medical authority, but rather to suggest alternative solutions to flaws in contemporary therapeutic practice, such as excessive bloodletting, which was repeatedly condemned by the Sardinian physician.[26] Porcell's treatise lacked an efficient preservation and cure for future plague epidemics, and offered no new insights, as admitted by the author in a short summary.[27] His tendency to fall back upon traditional remedies and intricate cold/warm and dry/wet systems of humoural interrelations was revealed in a later chapter, which emphasised the importance of drinking plenty of cold water, a recommendation later found in Francesc Micó's 1576 publication on the healing virtues of cooled drinks.[28] Porcell's work emphasised the limitations of anatomical practice in the early modern period, which often did not offer noticeable clinical and therapeutical improvements for the prevention and cure of diseases and epidemics. The emphasis on direct observation rather than a reliance on written

---

[25] 'La quinta y ultima Anathomia que hize fue en un mancebo, de edad de hasta veinte y cinco años de buena complexion. Tenia el tumor en la ingle derecha, de magnitud dun pigneon consascara, murio al tercero dia de su dolencia. Tenia grandes ascos, y vomitos antes que muriesse, y vomito unas coleras que no eran bien amarillas, ni bien verdes, sino entre mezcladas. Y porque se dezia que auia muerto desmastado por hauer tenido mucha conuersacion y trato con mugeres y yo nunca hauer hecho Anathomia en semejantes (aunque hartas, porque passan de cinquenta las que hasta hoy he hecho) la hize de major voluntad y gana que las otras, por ver al ojo lo que muchas vezes hauia leydo en muchas partes de Gal.' Porcell, *Información y curación de la peste de Caragoca*, f. 7r.

[26] 'Estas son las anathomías que en la peste de Zaragoza he hecho, y lo que en ellas he visto y hallado digno de consideracion. Los quales han sido causa que convalesciese tanto número de gente en el Hospital General de dicha ciuda, porque después de hechas he curado todo al contrario de como curaba antes de haberlas hecho, que no es sangrando, ni sajando, ni menos purgando por camera. Porque vea Vuestra Magestad, cuan necessarias son hacellas luego a los principios de semejantes enfermedades.' Ibid., f. 8r.

[27] 'La verdadera cura pues della consiste en tres cosas, la primera en quanto a corregir y templar el ayre de la casa y aposiento en donde habita y duerme el enfermo. La segunda en quanto alo que ha de comer y beuer, quanto, quando, quantas vezes y de que manera. La tercera en quanto alas medicinas y remedies, tanto en los que se han de dar y hazer para dentro, como en los que se han de applicar, y poner para fuera.' Ibid., f. 37v.

[28] 'En que tracta en que enfermedades conviene dar de beuer mucha quantidad de agua fria, y si conviene en toda peste, y si es conveniente y necessaria en esta peste de Zaragoza.' Ibid., f. 51r.

authority represented a progressive and important methodological advance in contemporary medicine. Yet the proposed cures based on this accumulated knowledge were still firmly rooted in Galenic humoral pathology and classical medicine – and its traditional prescriptions for fresh air and proper food as the best means to restore the disequilibrium of the diseased body.

Porcell's method and exercise seems to have been imitated during subsequent plague epidemics of Zaragoza, and the most devastating 1652 epidemic left behind another work on anatomical pathology based on recent autopsies on plague victims, carried out by the Professor of Anatomy Fransisco Huguet and the hospital surgeon (and author of the treatise) Joseph Estiche.[29] While the therapeutic measures had not made significant advances during the long time interval since Porcell's ordeal in 1564, one substantial improvement had appeared since then, namely the contruction of a *casa de anathomia* built in 1586 within the *Hospital de Nuestra Señora de Gracia,* which was described in unparallelled detail in a notarial agreement from the same year between its builder and the town council of Zaragoza.

The 'House of Anatomy' within the *Hospital de Nuestra Señora de Gracia*

As previously mentioned, the construction of Zaragoza's first anatomical theatre took place during one of the most turbulent periods in the early history of the university and after a lost lawsuit filed by the neighbouring University of Huesca. The construction of the anatomical theatre was the very first building programme launched by the new university during its early phase of consolidation and was initiated on 27 March 1586 at the expense of 3,240 sueldos.[30] The architectural plan for this building has survived in a written contract or *Capitulacion para hazer la casa de anathomia,* signed by the notary Martin Español, who kept an account of the agreement between the builder of the *casa de anathomia,* Andres de Capraneda, and representatives of the town council and medical faculty.[31] The exact time period devoted to construction is unknown, but the theatre was functioning by

[29]   J. Estiche, *Tratado de la peste en Zaragoza, en el año 1652* and J. Maiso González, *La peste de 1652 en Zaragoza* (Zaragoza: Separata facticia de Estudios del Departamento de Historia Moderna, 1973).

[30]   Angel San Vicente, *Monumentos diplomaticos sobre los edifiicios fundacionales de la Universidad de Zaragoza y sus constructores* (Zaragoza: Diputacion Provincial, Institucion Fernando el Catolico, 1981), p. 4.

[31]   'Los quales dixeron que sobre la fábrica y obra del aposento que se ha de hazer en él cimenterio del hospital de nta senora de grazia para hazer enel notomía de los cuerpos muertos, havian hecho y ordenado entre las dichas partes una capitulación y Concordia, la qual quisieron haber y tubieron por leyda y notificada debidamente y según fuero la dieron y libraron en poder de mí dicho notario.' Archivio Histórico Provincial de Zaragoza, Protocoles Notariales, Martin

the time of an inspection of the hospital in 1600, which recorded the presence of an anatomical theatre next to the hospital cemetery: 'there is a room with its lock and key and four windows with its wooden shutters that the surgeons and professors and physicians of the University use for performing anatomies'.[32] The four windows mentioned in this source were the result of an explicit requirement in the contract between the builder and the city council of Zaragoza, and indicate that the intended design was followed dutifully in this respect. Although the exact date of completion is unknown, the town council had placed a time limit of only two months' work on the construction of the entire theatre.[33]

The *Capitulacion para hazer la casa de anathomia* offered clear directives for the precise measures and materials to be used in the construction of the small theatre, and further instructions for the creation of the dissection table, which was to be surrounded by two circular rows of benches.[34] The meticulous approach to construction indicates an already-established matrix for such anatomical facilities, even though no similarly detailed descriptions exist in other contemporary Iberian contexts. The existence of a general model for such anatomical theatres was nonetheless indicated in Diego Antonio de Robledo's seventeenth-century textbook on surgery, entitled *Compendio chirúrgico útil y provechoso a sus profesores...* (Madrid, 1686). In a short treatise within this work, which consisted of a series of questions and answers regarding anatomy, Robledo answered the question: 'What facilities are needed to carry out this kind of work?' His response describing the ideal location for dissection overlaps in almost every respect with the stipulations of the 1583 contract made in Zaragoza: 'First of all, the place needs to be light and bright, and it should have open windows so the

---

Español, año 1586, fols 143–46, 'Capitulacion para hazer la casa de anathomia', 27 March 1586. Transcripción de Angel Sanvicente (1981), p. 167.

[32]   'A otro lado es el aposento que sirve de hazer lanotomia los cirujanos y Cathedraticos de la Universidad y medicos della con su cerreja y llave y quarto ventanas con sus belagoste de madera.' *Archivo de la Vicaría del hospital Provincial de Nuestra Señora de Gracia de Zaragoza. Libro de visitas del hospital año de 1600 en adelante (1600–08)*, p. 661, cited in José Pardo-Tomás and Àlvar Martinez-Vidal, 'Anatomical Theatres in Early Modern Spain', *Medical History*, 49 (2005), p. 273.

[33]   This is stressed in the surviving *Capitulación entre los jurados de Zaragoza y Andres de Capraneda obrero de Villa*, dated 27 March 1586: 'Item, que a de dar dicha obra hecha y acabada a contento de los señores jurados por todo el mes de mayo de este presente año 1586 y que la comencerá ... Item dásele por la dicha obra ciento y sesenta y dos libras pagaderas desta manera: las sesenta y dos luego y las demás el día que hubiere acabado la obra a contento de los señores Jurados.' *Capitulación...*, p. 169.

[34]   The most detailed scrutiny of this document has been carried out by Àlvar Martínez-Vidal and José Pardo-Tomas in their studies of early modern theatres of anatomy, and has led to an increased understanding of this particular phenomenon in an Iberian context. Their translation of the measuring system *palmos* into well-known systems of measurement will be used throughout this chapter.

air can be ventilated, and the fouls smells can leave the room. Secondly it should be adept, and the best shape is circular with many steps, and with the anatomical table in the middle, where the body is placed, and this should be able to circulate freely, so that everyone can see what is going on from their seats. Thirdly the surgeon who works there should have good hands and should only dissect the things he is teaching, using all the instruments necessary for this practice.'[35]

The overall design of the Zaragoza theatre seems very similar to Girolamo Fabrizi d'Acquapendente's extant theatre of anatomy at the University of Padua, which was finished in 1594.[36] Yet its immediate predecessor in Zaragoza eight years earlier was markedly smaller and certainly much less spectacular with its humble dimensions, less than half the size of its later Italian counterpart, and with only two circular rows for the audience, compared to six in the theatre in Padua.[37] The guidelines set out for the builder in Zaragoza specified that the small circular amphitheatre was to be constructed within a quadrangle building. Each side of this small square house measured only 30 *palmos* (or 5.79 m) and 24 *palmos* (or 4.63 m) in height, and was adorned with two small windows on each side of the building, as later confirmed in the previously mentioned *libro de visita* from 1600. This architectural plan was provided in the agreement drawn up by the notary, but contained no references to any of the previous considerations which had produced the meticulously contrived design.[38]

---

[35]   'Lo primero, es que el sitio donde se exerca ha de ser luminoso, y claro, y alrededor ha de tener ventanas abiertas para que se ventile el ayre ambiente, y el vapor cadaveroso salga a fuera. Lo Segundo el aposento ha de ser capaz, y la major forma es redonda, lleno de gradas, y en medio deve estar la tabla anatomica, en que se tienda el cuerpo, y esta se anda alrededor para que todos los circumstantes puedan ver qualquiera cosa desde su asiento. Lo tercero, que el cirujano que ha de obrar tenga buenas manos, y no corte mas de aquello que se ha de ir enseñando, y este prevenido de todos los instrumentos necessarios.' Diego Antonio de Robledo, *Compendio chirúrgico útil y provechoso a sus profesores escrito por el doctor D. Diego Antonio de Robledo (y Méndez), medico principal de la Real Casa de Nuestra Señora de Guadalupe y regente de la cathedra de cirugía de sus reales hospitales* (Madrid: Vicente Cabrera, 1686), p. 30.

[36]   The 1594 theatre of anatomy at the University of Padua was preceded by another smaller and more short-lived university theatre constructed the very same year as the one in Zaragoza but only active until 1590, as recorded in great detail in Cynthia Klestinec, *Theaters of Anatomy. Sudents, Teachers and Traditions of Dissection in Renaissance Venice.* (Baltimore: Johns Hopkins University Press, 2011), pp. 55-90

[37]   Giorgio Zanchin, *Il teatro anatomico di Padova, Rapprasentare il corpo, Arte e anatomia da Leonardo all' illuminismo* (Bologna. Bononia University Press, 2004) pp. 216–22.

[38]   'Prima se ha de hazer un aposento que tenga treinta palmos o mas en quadro contando el vacio del; tendran en alto estas paredes veinte y quarto palmos y mas; an de subir en quadro desta manera; que en cada esquina ternan un pilar de ladrillo y medio en quadro, y en medio de cada lienco otro pilar asimesmo de ladrillo y medio en quadro, y lo demás sera de tapia valenciana muy buena y provechosa; toda esta pared y pilares ternán fundamento de argamasa de cinco o seis palmos en ondo como esté más segura, y de dos ladrillos de ancho muy al provecho y segura.

The teaching floor was located one metre below the ground floor, from where a few steps led down to a small circular stage for the anatomy lectures. Here a rotating wooden dissection table measuring 1.73 m x 0.77 m rested on a central base and could be turned around during teaching. The building guidelines requested the construction of a locked cupboard for the instruments used in dissections, so that they might be stored safely outside teaching hours.[39] The theatre walls were made of *tapia valenciana*, interpreted by Pardo-Tomás and Martínez-Vidal as a type of white-plastered adobe. The floor was paved with brick and the roof of the building was allegedly covered with tiles in the 'Aragonese style'. The two-metre diameter of the lowered circular floor-level enabled the slightly shorter dissection table to rotate freely on its central base. This structure thus improved the didactic value of the intended dissections and enabled the assembled students to follow the dissections from numerous perspectives while seated on two rows of circular benches. An interesting reference to eight iron rings, four of which were placed on each side of the dissection table, was undoubtedly interpreted correctly by Martínez-Vidal and Pardo-Tomás as evidence that vivisection was carried out within the new theatre. This practice is mentioned occasionally in contemporary publications, but only vaguely in Iberian university statutes of the time.[40] References to small drains on one side of the dissection table and on the floor underneath was indicative of the untidiness of anatomical practice, which produced not only bodily fluids from the dissected cadavers but also large amounts of water from sponging and bathing the anatomised bodies.[41] Close attention to cleanliness and hygiene even in the construction of the theatre seemed to go beyond the requirement for the constructor, and suggested the additional future employment of a

---

Item en cada lienco de estas paredes ha de haber dos ventanas con sus sobreportales, con sus viaxes labados de algez, con sus ventanas de madera llanas y azepilladas, con sus picapuercos de hierro, con sus alguazas y sus aros; ternán de lumber cada ventana quarto palmos en ancho y cinco o más de caída, y de manera que se pueden poner encerados en ellas con sus aros.' AHZP, f. 144.

[39] 'Item, en una esquina de dicho aposento, atrás a esta postrera grada, ha de hazer un armario que tenga de lumber el aro quarto palmos en ancho y cinco en largo, con su ceraxa y llave, de manera que esté todo cerrado con una tabla por medio y con sus alguazas, y la parte que toca a la pared lavado de algez y muy pulido que esté y al provecho.' Ibid., f. 144.

[40] 'Item en medio de dicho aposento, más ondo que el suelo lo que pareziere que vastare, ha de haver una mesa recia de madera y lisa, firmada sobre un pie con su tornillo que pueda andarse alrededor, alta a la cintura; de largo terná dicha mesa nuebe y dies palmos, de ancho quadro palmos poco más; ésta ha de tener su guarnicion levantada alrededor y en cada lado por el largo della ocho cercillos de hierro, quarto en cada lado en la parte que parezerá, y en medio hazia una parte un aguxero grande como un real de a cuatro.' Ibid., f. 145.

[41] 'Devaxo la mesa, a un lado, ha de hazer un sumidero para hechar las agues y unas gradas desde la puerta para vaxar asta la mesa, como está en la traza.' Ibid., f. 144.

subordinate who could oversee the maintenance of the new anatomy theatre.[42] The design for this small amphitheatre contained two circular rows of benches surrounding the lowered central floor. The feet of the spectators sitting in the front row were thus placed on the same level as the surface of the dissection table, which was elevated one metre above the floor by its central rotating base. The seated rows of students could follow the dissections at a close range and certainly with a clearer view than the standing students crammed together in six tight rows at the later and much larger theatre in Padua, which could hold up to 240 spectators, and occasionally even housed a little orchestra of lute players during the demonstrations.[43]

There are no references to the actual teaching carried out within this 'house of anatomy' during the late sixteenth century. Later statutory orders, publications and salary lists from the early and mid-seventeenth century showed continued practice within the small theatre and presented anatomy as a core discipline within the training necessary for physicians and surgeons: 'Medicine and surgery are not mastered without hard work, diligent study, frequent lectures and rotation of books, knowledge of natural things, and an expertise in anatomy.'[44] The statutes of 1647 indicated that annual dissections at Zaragoza had reached the number known from the much larger University of Salamanca. These revised statutes seem to have imitated the larger institution in its division of the anatomical dissections into 'six systematic, and twelve regional', the same numbers found in the 1594 statutes from Salamanca. The 1647 statutes from the University of Zaragoza also imitated Salamanca's teaching programme in its term dates, which lasted from St Luke's Day (18 October) until St John's Day (1 March), and in its emphasis on Galen's *De usu patium* as one of the cornerstones of anatomical teaching.[45]

---

[42]   'El suelo de entre las gradas y dentro del antipecho donde estará la mesa ha de estar enladrillado y pulido todo ... Item a de lavar de alxez con plana muy pulido todo el aposento, las ventanas y gradas por una parte y por otra de manera que esté muy pulido y a contento de los señores.' Ibid., f. 144.

[43]   Cynthia Klestinec, *Theaters of Anatomy*, pp. 95 and 106-107

[44]   'La medicina, y la cirujia no se adquiren sin trabajo, vigilias estudio, frequente leccion, y revolucion de libros, conocimiento de cosas naturales, y sin pericia de la Anatomia.' Fernández, *Documentos para la historia de las profesiones sanitarias*, p. 180.

[45]   'El de Anotomia leerà cada año el libro de ossibus de Galeno, y despues las calidades del cuerpo humano; otros veces los libros de usu partium; otras los de motu musculorum, ó de disectione venarum, o de anatomicis administrationibus. Hara diez y ocho disecciones desde el Señor San Lucas, hasta el primero de Marzo, es a saber; seis universales, y doze particulares; y tenga por pena cada vez que las dexare de hazer, veinte reales; las dos partes para el Arca, y la tercera parte para el denunciador; y sean las tres calidades, de la cabeza, de los ojos, de las narizes, de las orejas, de la lengua, de la laringe, de brazos, de piernas, del utero; y finalmente de qualquire particulas que en el cuerpo humano se hallan; y las generales sean de las venas, tomandolas de su origen, que es

A novelty in the 1647 statutes and a testament to the importance attached to anatomical practice was a penalty fee of 20 reales for any dissection in the statutory quota of 'six systematic, and twelve regional dissections' which was not carried out, with one-third of the amount to be paid to the whistleblower. Developments in the study of anatomy at the University of Zaragoza and at the two other large Aragonese universities in Valencia and Barcelona contrasted with the waning of the discipline in Castile. The absence of an anatomical professorship at the University of Valladolid after 1550, the long period of vacancy in Alcalá from 1570 and the gradual decline of the theatre and chair of anatomy in Salamanca after 1600 coincided with construction and restoration programmes of the anatomical theatres in Aragon and a steady increase in the number of annual dissections at its universities. Anatomical teaching in mid-seventeenth-century Zaragoza followed the Aragonese tendency, but also had its own unique path, as evidenced by the 1649 petition for the city's midwives to attend the lectures of anatomy: 'We require and order the Professor of Anatomy to perform privately for the midwives the anatomies considered necessary on the appropriate parts, with suitable tact and discretion, and the women will be obliged to go to the hospital area or wherever they are told to go, on the date and time fixed by the Professor.'[46]

Another unique trait of Zaragoza was the mid-seventeenth-century influx of Italian medical doctors to the city, including Juan Bautista Juanini from the University of Pavia and the Sicilian physician Federico Bottoni, reminiscent of the similar transfer of Juan Tomas Porcell from Sardinia a century earlier. Other contemporary Iberian universities were seemingly unable to attract foreign anatomists, whereas the University and general hospital of Zaragoza seemed to have established a wider reputation as an active centre of anatomical practice and research. This status attracted several Italian scholars and was praised repeatedly in their later publications, such as Juanini's *Discurso político y phísico* (Madrid, 1679) and Bottoni's *Evidencia de la circulación de la sangre* (Lima, 1723). As previously emphasised, this exchange between the Iberian and Italian Peninsulas had formerly taken the form of a largely one-sided procession of Spanish anatomists into Italy; Andres Laguna, Pedro Jimeno and Juan Valverde are prominent examples. The medical institutions of Zaragoza offered an exception to this rule,

---

el higado, y sigiendolas por todas partes del cuerpo; en que se ramifican. Otro año de las arterias, comenzando desde el corazon, y prosiguiendo todas por las partes, en que van a parar. Otro año de todos los nervios, comenzando desde el celebro.' Ibid., p. 126.

[46] 'Ittem, estatuimos y ordenamos que el Catedrático de Anatomía que es, o, por tiempo será, sea tenido y obligado hacer las anathomias que parecieran necesarias de las partes convenientes al dicho exercicio de parteras a ellas privadamente, con el recato y compostura conveniente, y las dichas tengan obligación de acudir a la Camarilla del Hospital, o puesto que se les señalara, el día e hora señalado por dicho Catedrático.' AHPZ, Protocolos Notariales, Juan Francisco Sanchez de Castellar, 1649, f. 976, Colegio de Médicos y Cirujanos de Zaragoza. 'Ordinaciones para parteras', cited in Martínez-Vidal and Pardo-Tomas, 'Anatomical Theatres', p. 276.

with the arrival of Juan Tomás Porcell, Juan Bautista Juanini and Federico Bottoni in the sixteenth and seventeenth centuries, and the city was acclaimed by Bottoni as a European centre of medical research: 'In my view, the fame of the University of Zaragoza has been increased, since nobody in Europe is unaware that in this celebrated museum of the sciences, medicine flourishes at its highest level, due to the continual practice of anatomy that is performed twice a week in the theatre or room provided for this purpose in the famous General Hospital.'[47]

Such praise was in clear opposition to the decline of anatomy elsewhere in Spain and was far from the contemporary situation at the University of Alcalá, where the chair of anatomy ceased to exist from the late seventeenth century onwards. While the city of Zaragoza was without a fully developed university for most of the sixteenth century, it is noteworthy that it eventually managed to achieve renown outside Spain as a contributor to contemporary anatomical research. This could be seen in the works of Juanini and Bottoni, who both referred to previous studies performed at Zaragoza in their stern defences of Harvey's physiology, which was generally condemned in other Iberian universities, with the *Estudi General* of Barcelona being the only other exception. Bottoni's defence of Harvey's theory of the circulation of the blood referred to two weekly dissections carried out in Zaragoza during the late seventeenth and early eighteenth centuries. This seems to be something of an overstatement, as it would amount to almost 10 times the number of anatomical dissections carried out at the Universities of Valencia and Barcelona during the same period. Despite such exaggeration, the chair of anatomy at Zaragoza certainly seemed to have prospered throughout the seventeenth and early eighteenth centuries. The additional employment of a *disector* to assist the anatomical professorship during this period was only parallelled elsewhere in Iberia at the University of Salamanca, where the Professor of Anatomy Augustin Vazquez made a succesful request for an assistant in 1593. In Zaragoza the subordinate was responsible for cleaning and maintaining the anatomical theatre, and for the preparation of instruments and cadavers for the teaching.[48] This assistant who appeared in the 1681 statutes on anatomy represented a belated acknowledgement of the concerns for the maintenance of the anatomical theatre which had been presented a century earlier in the contract between its builder and the city council.

---

[47] 'Nadie de los europeos ignora, que en este celebrado Museo de la Sciencias, florece la medicina, en el más elebado crédito, debiendo este al continuo exercicio anathómico que dos vezes a la Semana se executa en el Theatro, o Salón, que para este efecto hay en aquel célebre Hospital General, concurriendo todos los Professores de esta sciencia a tan importante demostración.' Federico Bottoni, *Evidencia de la circulación de la sangre* (Lima: Igncio de Luna, 1723), p. 4.

[48] 'Cuidando de que el Teatro se halle con la limpieza y prolijidad correspondiente, teniendo a su custódia las llaves y lo que se requiere para hazer demostrables las partes, y cuidando del estuche y hierros anatómicos que a este fin tiene el Hospital de que es responsible dicho disector.' Asunción Fernández, *El Hospital Real y General de Nuestra Señora de Gracia de Zaragoza en el siglo XVIII* (Zaragoza, Institución Fernando el Católico, 1987), p. 330.

Later statutory orders from the eighteenth century indicated that the assistant post was maintained, but only with significant difficulties, as emphasised in the 1767 statutes, which showed signs of crisis in the anatomical teaching programme and an apparent end of the momentum gained by the professorship during the previous century.[49] In spite of this break with the the ambitious activities of the sixteenth and seventeenth centuries, José Maria Lopez Piñero's biography of the famous neurologist and Nobel laureate Santiago Ramón y Cajal presented an almost uninterrupted progression of later studies of anatomy and morphology at the medical faculty of Zaragoza, where the young physician from Zaragoza graduated in 1873.[50]

During the latter half of the sixteenth century, the combined information from Zaragoza relating to systematic autopsies and the detailed guidelines for the construction of an anatomy theatre provides a category of sources absent from contemporary Iberian university contexts – and invaluable supplements to the fragmented history of anatomy in late Renaissance Spain. The physical framework for anatomical teaching and research is not mentioned in other contemporary Spanish sources, and the guidelines for the construction of the anatomy theatre in Zaragoza therefore provide important glimpses into the complex construction and maintenance of such structures. Anatomical theatres have often appeared in the context of early modern universities such as Padua or Leyden, but in an Iberian context, contemporary anatomists instead carried out their work within existing hospital precincts that often pre-dated the establishment of nearby universities, as was the case in Zaragoza, Barcelona and Valencia. A rare example of a sixteenth-century hospital carrying out its own anatomical dissections without affiliation to a nearby university was found in the *Hospital del Monasterio de Nuestra Señora de Guadalupe.*

---

[49] 'Viniendo ahora a los vicios, y corruptelas de el Hospital digo que habiendo se fundado y heregido posteriormente la Uniberssidad, se el catedrathico de Anathomia el theatro Anatomico, con la renta, y con la Cathedra, y con la obligacion de hacer al año doze Anatomias: que aunque se hagan y junten todas, no baldran por media sino ay algun Zirujano que la esplique y con methodo y razon disseque. Este dissertor deve ser tan habil como el mismo Cathedratico; y la nombraba la Sitiada a uno de los principales mancebos, pero viendo estos que nada se utiliza, y hassi mismo la profession de este Reyno se halla avandonado, y que no pueden escapar de ser barberos, ninguno quiere aplicarse a hesta maniobra. Ynstrumento no hay ninguna de provecho, y quando los Cirujanos Maestros los necessitan los compran y gastan de su bolsillo.' Ibid., p. 256.

[50] 'Durante el siglo XVIII, el cultivo de la anatomía práctica se mantuvo en el Hospital de Nuestra Señora de Gracia a una estimable altura con trabajos como las investigaciones necropsies efectuadas por José Amar (1774) para diferenciar con precision entre pleuresía y pulmonía. Los que continúan diciendo que la obra de Cajal 'nació por generación espontánea' o fue un 'milagro' en un país sin tradición morfológico fingen desconocer hechos tan evidentes a nivel internacional como las aportaciones de Antonio de Gimbernat y de Juan Bautista Bru.' José Maria López Piñero, *Santiago Ramón y Cajal* (Valencia: University of Valencia, 2006), p. 132.

Chapter 8
Beyond the Universities

Anatomical Practice at the *Hospital del Monasterio de Nuestra Señora de Guadalupe*

Anatomical dissections performed at the *Hospital del Monasterio de Nuestra Señora de Guadalupe* maintained an autonomous character outside any nearby university context throughout the late 1500s. A few references from the *Hospital de la Santa Cruz* in Toledo and the *Hospital del Cardenal* in Seville indicate that similar activities took place in other Castilian hospitals. However, the Monastery Hospital of Guadalupe represents the only well-documented case within Spain where anatomists had routine experience with human dissection outside a university. Its tradition in this field is even said to pre-date the public anatomies introduced at universities in mid-sixteenth-century Castile. This notion remains speculative and rather dubious, but has nonetheless appeared in several historical accounts of the monastery. The modern-day visitor to the *Monasterio de Nuestra Señora de Guadalupe* is also reminded of the institution's pioneering status by a modern plaque placed at the entrance of the late medieval hospital buildings, which survive to this day: 'This building was erected by the protoprior of The Hieroninites Yañes de Figueroa in 1402 as a hospital of Saint John the Baptist and in this hospital the earliest dissections in the Spanish Kingdoms were carried out with papal privilege.'[1]

Similar misconceptions regarding 'the origin of Spanish anatomy' have been presented in other contexts as well, and some fallacies have been repeated to this day, despite the existence of convincing counter-evidence. Lopez Piñero's *Ciencia y tecnica...* (1979) looked critically at the attempts of the nineteenth-century historian Hernández Morejón to extrapolate the practice of anatomy in Guadalupe back to 1322, coinciding with the very foundation of the monastery. If confirmed, this date would precede even the earliest known privileges for dissections granted to the Universities of Montpellier and Lleida, but the claim has hitherto not been supported by any known sources. There is nonetheless rich documentary evidence to support the notion of a long tradition of medical practice existing at the monastery since the mid-fourteenth century, when

---

[1] 'Este edificio lo mandó construir el protoprior de la orden de Jerónimos Yañes de Figueroa, en 1402, como Hospital de San Juan Bautista y en este Hospital, se llevó a cabo por primera vez en los Reinos de España la dissección del cuerpo humano, por especial privilegio de Roma.'

Alfonso XI of Castile elevated the image of *Nuestra Señora de Guadalupe* to one of the most celebrated relics of his realms following his victory over Moorish forces at the Batalla del Salado (1340), one of the last decisive battles of the Reconquista. From this period onwards, the Monastery of Guadalupe became a distinguished centre of pilgrimage and served as residence for the Hieronymite Order for five centuries after they settled there in 1387.

The archival records of the monastery from the late fourteenth and early fifteenth centuries count a number of physicians, surgeons and pharmacists among the Hieronymite friars. During the mid-fifteenth century, the monastery was granted two pontifical privileges for its medical practices by Popes Eugenius IV and Nicholas V. The hospital's medical expertise was enhanced by two of its physicians, doctor maestre Juan de Guadalupe and Alfonso Fernández de Guadalupe, who were among the first medical doctors to be incorporated into the recently established *Tribunal del Real Protomedicato de Castilla*, founded by the Catholic Monarchs in 1477.[2] At that time, the hospitals near the monastery enjoyed rapid growth and by the end of the fifteenth century, they held a series of varied medical functions, with two separate hospitals for men and women and a third for the friars. Its organisational structure is related in detail in a chronicle of the monastery written by its later prior, Gabriel de Talavera. This *Historia de nuestra Señora de Guadalupe* was printed in Toledo in 1597 and included a chapter on the foundation of the hospitals entitled 'On the Famous Hospitals Raised and Supported by this Monastery'.[3]

The chronicle's reference to a section of the hospital which was devoted to ointments for syphilis victims consolidated an early mention from 1497 of a recently created infirmary for patients suffering from *las bubas* and showed a very early experience at the hospital with this recently discovered disease.[4] A

---

[2]   Guy Beaujouan, *La medicina y la cirugía en el Monasterio de Guadalupe* (Paris: Diana, 1966), p. 165.

[3]   'Ha edificado este monasterio en todo ilustre, dos hospitales: prendas certissimas su feruosa charidad, el uno diputado para remediar hombres, y el otra para curar mugeres Referire dellos loque todo el mundo conoce. Esta mas abaxo del seminario, un sumptuoso edificio, lauantado a honra del gran Baptista, a quien esta dedicada una iglesia dentro del, y assi mismo un quarto donde se hospedan y regalan los religiosos de otras ordenes, que visitan este santuario ... Ay en los quarto lienzos deste edificio quarto salas, o enfermerías, y en baxo otros aposentos, que sirue para buscar (según el tiempo) el regalo de los enfermos. La una destas enfermerias esta diputada para curar los capellanes, collegiales, y donados de nuestra casa: otra para los heridos: la tercera para reparo de las calenturas: y la ultima esta señelada para los que padecen enfermedades de mayor peligro. Deste claustro principal se passa a otro, donde se dan las unciones del mal Francés.' Gabriel de Talavera, *Historia de Nuestra Señora de Guadalupe: consagrada a la soberana magestad de la Reyna de las Angeles milagrosa patrona de este sanctuario* (Toledo: Thomas de Gúzman, 1597), f. 197–98.

[4]   'Si les plazia que para los enfermos de las buas se pusyese en un ospital ciertas camas a parte e les provyesen de lo necesario y concordose que se pusyesen doze camas o trece o las que

royal privilege was granted for the import of mercury to be used in ointments for syphilis and plague sufferers, and there are several references to a garden of medical plants existing within the hospital precincts during the late 1400s.[5] In 1502 a new state-of-the-art pharmacy was created for the manufacture of medicine, an innovation that was still subject to much praise in Talavera's late-sixteenth-century account.[6] Many varied medical activities, including work within different specialised fields, took place during the fifteenth and early sixteenth centuries, but we cannot be sure whether these practices also included anatomical dissections. Accounts written by visitors to the monastery, such as the physician Jeronimo Müntzer in the late fifteenth century and the Portuguese traveller Caspar Barreiros in 1542, both referred to the monastery as a centre and maybe even a school for medical and surgical treatment, but did not include references to anatomy studies within the hospitals of Guadalupe.

The Uncertain Origins of Anatomy Studies at the Hospital of Guadalupe

The uncertainty surrounding anatomical practice was dealt with briefly by Lopez Piñero, who concluded that the anatomy studies presented in Francisco Hernández's mid-sixteenth-century comments on Pliny were probably among the earliest dissections carried out in Guadalupe: 'When did the regular practice of dissection begin in Guadalupe? 'There is no evidence to show that it was introduced during the late medieval period. In fact, there is a dearth of direct references until the decade 1550–1560, although certain indications allow us to assume the existence of an earlier tradition.'[7] Lopez Piñero makes the reasonable assertion that Guadalupe came to practise anatomy at a very late stage, compared

---

fuesen menester en un hospital o lugar que a nuestro padre paresciere y que pongan en él personas diligentes que los syrvan y que les provehan de casa o del ospital, de fisicos de medicinas e otras cosas necesarias; e señalose el hospital de la pasyon donde se hazen las dichas cosas.' Arch. De Guadalupe, cod. 74, f. 1, cited in Beaujouan, *La medicina y la cirugía en el Monasterio de Guadalupe*, p. 161.

5   José Ignacio de Arana Amurrio, *Medicina en Guadalupe* (Badajoz: Diputación Provincial de Badajoz, 1990), p. 208.

6   'Ay en esta enfermería, por la parte que cae a los huertos una célebre y famosa botica, tan grande, tan limpia y bien acabada, tan abundante de medicina y muchedumbre de vasos que no creo tiene semejante officina toda España. Es tanto el cuydado que se tiene que no huela a lo que es, siendo las medicinas perfectíssimas, que quitan aquel común enfado y aborrecimiento que suelen tener los enfermos.' Ibid., p. 197–98.

7   'Quando se introdujo en Guadalupe la práctica regular de disecciones anatómicas? Nada autoriza a pensar en tempranas fechas bajomedievales. De hecho, carecemos de noticias directas hasta la década 1550-1560, aunque algunos indicios permiten suponer una tradición anterior.' José Maria López Piñero, *Ciencia y tecnica en la sociedad Española de los siglos XVI y XVII* (Barcelona: Labor, 1979), p. 322.

to Morejón's prior claims of a tradition preceding even the records of early fourteenth-century dissections within the Crown of Aragon. Both Piñero's *Ciencia y tecnica* and Luis Granjel's later *La medicina española renacentista* furthermore doubted the existence of a papal privilege for dissection granted in the fourteenth or fifteenth century. The alleged Guadalupe privilege has nevertheless been cited repeatedly since the late sixteenth century in the wake of Talavera's explicit reference to 'the anatomies that the surgeons carry out here with a privilege granted by His Holiness'.[8] No other known contemporary sources uphold this claim of papal support for the practice of anatomy in Guadalupe. If found, it would be the only of its kind in an Iberian context, and it may have been confused with the other papal privileges for medical practice granted to the Hieronymite friars in 1442 and 1451.[9] Pontifical interference in this matter would conflict with the evidence of preceding privileges for dissections, granted only by the monarchs, such as those bestowed upon a number of Aragonese and Castilian university towns between the fourteenth and sixteenth centuries.

This unsupported claim of papal involvement has been cast into doubt by Piñero and Granjel, but their outlines of famous physicians trained in Guadalupe nonetheless maintained equally undocumented references to the presence of the famous surgeon and physician Fransisco Arceo at the Hospital of Guadalupe. Piñero's abstract on Arceo in *Diccionario histórico de la ciencia moderna en España* presented the early training undergone by Arceo at Guadalupe as a fact, but offered no reference to the source of this information.[10] Theories on the practice of dissection in Guadalupe prior to Fransisco Hernández's well-documented anatomical training in the 1550s might have been supported by the presence of Arceo – a physician born in 1493, educated in the first decades of the 1500s and renowned for a much later treatise on plastic surgery included in his *De recta curandorum vulnerum ratione* (Antwerp, 1574). This publication

---

8　　'Asisten en estos hospitales ordinariamente quarto cirujanos, y otro mas doctor que ellos, para las enfermedades mas graues. Lee este la facultad, y lo mesmo haze el Dotor de medicina. A estas lecciones acuden con gran prouecho, por la experiencia ordinaria que se haze de lo que se enseña, y anotomias que pueden hazer los cirujanos, por indulto de su Santidad. De aqui ha nacido salir desta casa tan grandes medicos, que sus partes y fama los ha llevado a las de los Reyes.' De Talavera, *Historia de Nuestra Señora de Guadalupe*, f. 221.

9　　This is discussed at great length in de Arana Amurrio's *Medicina en Guadalupe*, in which the author reintroduces the notion of a late medieval papal grant for dissections in Guadalupe without offering new evidence to support his theory.

10　　'Estudió medicina en la Universidad de Alcalá y trabajó varios años en los hospitales del Monasterio de Guadalupe, que eran entonces un prestigioso centro de perfeccionamiento clínico para médicos que ya habían obtenido su título, además de tener una escuela para cirujanos.' José Maria López Piñero, 'El saber anatomico y la dissección de cadaveres humanos en la España en la primera mitad del siglo XVI', *Cuadernos de la historia de la medicina Española*, 13 (1974), pp. 64–65.

makes no references to the author's own education in Guadalupe, a background also disputed by an early biographical article from 1913.[11]

Talavera's 1597 account of famous physicians at the monastery hospital did not refer to Fransisco Arceo either. Notions of his practice at the hospital in Guadalupe remain purely speculative, in line with the often repeated yet equally undocumented claims of Miguel Servetus' early education in Zaragoza. Although Arceo's sole publication on plastic and orthopaedic surgery, cancer and syphilis indicates that the author possessed detailed anatomical knowledge, there is no evidence to support the notion that his prior training had been obtained during hospital practice in Guadalupe. We do know that the monastery hospital did house a number of other Spanish physicians, such as Benito Bustamente Paz, whose death there was documented in 1555, and Fransisco Arias Lopez, who had also died during his medical practice at the hospital in 1551. According to the archival records, Arias Lopez 'was here in order to gain experience'.[12] This latter information supports Fransisco Hernández's account from the subsequent decade of physicians who came to the monastery as part of their concluding two-year practicum. Perhaps a formal medical school was developing within the hospital precincts during the mid-sixteenth century; Caspar Barreiros' 1542 account mentioned schools of both grammar and surgery in Guadalupe, and thus indicated a Castilian professionalisation and institutionalisation of surgery decades before its introduction at the Universities of Salamanca (1566), Alcalá de Henares (1593) and Valladolid (1594).[13] It is possible that this alleged school of surgery carried out dissections as part of the training undergone by its students, but this has not been supported by the surviving sources. The first definitive evidence of anatomy studies taking

---

[11]   'No se tienen certeras noticias del sitio en que hizo sus primeros estudios, siendo, según unos, la Universidad complutense, y en concepto de otros, relizó su aprendizaje en la escuela práctica del Monasterio de Guadalupe, donde existía un gran hospital, en el que podían adquirirse utilísimos conocimientos. Es indudable que este hospital era una verdadera policlinica, de gran provecho para que el estudioso y observador que recibía su enseñanza, llegara a ser un buen medico; lo qual unido a la circumstancia de poseer un privilegio del Papa para apertura de cadaveres, cuyo hecho no era en todas partes permitido entonces, constituyó el motivo de que muchos elegiesen este hospital para verificar con aprovechamiento sus estudios.' Joaquin Olmedilla y Puig, *Francisco Arceo, illustre medico y escritor español del siglo XVI (*Madrid: Real Academia de Medicina, 1913), p. 6.

[12]   'Estava aqui tomando experiencia.' Cited in de Arana Amurrio's *Medicina en Guadalupe*, p. 159.

[13]   'Guadalupe tiene dos colegios, uno de gramática y otro de cirugía ... los de cirugía son cuatro y se hacen buenos letrados' en esta facultad porque, aparte de sus lecciones y conferencias de letras, adquiren mucha práctica en las curas del hospital, donde siempre hay heridos y enfermos de diversas enfermedades.' Caspar Barreiro, *Chorographia de algums lugares que stam em um caminho que fez G.B ó anno de MDXXXVI, comencando na cidade de Badajoz em Castella te á Milam em Italia* (Coimbra: Joao Alvarez 1561), f. 38.

place in Guadalupe appeared in Fransisco Hernández's later accounts from the 1560s, which did not therefore precede the introduction of anatomy at the leading Castilian universities.

First Documentation of Anatomy at the *Monasterio de Nuestra Señora de Guadalupe*: Fransisco Hernández's *Historia natural* (1565–69)

The first descriptions of a systematic and extensive practice of dissection in Guadalupe emerged in biographical comments of Fransisco Hernández's unpublished manuscript *Historia natural de Cayo Plinio Segundo – traslada y anotada por el Doctor Francisco Hernández*, which was written between 1565 and 1569. This Castilian translation and commentary was inspired by the work of another Spanish anatomist/botanist, Andrés Laguna, whose contemporary translation and deluxe two-volume binding of Dioscorides' botany, *Pedacio Dioscorides Anazarbeo, acerca de la material medicinal, y de los venenos mortiferos*, was printed in Madrid in 1555. Unlike Laguna's masterly and beautifully illustrated magnum opus, Hernández's similarly ambitious work on Pliny unfortunately shared the faith of his residual vast and ill-fated written production; his *Historia natural* never reached a contemporary printing press and remained unpublished in its totality until the late twentieth century. The text includes much valuable information about Hernández's own experience as a physician and an active anatomist at the Hospital of Guadalupe. The manuscript published in German Somolinos d'Ardois' seven-volume *Fransisco Hernández. Obras completas* (Mexico City 1959–84) nonetheless leaves important questions unanswered, such as the uncertainties regarding anatomical practice in Guadalupe prior to Hernández's own arrival in the mid-1550s. Hernández wrote his comments on Pliny after transferring from Guadalupe to the Hospital de Santa Cruz in Toledo in the early 1560s, and at the Madrid court of Philip II, where he was employed as *Médico de la Casa Real de su Majestad* from 1567.[14] During his professional life, Hernández was affiliated with a broad network of some of the most eminent scholars of his time – not only colleagues in the fields of medicine and botany, but also scientific circles outside his own areas of expertise. He was engaged in botanical excursions in Extremadura and Andalucia with his colleague from Guadalupe, Francesc Micó, and the anatomist/botanist Juan Fragoso, and during his years in Toledo, he maintained personal friendships with the renowned Italian engineer and clockmaker Juanelo Turriano, and the Toledan astronomer Bernardo Pérez de Vargas.[15]

---

[14]   Enrique Alvarez Lopez, 'El Dr. Fransisco Hernández y sus comentarios a Plinio', *Revista de Indias*, 3 (1942), p. 254.

[15]   José Pardo Tomás and José Maria López Piñero, *La influencia de Fransisco Hernández (1515–87) en la constitución de la botánica y la material médica modernas.* (Valencia: University

Hernández's 1567 employment at the nearby Madrid court of Philip II expanded this social and professional network, which came to include the architect of the Escorial, Juan de Herrera, and the organiser of its famous library, the humanist Benito Arias Montano, who was praised by Hernández in a long eulogy that still survives. Even though his employment as court physician in Madrid did not overlap with Vesalius' similar appointment a few years earlier, Hernández's *Historia natural* included references to the author's friendship with the famous anatomist: 'Andreas Vesalius, an excellent man in the field of anatomy, and a very good friend of mine, while he was still alive.'[16] Vesalius himself was not similarly informative, and so the only confirmation of the alleged friendship between the two renowned physicians comes from Hernández himself. After a period of almost 20 years since his early anatomical training in Guadalupe, Hernández put his expertise to use on the other side of the Atlantic, cooperating with the surgeon Alonzo López de Hinojosos on autopsies carried out on victims of a 1576 epidemic of *cocoliztli* (Great Pestilence) in Mexico City. Hernández's account of his experience as an anatomist at the hospital of the monastery of Guadalupe provides valuable information about the study of anatomy outside the university faculties of sixteenth-century Spain. It also includes detailed knowledge of the writings of contemporary anatomists, such as Vesalius and Realdo Colombo. Hernández's *Historia natural* referred to his cooperation with Francesc Micó in Guadalupe – the only colleague mentioned by name – who was seemingly in charge of the dissections carried out within the monastery hospital. The joint efforts of Hernández and Micó in Guadalupe during the late 1550s, and their explicit aim to bring anatomy up to date were touched upon repeatedly in the *Historia natural.*[17]

However, Hernández did not clarify whether this allegedly improved anatomical method of hands-on dissection was a reform of existing practices or rather the very first introduction of anatomy studies at Guadalupe. Nor is it clear if it was a new experience for the physicians involved or a continuation of a familiarity with dissection established at Hernández's and Micó's former Universities of Alcalá de Henares and Salamanca. A supplementary comment suggests that Hernández and his colleagues were following in the footsteps of other physicians who came to Guadalupe to gain experience and to improve their practical skills as physicians and surgeons: 'We gave ourselves with the greatest care and will to understand all

---

of Valencia, 1996), p. 40.

[16]   'Andreas Vesalio, varón excellente en anatomía y mientras vivía amigo nuestro.' Cited in Germán Somolinos d'Ardois, *Vida y obra de Francisco Hernández* (Mexico City: Universidad Nacional de Mexico, 1960), p. 132.

[17]   'En Guadalupe dexamos puesta por la bondad de Dios, el anatomia de su punto, como hasta alli no se hubiese cortado todo, sino los miembros interiores solo.' Germán Somolinos d'Ardois, *Fransisco Hernández. Obras completas* (Mexico City: Universidad Nacional de Mexico, 1958–84), vol. 5, p. 99.

that was necessary for the true and consummate physician and the well instructed surgeon.'[18] Hernández indicated a possible awareness of Pedro Jimeno's and Juan Valverde's contemporary discovery and descriptions of a third auditory ossicle, unnoticed in Vesalius' *Fabrica*, in a reference to his own studies of the auricular bones: 'I shall not here describe the little bones of the ear ... which I, not without great delight, observed in the anatomies and dissections that I performed or attended during my sojourn at Guadalupe.'[19] Hernández also revealed a detailed knowledge of the even more recent discovery of the pulmonary system, which had been described by Miguel Servetus in his *Christianismi Restitutio* (Vienna, 1553) and later in Juan Valverde and Realdo Colombo's anatomical textbooks *Historia de la composición del cuerpo humano* (Rome, 1556) and *De re anatomica* (Venice, 1559). This contemporary recognition in Guadalupe of these recent insights into the form and functions of the human body offers a valuable example of the rapid European and Iberian distribution of anatomical knowledge. While Lopez Piñero's *Ciencia y tecnica* included Guadalupe in its references to a Vesalian movement in sixteenth-century Spain, it is difficult to conclude with any certainty whether the studies of Hernández and Micó in Guadalupe had particular roots in the anatomical guidelines set forth by Vesalius. A copy of *Fabrica* was recorded in the 1555 inventory of the monastery, thus indicating a familiarity with Vesalius' works within the hospital. Furthermore, a few examples of explicit criticism of Galenic descriptions based on animal dissections were found in Hernández's work, such as the notion of the difference between the human uterus and those found in cows, goats and sheep: 'Its form (according to what I saw at Guadalupe in the corpse of a pregnant woman that we anatomized) is of an elongated roundness, quite different from that of cows, goats, and sheep, as opposed to the opinion of Galen.'[20]

Hernández's familiarity with the latest publications on anatomy could be seen in his appraisals of Realdo Colombo's 1559 publication *De re anatomica*, which was understood to be a significant step forward from the works of both Galen and Vesalius.[21] Since Hernández did not mention Miguel Servet or Juan Valverde, his knowledge of the pulmonary system must have reached him through Colombo's 1559 textbook, which described this physiological discovery in detail. Hernández's description of the passage of the blood (from the right ventricle of the heart to the left through the lungs and the venous artery) was

---

[18]　'A la qual nos dimos con mayor cuidado y voluntad por entender quanta necessidad tuviese della el consumado y verdadero medico y el bien instruido cirujano.' Ibid., p. 99.

[19]　'No cuento aqui los ossequelos de los oydo..., que yo no sin grande deleite, en las anatomias o disecciones que hice estando en Guadalupe y con los que a ellas assistían, consyderava.' Ibid., p. 120.

[20]　'Su figura (según lo vi en Guadalupe en una preñada que anatomizamos) es de un redondo prolongado, harto diferente de las de las vacas, cabras y ovejas, contra el parecer de Galeno.' Ibid., p. 143.

[21]　'Realdo Colombo: varón muy excellente en anatomía, y que no tuvo tanta cuenta con seguir las pisadas de Galeno o de Vesalio, aunque también fue en esta cosa diligencia admirable, como en inquirir y averiguar la verdad.' Ibid., p. 130.

short and simplified, but nonetheless showed his immediate understanding of 'the smaller circulation'.[22] Other references to different anatomical observations indicated that a significant number of dissections had been carried out by the author and his colleagues. Hernández's descriptions emphasised an awareness of variations in anatomy from case to case, as seen in his account of a male body with unusually long and numerous ribs: 'In a man that we dissected in Guadalupe I saw how his whole stomach and also the liver and spleen were covered by his ribs, and there were much longer and more numerous than normal ones.'[23] A short reference to dissections of brains, 'which I have seen and dissected many times in both humans and animals', described the cerebral membranes, thus indicating Hernández's interest in neurological studies carried out through dissections and vivisection in Guadalupe and later in Toledo.[24] His *Historia natural* also referred to his experiences beyond Guadalupe and Toledo, in mentions of a female patient at the *Hospital del Cardenal* in Seville, whose cerebral membrane was visible to the naked eye, as observed during his practice at this hospital.[25] Hernández's anatomical experience in the late 1550s also included a series of documented animal dissections, as recorded in his account of the inner structure of a chameleon dissected in Guadalupe: 'While I was in Guadalupe as a physician at that monastery and hospital I remember having seen

---

[22] 'Y despues de ser alli (según que dije) preparada y adelgazada, vuelve juntamente con el aire por la arteria venal al ventrículo izquierdo ... Largo sería contra sus dos orejas, el discurso de las venas cava y arterial, que salen de su concavidad derecha y de las arterias venal y aorta, que salen de la izquierda. Item las once membranas que se llaman tricuspids, y la arteria coronal que sale de la aorta y sus usos.' Ibid., p. 114.

[23] 'Yo vi un hombre que abrimos en Guadalupe tener cubierto todo el estómago y tambien el hígado y bazo con las costillas, las quales eran las más largas y en mayor número que las ordinarias.' Ibid., p. 99. His subsequent description may well have resulted from a consultation of a textbook on osteology; it outlined the usual structure of the human thorax and the normal number and character of the ribs: 'catorce verdaderas, diez mendosas, pero a veces se hallan a cada lado once o trece, y es más frequente lo segundo'.

[24] 'Tiene el cerebro dos telas que le amparan, defienden y cubren allende de otros usos, la una llamada dura, aunque bien mirado son dos, y ésta apartada del cerebro para que se dé lugar a su pulsar y latir, porque lo hace como el pulso, según que en mi juventud, y casos de necisidad, que ejercité el arte de quirúrgica, o siendo acompañado de cirujanos lo observe no pocas veces.' A short reference to the eyesight, 'Se forma de los nervios ópticos por donde viene la vista del cerebro a los ojos o espíritus visivos', once again highlighted his studies of nerves and brains, which he claimed to have studied several times in both humans and animals. Ibid., pp. 64–66.

[25] Anatomy studies at the Hospital del Cardenal were not formally carried out until the following century and were allegedly initiated by the Pisan physician Tiberio Damián. Joaquin Herrera Dávila, *El hospital del Cardenal de Sevilla y el Doctor Hidalgo de Agüero* (Seville: Edición de la fundación de Cultura Andaluza, 2010), p. 222.

a chameleon owned by one of the friars and described by Pliny. After its death we dissected it, and I saw within its belly a long string of eggs.'[26]

Hernández's varied training in dissection advanced following his transfer to the *Hospital de Santa Cruz* in Toledo, where he allegedly performed vivisections on animals. One was carried out on a dog, which was deprived of its bark during the exercise: 'Some years ago the excellent Toledan architect, painter, and sculptor Nicolas de Vergara and I experimented with the vivisection of the recurrent nerves of a dog, which was deprived of its bark and voice. When you cut the nine pairs of nerves which exit from the brain, (the encephalon), and from which the recurrent nerves derive, it removes the ability to speak, as shown in the experiments by me and Nicolas de Vergara.'[27] The cooperation with a local artist in these experiments is an important supplement to other contemporary descriptions of collaboration between anatomists and artists in the late Spanish Renaissance. The vivisection of animals at the *Hospital de Santa Cruz* is furthermore evidence of similar practice in Toledo to that carried out in the contemporary university and hospital contexts of Valencia, Alcalá, and Zaragoza. Hernández's experiment was an almost complete imitation of Vesalius' vivisection of a dog recorded by Baldassar Heseler in Bologna in 1540.[28] These experiments were probably inspired by Galen's famous vivisections of porcine spinal columns carried out for a Roman audience in the second century AD, during which the pigs' limbs and ability to squeal were systematically paralysed.

After Philip II nominated Francisco Hernández to *Protomédico de nuestras Indias, islas y tierra firme del mar Océano* in 1570, the anatomist made use of his training in the New World. Hernández worked at the *Hospital Real de Indios* in Mexico City, where he supervised the surgeon Alonzo Lopez de Hinojosos during autopsies of victims of the epidemic disease *cocolitze* which engulfed the Mexican capital in 1576. Hernández described this exercise with anatomical pathology in New Spain the next year in a short relation entitled *De morbo*

---

[26]   'Acuérdome haver visto un chamaleon en Guadalupe, siendo médico de aquel monasterio y hospital, en poder de un fraile, qual Plino lo pinta; y después de muerto, le anatomizamos yo y algunos amigos medicos que estavan allí asistiendo a la práctica de la medicina, chirurgía y dissenctión, y entre otras cosas miramos, no sin grande meravilla, una sarta de huevos que tenía tan larga.' P. Barreiro, 'El testamento del Dr. Fransisco Hernández', *Boletín de la Real Academia de la Historia*, XCIV (1929), p. 478.

[27]   'Los años pasados por causa de experiencia cortamos yo y Nicolas de Vergara, arquitecto, pintor y escultor toledano excelente, a un perro los nervios reversivos y ansí le privamos totalmente del ladrido y la voz (...) Item los nueve pares de nervios que de él salen (del encéfalo), de cuyo nono par se deriven los reversivos, los quales cortados se quita la voz, según que Nicolas de Vergara y yo lo experimentamos, varón en escultura y pintura excelente...' Somolinos d'Ardois, *Vida y obra de Francisco Hernández*, p. 121.

[28]   Baldassar Heseler, *Andreas Vesalius' First Public Anatomy at Bologna, 1540*, translation and commentary by Ruben Erikson, p. 290

Novae Hispaniae anni 1576 vocato ab indis cocolitztli.[29] The epidemic was dealt with in more detail in the first Mexican textbook on anatomy and surgery, Lopez de Hinojosos' *Summa y recopilación de cirurgía* (Mexico City, 1578). Both texts will be discussed in the next chapter.

Hernández's training in both anatomy and botany in Guadalupe was to become routine in contemporary university contexts, where the two branches of medical learning became increasingly interrelated during the sixteenth century. Hernández's colleague in Guadalupe, Francesc Micó, who went on to become professor of *anatomía y simples y coses de apothecaris y cirurgia* at Barcelona, was an obvious example of the burgeoning relationship between anatomy and medical botany. This affiliation was seen at a number of Iberian universities, particularly at Valencia, Barcelona and Zaragoza, where the two disciplines were eventually united in joint chairs of *anatomia y simples, anatomia y materia medica* or *anatomia y hierbas.* Talavera's appraisals of the medical garden and the developed facilities for the manufacture of medicines in Guadalupe was an indication of the essential nature of medical botany: 'I don't think there is any similar facility in all of Spain.'[30] In Guadalupe this pharmacological expertise coincided with the practice of human dissection, and Francesc Micó and Fransisco Hernández were occupied in both fields during their employment at the monastery hospital. Besides his anatomical training in Guadalupe, Hernández made a number of botanical excursions during this period and his commentary on Pliny was full of references to botanical findings in the vicinity of the monastery: 'I have seen the Greek oregano described by Dioscorides in some parts of Spain, and especially in the mountains close to the town of Guadalupe.'[31] Hernández's anatomical experience was certainly surpassed by his later natural studies, and his reputation as the 'Pliny of the New World' was mainly due to his treatises on the botany and zoology of New Spain. Only parts of this richly illustrated and multi-volumed work were published, and only incomplete copies made, before the original manuscripts were lost in the 1671 fire of the Escorial library.[32]

Anatomical training was arguably only a minor part of Hernández and Micó's broader activities as naturalists, and indeed both were later credited for their contributions to botanical knowledge rather than for any important studies of anatomy. Similar developments were seen in the careers of Andrés Laguna and Juan Fragoso, who gradually replaced their anatomical activities

---

[29] Unpublished manuscript dated 1577. Published in a Castilian translation in Germán Somolinos d'Ardois, *Fransisco Hernández: Obras completas.*

[30] 'Que no creo tiene semejante officina toda España.' De Talavera, *Historia de Nuestra Señora de Guadalupe,* f. 197v.

[31] 'El oregano heracleótico de Dioscórides, el qual he visto en algunas partes de Hespaña y principalmente en las montes de cercanos al pueblo de Guadalupe.' Somolinos d'Ardois, *Vida y obra de Francisco Hernández,* p. 127.

[32] Pardo Tomás and López Piñero, *La influencia de Francisco Hernández,* p. 45.

and publications with studies on botanical medicine. We might also consider Andreas Vesalius here, whose *Fabrica* was followed by a botanical treatise on the healing properties of the China root. This differing medical focus probably represented an advantage in the attempts to obtain a position at the court of Philip II, where neither Vesalius nor Hernández seem to have been engaged in anatomical activities after their respective appointments in 1559 and 1567. In his *Historia natural* from that period, Hernández nonetheless indicated his involvement in a large written work of commentaries on Galenic anatomy: 'It would be outside the scope of our work here to repeat what we are writing with more length, elegance, and clarity in our medical studies and commentaries on the anatomical books of Galen.'[33] A consultation of this unfortunately lost work would probably clarify the many uncertainties that still surround the studies of anatomy at the monastery hospital of Guadalupe.

Anatomy in Guadalupe During the Late Sixteenth Century

Even though Hernández's descriptions of anatomical studies within the *Monasterio de Nuestra Señora de Guadalupe* were not consolidated by similarly detailed sources from the late sixteenth century, Gabriel de Talavera's 1597 chronicle emphasised a thriving practice within this field in a reference to 'the anatomies that the surgeons carry out here'.[34] The anatomical training provided for surgeons at Guadalupe is comparable to the contemporary university programmes known from Salamanca and Alcalá de Henares, but did not produce similarly detailed statutes regarding procedure and frequency. While dissections were often carried out on recently executed criminals provided by the local authorities in Salamanca and Alcalá de Henares, the anatomists at Guadalupe tended to acquire their subjects from among recently deceased hospital patients, although the evidence to support this is very weak. An unexpected allusion to the anatomical activities within this particular hospital context was made by the contemporary author Eugenio de Salazar in his *Silva de poesía*, a compilation of satirical poems and prose from the early 1580s. This work included a reference to the author's encounter with a group of sailors who allegedly chewed the meat off their bones with great skill and diligence, as if they had been trained in anatomical dissection: 'They take the poor bones in their hands, and rip and

33   'Sería fuera de nuestro intento hacernos ya repetir lo que largamente y con distinción y claridad scribimos en nuestra medicina y comentarios sobre los libros anatómicos de Galeno.' Somolinos d'Ardois, *Vida y obra de Fransisco Hernández*, p. 129.

34   De Talavera, *Historia de nuestra Senora de Guadalupe*, f. 221v.

rend off their nerves and cords, as if all their lives they had practised anatomy at Guadalupe or Valencia.'[35]

It is remarkable that Salazar, a poet and legal administrator of Guatemala and Mexico who was educated in law at the Universities of Salamanca and Alcalá de Henares, did not refer instead to these two institutions as well-known Castilian centres of anatomical training. His attention to the studies carried out within the monastery hospital of Guadalupe is an indication of the contemporary reputation of this training ground for anatomists, acknowledged even by laymen with no medical background or expertise. Salazar's allusion to the anatomists of Guadalupe and Valencia is furthermore a rare literary supplement to the history of Spanish Renaissance anatomy, which has most often been passed down in medical textbooks and administrative sources. The author's educational background from the Universities of Alcalá de Henares and Salamanca was equivalent to the education of Fransisco Hernández and Francesc Micó at these two prominent institutions. In all likelihood, the influx of these and other recently educated physicians from the large Castilian universities into the hospitals of Guadalupe was less ad hoc and informal than the surviving evidence indicates in the vague references to physicians who went there to gain experience. The anatomical training for young medical graduates from Salamanca and Alcalá de Henares could instead be interpreted as a fairly formal collaboration between the monastery hospital and the most prominent medical faculties of Castile. Evidence of such agreements between universities and hospitals can be found repeatedly in an Iberian context, where university statutes often stipulated that anatomical studies were to be carried out within nearby hospital precincts. This was certainly the case in Valencia, Barcelona, Zaragoza and Alcalá de Henares.

The human dissections documented at Guadalupe may even have had parallels elsewhere, as indicated in Hernández's short reference to a vivisection carried out within the *Hospital de Santa Cruz* in Toledo, and a passing remark on his anatomical observations at the *Hospital del Cardenal* in Seville. Bartolomé Hidalgo de Agüero's short anatomical treatise, *Anatome del cuerpo humano*, which was included in his *Thesoro de la verdadera Cirugia y via particular contra la comun* (Seville, 1604), suggested a familiarity with anatomical practice at this hospital, where Hidalgo was employed during the late sixteenth century. Unfortunately, this textbook on surgery provided no accounts of the author's personal anatomical experience. Only Hernández's account from Guadalupe provided a detailed testimony of the practice of Spanish Renaissance anatomy beyond the universities. The unclear prior references to a medical school for physicians and surgeons in Guadalupe during the fifteenth and sixteenth

---

[35] 'Cogen entre manos los pobres huesos, y así los van desforneciendo de sus nervios y cuerdas, como si toda su vida hubiesen andado a la práctica de la anatomía en Guadalupe o Valencia.' Cited in Beaujouan, *La medicina y la cirugía en el Monasterio de Guadalupe*, p. 168.

centuries can be supplemented with more substantial evidence from the seventeenth century. Contemporary textbooks referred to 'the chair of surgery of the royal hospital' and also included mentions of a formalised professorship of surgery, as presented by Pedro Gago de Vadillo in his *Discurso de la verdadera cirugia* (Madrid, 1632): 'having carefully and for a long time (more than 40 years) practised surgery at the Hospital of Guadalupe in Spain under my master doctor Agustin Ollés, and thereafter in Peru for 16 years, in the city and hospital of Huamanga. Thereafter nine years as physician and surgeon at the mining hospital of Castrovirreyna following a provision from the Viceroy, and then three years in this famous city of Lima.'[36]

Vadillo's 40 years of surgical experience locates his training in Guadalupe under 'maestro Agustin Ollés' in the last decade of the sixteenth and the first years of the seventeenth century before he transferred to Peru. This period coincides with the references made by Talavera in 1597 to anatomical training for surgeons at the Guadalupe Hospital. Yet Vadillo's treatise offered no further details of his early apprenticeship in Guadalupe and his few, fairly esoteric references to anatomy did not indicate any personal practice in this field.[37] A more comprehensive treatise on anatomy, *De Anotomia de la parte natura*, appeared a few decades later in Diego Antonio de Robledo's *Compendio chirúrgico útil y provechoso a sus profesores escrito por el doctor D. Diego Antonio de Robledo (y Méndez), medico principal de la Real Casa de Nuestra Señora de Guadalupe y regente de la cathedra de cirugía de sus reales hospitales* (Madrid, 1686). In this work, the self-professed professor of surgery at the monastery hospital dealt with several questions relating to human anatomy and also outlined the ideal anatomical theatre, as described in the previous chapter. Unfortunately, Robledo failed to reveal whether such model facilities for dissections existed in Guadalupe at the time of his *Compendio*.[38] Another question – 'What kind of

---

[36] 'Aviendo por largo tiempo (mas de quarenta años) considerado atentamente con la experiencia que tengo curando cirugía desde el Hospital de Guadalupe en España, en tiempo del doctor Agustin Ollés, mi maestro, y despues en este Reyno del Piru diez y seis años en la Ciudad de Guamanga, y hospital della; y despues nueve en la de Castro-Virreyna, y Hospitales de sus minas, como Medico, y Cirujano, por provisiones de los señores Virreyes; y tres años en este insigne Ciudad de Lima.' Pedro Gago de Vadillo, *Discurso de la verdadera cirugia* (Madrid: Juan Micòl, 1632), *Al lector*, n.p.

[37] 'Los anatómicos Christianos, que haziendo anatomia, y tratando de la disección del cuerpo humano, viendo y tocando con sus manos esta admirable compustura, y miembros, vienen a confessar, y dezir todos, que entre los grandes provechos que se siguen de la anatomia. El primero es alabar a Dios nuestro Señor de ver la compostura, y fabrica del hombre, y tanta diversidad de miembros, y tantas facultades, y virtudes, que les dio.' Ibid., p. 143.

[38] Diego Antonio de Robledo, *Compendio chirúrgico útil y provechoso a sus profesores escrito por el doctor D. Diego Antonio de Robledo (y Méndez), medico principal de la Real Casa de Nuestra Señora de Guadalupe y regente de la cathedra de cirugía de sus reales hospitales*, p. 30.

bodies should be dissected?' – referred to the studies of comparative anatomy and vivisections of animals, as presented in Hernández's sixteenth-century accounts from Guadalupe: 'The dissection of dead bodies is always carried out on humans, and vivisections on pigs or dogs because of the resemblance which both have with human bodies.'[39] Before moving on to a summary of different anatomical categories, the author placed himself firmly within a theoretical framework of humoral physiology, which seemed unfamiliar with the progressive contemporary works on physiology presented elsewhere in Spain in the wake of Harvey's theory of the larger circulation: 'Even though most of the authors who have written about anatomy have begun their accounts with the regions where their dissection of the body commences, I have nonetheless varied the order, because it seemed to me more obvious to start with the anatomy of the mouth, which is the first part, which is nourished by the air, and through which the heart is refrigerated.'[40] As seen in this repository of dated Aristotelian-Galenic physiology, the attempts made by Hernández and his colleagues during the mid-1500s to bring anatomy at the hospitals of Guadalupe 'up to date' was obviously not imitated with similar success in the late seventeenth century.

---

[39] 'La anatomia del cuerpo muerto se haze en el humano cada vez; y la del vivo en un lechon, o perro, que entrambos con a proposito para la similitude que tiene con el hombre.' Ibid.

[40] 'Aunque los mas autores que han escrito de anotomía han comencado por donde se deve dar principio a la obra quando se exerce en un cadaver; no obstante, variando el orden me ha parecido dar principio a la anotomia por la boca, por ser la primera puerte por donde se allimenta el vientre, y por donde se refrigera el Corazon.' Ibid.

Chapter 9
Beyond Iberia

Anatomy and Anatomists in Sixteenth-Century 'New Spain'

After the Spanish conquest of Mexico in 1521, a remarkably swift institutionalisation and dispersal of European medicine followed the arrivals of the first physicians and printing presses in *Nueva España*. Yet the earliest known publications on New World medicine were produced in Spain, and were not based on direct personal experience, but on recently compiled material and secondary information from the Americas. The first printed work by a Spanish physician with direct experience from the newly conquered territories was Cristóbal Méndez's *Libro del exercicio corporal y sus provechos* (1553). This comprised a small treatise on the importance of exercise and hygiene with only scattered anecdotes of Mendez's own medical training in Guatemala and Mexico between 1534 and 1545. This 1553 publication coincided with the foundation of the University of Mexico, which was established even before the similarly developed Universities of Barcelona and Zaragoza. While medicine was taught at this institution from the very beginning, a formal medical chair was not created until 1578. This first university chair of medicine in the New World was held by Juan de la Fuente, a medical laureate from the University of Alcalá de Henares, who retained the new professorship until his death in 1595. An inventory of the medical books brought to Mexico by de la Fuente in 1561 has survived and includes Vesalius' *Fabrica*. This first appearance of Vesalius in a New World context is another indication of the spread of Vesalian anatomy throughout the Spanish kingdoms, independent of the anatomist's own contemporary residence in Spain.[1]

Explicit references to Vesalius appeared soon after in the second known work by a Spanish physician with medical experience from the Americas, Pedro Arias de Benavides' *Secretos de Chirurgia, especial de las enfermedades de Morbo galico y Lamparones Mirrarchia, y assi mismo la manera como se curan los Indios de llagas y heridas y otras pasiones en las Indias, muy util y provechoso para en España y otros muchos secretos de chirurgia hasta agora no escriptos* (Valladolid, 1567). This first systematic therapeutic and surgical treatise from the Americas was primarily based on Benavides' personal experience in Mexico City at the *Hospital del Amor de*

---

[1]   Enrique Gonzalez Gonzalez, 'La enseñanza médica en la Ciudad de Mexico durante el siglo XVI' in *El mestizaje cultural y la medicina novohispana del siglo XVI* (Valencia: Instituto de Estudios Documentales e Históricos sobre la Ciencia, 1995), p. 136.

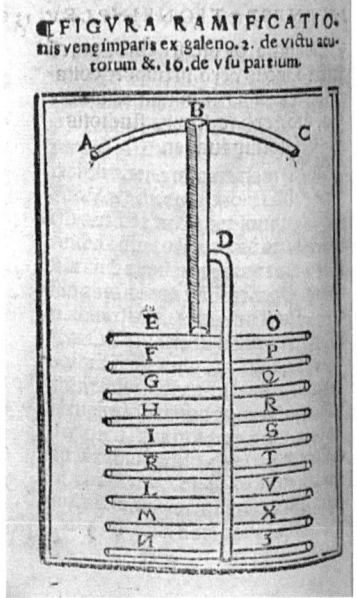

❰FIGVRA RAMIFICATIO-
nis venę imparis ex galeno. 2. de victu acu-
torum &. 10. de v fu partium.

Figure 9.1   The first book on medicine to be printed in the Americas – Francisco Bravo's *Opera Medicinalia* (Mexico City, 1570) – included a brief appraisal of Andreas Vesalius and the earliest known anatomical illustration from a New World textbook; a crude and schematic representation of Jan Stephan van Calcar's drawn image of the veins of the thorax, which he had made for Vesalius' 1539 *Venesection Letter.* The representation in Bravo's book depicted the veins wrongly, in mirrored view. Courtesy of 'Biblioteca Histórica José María Lafragua de la BUAP'

Dios, an institution more commonly known as *Hospital de las bubas*. As indicated in the title of his textbook, the author's experience with syphilitic patients and those suffering from *lamparones* or scrofula/struma comprised the larger part of the narrative. In a short chapter on eye disease or *oftalmía*, the author praised 'El Vessalio famoso cirurgiano y anathomista', but disregarded his proposed surgical procedure for this category of illnesses: 'The cure proposed by Vesalius is very laborious ... it would take too long and be too arduous for the patient.'[2] Only three years after Benavides' *Secretos de Chirurgia*, the Andalusian physician Francisco Bravo published the first textbook on medicine to be printed in the Americas, a work entitled *Opera medicinalia* (Mexico City, 1570), which also paid a brief tribute to the recently deceased anatomist: 'Vesalius, a man unanimously skilled and well trained in anatomical dissections and investigations.'[3] Bravo's work also included the earliest known anatomical illustration from a New World textbook: a crude and schematic representation of Jan Stephan van Calcar's engraving of the veins of the thorax, which had been made as the only diagram for Vesalius' 1539 Venesection Letter.[4]

---

2   'Muy trabajosa cura la del Vessalio ... Tardase en esta obra mucho, y paraceme a mi que es mas trabajosa para el enfermo.' Pedro Arias de Benavides, *Secretos de Chirurgia, especial de las enfermedades de Morbo galico y Lamparones Mirrarchia, y assi mismo la manera como se curan los Indios de llagas y heridas y otras pasiones en las Indias, muy util y provechoso para en Espana y otros muchos secretos de chirurgia hasta agora no escriptos* (Valladolid: Francisco Fernandez de Córdoba, 1567), f. 138r and 139v.

3   'Vesalio viro in anatomicis indagantis sectionibus omnium consensu peritissimo ac exercitatissimo.' Francisco Bravo, *Opera medicinalia* (Mexico City: Pedro Ocharte, 1570), f. 135r.

4   Andreae Vesalius, *Epistola, docens venam auxiliarem dextri cubiti in dolore laterali secandam & melancholicum succum ex venae portae ramis ad sedem pertinentibus, purgari* (Basle: Robert Winter, 1539), f. 139v.

The *Cocoliztli* Epidemic of 1576: Francisco Hernández's and Alonso López de Hinojosos' Accounts of the Autopsies Carried out at the *Hospital Real de Indios*

These isolated acknowledgements of Vesalian anatomy in a New World setting were not indicative of subsequent medical practices of late sixteenth-century Mexico. While Galenic doctrines were refuted in several Iberian centres of medicine during the late sixteenth century, contemporary New World physicians actively consolidated this waning tradition. A corresponding reliance on traditional medicine was seen in the first textbooks on anatomy and surgery written and published in the Mexican capital, namely Alonso Lopez de Hinojosos' *Summa y recopilacion de chirurgia* (1578) and Augustín Farfán's *Tractado breve de anothomia y chirugia, y de algunas enfermedades que mas comunmente suelen hauer en esta Nueva España* (1579). These remarkably early New World publications were almost entirely based on classical and medieval authorities, and were practically devoid of references to current contributors to the fields of anatomy and surgery. Despite their lack of proposed improvements and corrections to existing anatomical knowledge, these works are nonetheless noticeable as the first American textbooks on anatomy and as attempts to amalgamate well-known Galenic doctrines with previously unknown diseases and remedies of the New World.

An epidemic of *huey cocoliztli* or the 'Great Pestilence' laid large parts of *Nueva España* to waste after its first documented appearance in the summer of 1576. *Cocoliztli* was the native Nahuatl term for this pandemic disease, which has since been subject to hesitant diagnoses of both typhus and smallpox, and the Nahuatl classification has even been explained as a common name for a series of endemic diseases which swept through Mexico and Central America during the late sixteenth century. The introduction of Hinojosos' treatise referred to a similarly widespread uncertainty within learned circles regarding the nature of the illness following its first outbreak in 1576: 'At the end of August of 1576 we heard the first news in Mexico City of a terrible disease, from which many native Indians had died ... The astrologers said that it was caused by the conjunction of certain stars. The physicians said it was the plague.'[5] Hernández's description of the epidemic placed the outbreak two months earlier and emphasised that it was still raging by the time of his brief report. It had by then begun to attack the Spanish colonists, having already dispensed with large numbers of the native

---

5   'En fin del mes de agosto de mil quinientos setenta y seis años se comenzó a sentir en esta ciudad de México una muy terrible enfermedad de la cual morían muchos de los indios naturales ... Los astrólogos dijeron que la causa era conjunción de ciertas estrellas. Los médicos decían que era pestilencia.' Alonso Lopez de Hinojosos, *Summa y recopilacion de chirurgia* (Mexico City: Antonio Ricardo, 1578), p. 207.

and black populations: 'This plague began in the month of June 1576, and was still not over in January, when we wrote this description. From New Spain it invaded all the cold regions in a circle of about 400 miles, and was somewhat easier on the warmer regions (that is, it attacked rather less) infecting different areas in turn, beginning with those occupied by Indian tribes, then places where Indians and Africans lived, then those with a mixed population of Indian and Spanish, and later still, those areas occupied by Africans, and now finally it is attacking the Spanish.'[6] Tentative estimates assess that the 'Great Pestilence' may have reduced the population of Mexico and Central America by more than half, with a devastating impact on the native Indian and black populations of *Nueva España*. According to Francisco Hernández's contemporary account, the European colonisers also suffered, but were allegedly far less affected than their Indian and African subjects.

This racial imbalance caused instant demographic changes within the colony, which were soon anxiously noticed by Hinojosos in his printed account from 1578: 'During that period many Blacks and Chichimeca Indians died, which left New Spain and the mines almost without any workers.'[7] Systematic autopsies of victims of the 'Great Pestilence' were documented in a final section of Alonso López de Hinojosos' *Summa y recopilacion de chirurgia* and in a two-folio letter by Francisco Hernández, entitled *De morbo Novae Hispaniae anni 1576 vocato ab indis cocolitztli* and dated January 1577, thus preceding the publication by Hinojosos. Hernández produced this brief account shortly before he returned to Spain the following month, and six years after he had arrived in Veracruz in February 1571. He spent his first four years in New Spain on scientific excursions throughout the colony and the last two in the Mexican capital, where he compiled the vast material gathered from his expeditions and from his occasional medical practice at the local *Hospital Real de Indios*.[8]

Hinojosos' account of the *cocolitztli* epidemic was included in his textbook for practising surgeons of Mexico and Central America, and documented his experience in human dissection, which he had carried out during his employment as surgeon both within and beyond the *Hospital Real de Indios* of Mexico City: 'And this I have witnessed many times in the anatomical dissections, which

---

6   Unpublished manuscript dated January 1577. Published in a Castilian translation in Germán d'Ardois Somolinos, *Fransisco Hernández: Obras completas*. The version used for this chapter is an English translation in Simon Varey (ed), *The Mexican Treasury: The Writings of Francisco Hernández* (Stanford, Stanford University Press, 2000), p. 84.

7   'En el propio tiempo murieron muchos negros y indios chichimecos, que quedó México y las minas y toda la Nueva España, casi sin servicio.' Hinojosos, *Summa y recopilacion de chirurgia*, p. 210.

8   José Pardo Tomás and José Maria López Piñero, *La influencia de Fransisco Hernández (1515–87) en la constitución de la botánica y la material médica modernas*. (Valencia: University of Valencia, 1996), p. 44.

I have carried out with my own hands in the Hospital of the indies.'[9] The most original contribution of his work was found in the final and seventh *Tratado*, 'De pestilencia', which was devoted to his experience of the horrific epidemic, which had no European precedent. Hernández's and Hinojosos' accounts represented the first written and published records of systematic studies of anatomical pathology in the Western hemisphere – although similar practices appear to have taken place elsewhere in the Americas in a few poorly documented cases.[10] Hernández's manuscript from 1577 and Hinojosos' printed account from the following year both presented human dissection as a formal exercise organised within the *Hospital Real de Indios*, allegedly instigated by an official petition from the Viceroy of *Nueva España*, Don Martin Enriquez.[11] Similar to Hernández's account from Guadalupe in the company of Francesc Micó almost 20 years earlier, Hinojosos' eyewitness report from the *Hospital Real de Indios* indicated that pathological studies were carried out with Hernández in a somewhat secondary role. Even though Hernández's account did not refer to Hinojosos, whose later textbook

---

[9]   'Y esto lo e visto muchas veces en Anatomías que he hecho por mis propias manos en el Hospital de los indios desta ciudad.' Hinojosos, *Summa y recopilacion de chirurgia*, p. 99.

[10]   Fernando Bouza has written about the autopsy of a deformed girl in Lima, Peru during the late seventeenth century and also referred to a preceding case in 1533, where an autopsy of a newborn child was carried out by a Cuban surgeon. Bouza admitted the exceptionality of this second incident: 'La primera autopsia realizada a un neonato en América data de 1533 y fue ejecutada en la ciudad de Santo Domingo por el cirujano Juan Camacho, es sabido que sobre las autopsies a cadaveres pesaba una serie de prejuicios que hacía que no fueran una práctica habitual en el Siglo de Oro.' Fernando Bouza, *Enanos, bufones, monstrous, brujos y hechiceros* (Barcelona: DeBolsillo, 2005), p. 34. The autopsy of the deformed child from Lima was recorded in a contemporary account by the dissector and physician Francisco Bermejo in *Desvíos de la naturaleza o tratado de origin de los monstruos* (Lima: José de Contreras y Alvarado, 1695). Pedro Gago de Vadillo's *Discurso de la verdadera cirugia* (Madrid, 1632) was based on a previous experience with anatomy and surgery in the Peruvian capital. Federico Bottoni's defence of Harvey's theory, *Evidencia de la circulación de la sangre* (Lima, 1727), is further testimony of the 'Ciudad de los Reyes' as an American centre for anatomical studies during the seventeenth and eighteenth centuries, but without any previously known tradition that coincides with the late sixteenth-century publications by Hinjosos and Farfán. The first South American professorship of anatomy was established in Lima in 1729 and was imitated much later in the residual colonies of Spanish America, Chile (1773), Cuba (1797), Buenos Aires (1801), Guatemala (1809) and Caracas (1811), as listed in T.V.N Persaud's *A History of Anatomy, The Post-Vesalian Era* (Springfield: Charles C. Thomas Ltd, 1997), pp. 287–88.

[11]   'Sabido por el muy excellente señor Vissorey, que los remedios de tan famosos medicos, y sus pareceres no aprovechan, mandó que se hiciesen anatomias. Y por ser el Hospital Real el más acomodado, y adonde hay mayor refrigerio que en toda la nueva España por favorecerlo tan ampliamente como siempre lo favorece se Excelencia por respeto de ser este bien para los naturales, y aver en el dicho Hospital, en el dicho tiempo, más de doscientos enfermos de ordinario.' Hinojosos, *Summa y recopilacion de chirurgia*, p. 209.

instead emphasised that Hinojosos himself had carried out the autopsies single-
handedly and that Hernández had only been present at the hospital in his function
as a superintendent and envoy of the viceroy: 'And so the anatomies were carried
out, and I with my own hands performed these in the company of Doctor Francisco
Hernández Protomedico to his majesty, who was conducting experiments with
medical plants, purgatives, and other remedies from New Spain.'[12]

Hernández and Hinojosos both referred in their reports to the green blood
of the infected patients obtained by venesection and the exceptionally swollen
livers of recently deceased victims as observed through autopsies. Both authors
also noticed the infected matter found in the dissected inner organs of the
deceased. Hernández's two-folio account comprised a short and matter-of-fact
description of the pathological changes to central bodily organs: 'Autopsies
showed that the dead had very swollen liver, blackened heart emitting a pale
yellow liquid, and later black blood, black and semiputrefied spleen and lungs;
atrabilousness could be seen in their blood vessels, the dry stomach and the rest
of the body, wherever it was dissected was extremely pale.'[13] These observations
were based on autopsies performed by Alonso López de Hinojosos, whose
Summa y recopilacion de chirurgia gave a much more detailed account of the
surgeon's work overseen by Hernández, Philip II`s renowned *Protomédico de
nuestras Indias, islas y tierra firme del mar Océano*: 'And even though we tried
to cure with theriac and Spanish lime, it did not help. Because the disease was
immense and very dangerous, and the sick had a bad and very hard liver, which
looked so deformed that it appeared to be the liver of an ox, and it raised the ribs
upwards, and deformed the chest.'[14]

After the laborious efforts of Hinojosos, who was at that time 40 years old and
without a university medical degree, the diagnosis emerging from these autopsies
was offered by the learned Hernández, who was 20 years the surgeon's senior. The
verdict of the nature of the disease was rooted in Hernández's lengthy medical
expertise and recently obtained status as *Protomédico de nuestras Indias* – and
it was later reported to the viceroy that: 'He said that it was a poison, for which
they would have to make and use an antidote against poison. And the Viceroy
sent the said protophysician, so that he could assist in the cure of the patients of

---

[12] 'Y assi se hicieron las Anatomias y yo propio, por mis manos las hize, estando presente
el doctor Francisco Hernández, Prothomédico de su Magestad, que al presente estava haciendo
experiencia de las yervas medicinales, purgativas y otras cosas de esta Nueva España, las quales
hacia anatomias que se hicieron, dio noticia de ello a su excelencia.' Ibid., p. 209.

[13] Varey, *The Mexican Treasury*, p. 84.

[14] 'Y aun se aplicava el atriaca y quanepile, no aprovechava; por ser la enfermedad
grande y muy peligrosa, porque tenian los enfermos el hígado acirrado y muy dura, que se les
paraba tan dissorme, que parecia hígado de toro, y alzaba las costillas hacia arriba y hacia el
pecho muy deforme.' Hinojosos, *Summa y recopilacion de chirurgia*, pp. 209–10

the hospital.'[15] While Juan Tomas Porcell's previous autopsies of plague victims in Zaragoza had advocated improved diet and fresh air to combat future epidemics, Hernández instead proposed curing the Mexican plague by offering an antidote to its infected victims. Both cases presented a discrepancy between pioneering pathological observations carried out through systematic autopsies and traditional cures of vegetal medications, ointments and purgatives with supposedly very limited therapeutic effects: 'The patient would drink barley broth, the pith of a cultivated celeriac, the root of coanenepilli, and fennel seeds, and from time to time cocotlacotl, chipaoac and atochietl would be used as well, which we have described in more detail in our history of the plants of New Spain, in order to open all the orifices through which the poison, which was also evacuated by the urine could be expelled. Those unnatural swellings, even the immature ones, behind the ears could be stopped with an application of red-hot iron, and the pus that flowed from the ears was cleaned up with cotton and rose honey.'[16]

The amalgamation of Old and New World *materia medica* was nonetheless a noticeable indication of a swift incorporation of native medical practices and remedies. Hernández's own presence in the New World was a result of the often far-reaching politics of the proto-absolutist state of Philip II. The transfer of renowned naturalists to distant corners of the Spanish Empire served the combined purpose of improving the health in new colonies and strengthening the imperial centre through a monopoly of recently discovered medical practices and products. Yet the autopsies performed in a New World context may not have met the expectations of the King's representative in New Spain, Viceroy Don Martin Enriquez, who was accredited as the instigator of this practice at the *Hospital Real de Indios* by Hinojosos. Beyond the interesting historical accounts of the observations of dissected victims of *cocolitztli*, not much applicable knowledge was obtained from this empirical scrutiny; it produced no immediate improvements of contemporary cures and only a vague comprehension of the exact nature of the epidemic. Hernández's account of the *cocolitztli* was never published and remained unknown until it was rediscovered in the 1780s. It may have been only a preparatory sketch for a more systemic and now-lost work, as has been suggested by the systematic compiler of Hernández's work, the Spanish-Mexican historian German Somolinos D'Ardois. The printed account from 1578 appears to have been widely distributed among contemporary Mexican surgeons and physicians. The first impression of Hinojosos' textbook sold out and was reprinted in a second edition (1595), which differed slightly from the first edition in that it was missing the original *cocolitztli* account.

---

[15] 'El dixo que era veneno. Para lo qual convenia que se truxesen, y ussasen de cosas contra veneno. Y assi lo mando su Excelencia al dicho Prothomédico, assistiesen y curase los enfermos en el dicho Hospital.' Ibid., p. 209.

[16] Cited in Varey, *The Mexican Treasury*, p. 85.

Hernández's and Hinojosos' descriptions of the 'Great Pestilence' were ignored altogether in a subsequent chronicle of the epidemic, which was presented in Fray Augustín Dávila Padilla's *Historia de la fundación y discurso de la Provincia de Santiago de Mexico* (Mexico City, 1596). In this later – and arguably much less reliable – account, the author credited the medical professor Juan de la Fuente as being the true instigator of autopsy in 1576 and described how de la Fuente allegedly 'made an anatomical dissection of an Indian in the Royal Hospital of Mexico'.[17] Padilla's claim has not been supported by further evidence and may have been a loosely founded eulogy of the recently deceased de la Fuente, holder of the first New World university chair of medicine, who died in 1595. Padilla's textbook coincided with the publication of the second edition of Hinojosos' *Summa y recopilacion de chirurgia*, which lacked the treatise *De pestilencia* that had concluded the 1578 edition. Perhaps the report on *cocolitztli* had by then lost its value as a medical news bulletin; the initial chapters on anatomy were reprinted almost verbatim from the 1578 edition.

Alonso López de Hinojosos' *Summa y recopilacion de chirurgia*

Hinojosos' textbook on surgery included the first text on anatomy to be printed in the New World; this anatomical treatise comprised the first section of both the original book, published in 1578, and the second 1595 edition. In spite of its pioneering status as the first of its kind in New Spain, the treatise was almost entirely based on traditional medical knowledge from the Old World. This was also the case with Augustín Farfán's more comprehensive textbook on anatomy which was published the following year. Mexican textbooks on anatomy and surgery which appeared in the late 1570s had no equivalents elsewhere in the Spanish-American colonies and would not be imitated in English and French America for more than two centuries. The two unique publications appear to have been the product of a relatively small group of physicians active at the hospitals and university of the Mexican capital.[18] An indication of this intimate circle was presented in the initial approval of Alonso López de Hinojosos' *Summa y recopilacion de chirurgia*.[19] It opened with brief endorsements written by Francisco Bravo and Augustín Farfán – respectively authors of the two earliest

---

[17]   'Hizo Anatomia de un Indio en el Hospital Real de Mexico.' Dávila Padilla, *Historia de la fundación y discurso de la Provincia de Santiago de Mexico* (Mexico City, 1596; Facsimile, Mexico City: Editorial Academia Literaria, 1955), p. 33.

[18]   Fernando Chico-Ponce de León, 'The First Neuroanatomical Text Published on the American Continent: Mexico City, 1579', *Childs Nerv Syst* (2004), p. 10.

[19]   The full title of Hinojosos' work is *Summa y recopilacion de chirurgia. Con un arte para sangrar muy util y prouechosa. Compuesta por Maestre Alonso Lopez, natural de los Inojosos. Chirurjano y enfermero del Ospital de San Joseph de los Yndios, desta muy insigne*

textbooks on medicine and anatomy published in the Americas – and by Juan de la Fuente, the first professor of medicine at the University of Mexico City.

A further official endorsement of Hinojosos' treatise was given by the highest authorities of *Nueva España*, Viceroy Don Martín Enríquez and Don Pedro Moya de Contreras, *por la gracia de Dios Arzobispo de México, del consejo de su Majestad*, whose written approval of the author offered additional praise of his three renowned medical colleagues: 'Having heard the opinions of the doctors Bravo, Fuente, and Farfán, and their expressions of its usefulness we have continued with the edition of this book entitled *Suma y recopilación de cirugía*, written by master Alonzo Lopez, superintendant and surgeon at the Royal Hospital of The Indies in this city.'[20] These three physicians had all received medical training at universities in their native Castile: both de la Fuente and Farfán had previously studied at the medical faculty of Alcalá de Henares during the 1550s, while Francisco Bravo had been educated at the less prominent University of Osuna, as referenced in his *Opera medicinalia*. Interestingly, Juan de la Fuente's introductory lines touched upon his own collaboration with Hinojosos in both Spain and Mexico City: 'For many years I have seen Master Alonzo cure and experiment in his art of surgery, both in Spain and in this city.'[21] The acquaintanceship of the two medical practitioners thus preceded de la Fuente's departure from Spain in 1561. Alonzo López de Hinojosos followed a few years later, as seen in a recent discovery in Archivo General de Indias, where Hinojosos and his family appeared in the records of passengers who departed from Seville to Veracruz on 8 February 1564.[22] Among this entangled group of Castilian physicians, only Hinojosos (who was born in Villaescusa, but instead took his name from the larger neighbouring municipality of Hinojoso near Cuenca) did not refer to any prior university training received in Castile. The

---

Ciudad de Mexico. Dirigido al ill. y R. S. Don P. Moya de Contreras, Arzobispo de Mexico y del Consejo de su Majestad (Mexico City: Antonio Ricardo, 1578).

[20] 'Habiendo visto los pareceres de los doctores Bravo, Fuente y Farfán y la utilidad que significan se conseguirá de la impresión de este libro intitulado Suma y recopilación de cirugía compuesto por el Maestre Alonso López, mayordomo y cirujano del Hospital Real de los Indios de esta ciudad y porque en él no hay cosa contra la Santa Fe Católica, damos licencia al dicho Maestre Alonso para que lo pueda imprimir, poniendo al principio de él esta licencia y los dichos pareceres.' Ibid., p. 71.

[21] 'Hace muchos años que he visto curar y experimentar a Maestre Alonso en su arte de Cirugia en España y en esta dicha ciudad.' Ibid., p. 74. A visit to the *Archivio de las Indias* in Seville did not establish the departure date of Bravo or Farfán.

[22] 'Alonso López, barbero, natural de Villaescusa, hijo de Pedro Grande y de Inés López, con su mujer, Juana Hernández, con sus hijas Ana e Inés, a Nueva España.' Archivo General de Indias, Pasajeros, L. 4, E. 3133, 1800. Cited in Gerardo Martínez-Hernández, 'La llegada del cirujano Alonso López de Hinojosos a la Nueva España', *Revista médica del Instituto Mexicano del Seguro Social*, 49(4) (2011), p. 460.

recently discovered records of his travel to New Spain referred to him as *barbero*, and the initial *Licencia* of his textbook, signed by Juan de Cuevas on behalf of Viceroy Don Martin Enriquez, also emphasised his inferior rank. Hinojosos was not referred to as *Doctor* like his three aforementioned colleagues, but instead as *Maestre Alonso cirujano y enfermero del Hospital Real de los Indios en esta ciudad*. The following approval of his work highlighted its humble and merely practical use as a medical guide for remote areas with no trained physicians or surgeons: 'It is very useful and beneficial for those who live where there are no physicians and surgeons. They can take advantage of the news presented in it, and use it to cure their diseases.'[23] The author recognised the unassuming character of his work, aimed at 'anyone who knows how to read'.

The first of the seven chapters of Hinojosos' textbook was a short introduction to anatomy entitled *De la anatomia y de las partes del cuerpo*, noticeable as the first American treatise of its kind. This brief account was less conspicuous as a contribution to contemporary anatomical knowledge in that it was seemingly unaware of recent corrections of human anatomy and physiology. Hinojosos seemed to rely on late medieval surgery, humoral physiology and a tentative comparative osteology embedded in the tradition of *galenismo arabizado*. This was clearly emphasised in his hypothetical account of the components and estimated number of bones of the human skeleton: 'Bones are simple members of spermatic essence, of cold and dry complexion, and hard and earthly substance, insensitive and incapable of feeling pain. According to Avicenna the number of bones is 248.'[24]

Apart from a few references to Juan de Vigo – an Italian Renaissance surgeon from the late fifteenth and early sixteenth centuries – and a sole appearance by Juan Fragoso, no recent medical authorities appeared in Hinojosos' account. The author relied for the most part on the mid-fourteenth-century *Chirugia magna* by Guy de Chauliac, whose surgical textbook may have reached Hinojosos in one of the five Castilian translations published during the late fifteenth and early sixteenth centuries. Hinojosos' initial short treatise on anatomy opened with the rhetorical question *Qué cosa es Anatomía?*, which the author answered through the authority of Guido de Cauliaco (whose name he Hispanicised): 'Anatomy according to Guido is a dissection or division of any dead body, and principally of the human body, which is the subject of this art of surgery. It is a part of medicine, an experience which comes with the knowledge of the disease and health of the

---

[23]   'Muy útil y provechosa para gentes que estando donde no hay médicos ni cirujanos, se podrán aprovechar de los avisos que en ella hay y curarse de sus enfermedades.' Hinojosos, *Summa y recopilacion de chirurgia*, p. 77.

[24]   'Huesos son miembros simples, de esencia espermática, de complexión fría y seca, de substancia dura y terrestre, insensibles e incapaces de sentir dolor. Según Avicena son doscientos cuarenta y ocho todos los huesos.' Ibid., p. 19. The correct number of bones in the human skeleton is 206.

human body.'[25] Hinojosos' initial treatise described the four virtues of anatomy through which an appreciation of the human body and its morphology and physiology could allegedly be obtained, together with rare pathological insights for the prognosis of different illnesses suffered by this divine creation.[26]

Hinojosos entered the Companía de Jesús a few years later in 1585, 'now widowed and leaving two sons and a daughter to the church', but spending his final decade among the Jesuits did not put an end to his surgical practice and writings. A second revised version of *Summa y recopilacion de chirurgia* was published in 1595, two years before his death, and while it did not include the *cocolitztli* account from the 1578 edition, it was supplemented by new chapters on rheumatism, paediatrics, and a joint chapter on gynaecology and obstetrics.[27] It also included a reference to a recent public dissection carried out by Hinojosos two years earlier in Oaxaca on an executed and quartered criminal delivered by the city mayor. The demonstration served to trace the exact location of the human liver, after an apparent dispute over this issue between Hinojosos and his medical colleagues: 'I dissected a man who had been quartered by the authorities on this public square in Mexico in front of many spectators, whom I called in as eyewitnesses even though I had carried out many anatomies before.'[28] The new 1595 edition included a sole reference to Bernardino Montaña, thus accrediting another Spanish anatomist whose work was similarly dependent on classical and late medieval doctrines – besides its use of the Vesalian plates for merely decorative purposes. The reprinting of Hinojosos' and Farfán's publications in new editions indicates the relative success of these early Mexican textbooks, and their seemingly successful distribution of medical knowledge to remote corners of New Spain. This objective was accentuated in the 1595 edition in a lament of

---

[25] 'Anatomía según Guido es una derecha disección o división de cualquier cuerpo muerto principalmente del cuerpo humano, el cual es el sujeto de este arte de Cirugía la cual es una parte de la medicina, experiencia que hace venir en conocimiento de las partículas enfermas o sanas de todo el cuerpo.' Ibid., p. 84.

[26] 'El primero es la grande admiración que causa a quien considera cómo nos hizo Dios a su imagen y semejanza en cuanto al alma y en cuanto al cuerpo, una composición con tanta diversidad de miembros, unos fríos y otros calientes, otros húmedos y secos, con unidad y coligancia los unos con los otros que cuando un miembro es dañado o corrompido, todos los demás lo sienten... El segundo provecho es el conocimiento de cada miembro ... El tercer provecho es el saber pronosticar de las enfermedades y miembros, si quedarán amancos o sanos, o tuertos o derechos. Y el cuarto, pronosticar las enfermedades que en cada miembros pueden sobrevenir.' Ibid., p. 84.

[27] Alonso Lopez de Hinojosos, *Summa y recopilacion de chirurgia* (Mexico City: Pedro Balli, 1595), f. 166.

[28] 'Hize abrir un hombre que por justicia mandaron hacer cuartos en esta plaza de Mexico y delante de muchos testigos lo pedí por testimonio aunque yo había hecho muchas anatomias.' Ibid., f. 85.

the sad state of medicine in these peripheries: 'I wrote this book because I worry about those who work outside the city in mines and farms, and remote villages and regions which do not have proper medical remedies.'[29] The author was aware that medical knowledge and training in fields such as anatomy and surgery was still not widely available in the Mexican and Central American colonies, where Spanish physicians were scarce in number and Old World medical theories and practices were largely unknown. Yet his *Summa y recopilacion de chirurgia* offered no significant contributions to contemporary medical knowledge, given its reliance on late medieval anatomy and surgery. However, the fact that this work now exists in only a handful of surviving copies indicates that the objective of the author was met and that his manual was literally thumbed out of existence in remote medical workshops of New Spain.

Augustin Farfán's *Tractado breve de anothomia y chirugia, y de algunas enfermedades que mas comunmente suelen hauer en esta Nueva España*

Alonso López de Hinojosos' late decision to affiliate himself with a religious order was paralleled by his contemporary New World author, Augustín Farfán, who entered the Order of Saint Augustine in 1569. Farfán's *Tractado breve de anothomia y chirugia, y de algunas enfermedades que mas comunmente suelen hauer en esta Nueva España* from 1579 was the first New World publication to deal primarily with anatomy. It was published only a year after Hinojosos' textbook on surgery and was also printed by Antonio Ricardi, a Piemontese publisher who was at that time active in the Mexican capital. Farfán's *Tractado breve de anothomia y chirugia* was later included in an expanded medical textbook entitled *Tractado breve de medicina y de todas las enfermedades* (Mexico City, 1592 and 1610). In spite of the alterations made to these two later treatises, the 1579 text on anatomy was printed almost verbatim in both editions.[30]

---

[29] 'Doliéndome yo desto y por los que estan fuera desta ciudad en minas y estancias, pueblos y partes remotas que carecen de los remedios convenientes, hize este libro'. Ibid., *Al lector*, n.p.

[30] Farfán's 1579 publication is exceedingly rare and I have only been able to study the more common 1592 and later 1610 editions. It is therefore reassuring when the Spanish-Mexican historian German Somolinos d'Ardois maintains that the 1579 edition is included almost verbatim in the later editions: 'Tal vez los capitulos más similares en ambas obras sean los que constituyen el tratado anatómico, con el qual comienza la edición de 1579 y la termina la de 1592. El número y título de los capítulos es el mismo, pues aunque en la segunda edición som diez y seis contra quince en la primera, la diferencia se debe a que el capítulo preliminar y generalidades, no esta numerado en 1579 y en cambio sí lo está en 1592. No hay ninguna variación de fondo, tal vez los capítulos sean algo más detallados y con mayor nomenclatura en la edicion primera pero el contenido de ambos, sus ideas y las descripciones

TRACTADO BREBE DE MEDICI
na, y de todas las enfermedades, hecho por el
padre fray Auguſtin Farfan Doctor en Mediçi
na, y religioſo indigno de la orden de ſanc
Auguſtin, en la nueua Eſpaña. A hora
nueua mente añadido.
(*)
DIRIGIDO A DON LVYS DE VE
laſco cauallero del habito de Sādtiago,
y Virrey de eſta nueua Eſpaña.

En Mexico, Con Priuilegio en caſa de Pedro
Ocharte De 1592. Años.

Figure 9.2   Augustín Farfán's *Tractado breve de anothomia y chirurgia* (Mexico City, 1579) was the first New World publication to deal primarily with anatomy and was published only a year after Alonzo López de Hinojosos' *Summa y recopilación de cirugia* (Mexico City, 1578). Title page of a revised edition of Farfán's textbook, whose 1592 title was changed to *Tractado breve de medicina, y de todas las enfermedades ... ahora nuevamente añadido* (México, Casa de Pedro Ocharte, 1592). Courtesy of the Wellcome Library, London

An illustration on the title page showed the author in profile, dressed as an Augustinian friar with his head shaved, and presented Farfán as 46 years old at the time of publication, thereby establishing his year of birth as either 1532 or 1533. Farfán was born and educated in Seville, and his text briefly referred to his prior medical studies at the University of Alcalá de Henares during an unknown period. This information once again shows that the *Complutense* outside Madrid was a leading breeding ground for the physicians and naturalists of New Spain. Little else is known about Farfán's life prior to his arrival in Mexico City, where he allegedly obtained a position as *Protomédico* immediately before his clerical vocation.[31] Later accounts referred to Farfán's employment as a court physician to Philip II, but his name does not appear among the physicians of the Spanish court and it is dubious whether Farfán really obtained any of these formal positions.[32] He did not conclude his medical studies until some years after his transfer to Mexico, a few years before he joined the Order of Saint Augustine in 1569. Some rather tentative accounts of his life present a developing deafness as the main reason for his gradual abandonment of a medical career in favour of a monastic life, but offer no

---

anatómicas son análogas.' German Somolinos d'Ardois, 'Los impresos medicos mexicanos (1553–1618)' in *El mestizaje cultural y la medicina novohispana del siglo XVI* (Valencia: Instituto de Estudios Documentales e Históricos sobre la Ciencia, 1995), p. 213.

31   Chico-Ponce de León, 'The First Neuroanatomical Text Published on the American Continent: Mexico City, 1579', p. 11.

32   On the physicians of Philip II, see E. Subiza, 'Los médicos de Felipe II, Aportación a su studio', *Archivo Iberoamericano de Historia de la medicina y la antropología médica*, 6 (1954), pp. 377–90 and Javier Puerto, *La leyenda verde* (Salamanca: Junta de Castilla y León, 2003).

source for this information.[33] Farfán's anatomical training is similarly hard to trace and is not documented in any clear and unambiguous accounts.

The opening of his anatomical textbook nonetheless presented a few words of warning regarding the irreparable errors caused by surgeons with no anatomical skill, thus indicating his own experience in the field: 'As the human body (as it is) is the subject of the physician and the surgeon it is fitting here to present a brief text on anatomy ... for the surgeon it is useful to know where to cut and cauterise. And many of those who do not know anatomy make irredeemable errors every day.'[34] This statement was in many ways a repetition of previous claims made by the fourteenth-century surgeon Guy de Chauliac, which were repeated by later Spanish translators and commentators such as the Valencian anatomists Miguel Juan Pascual and Juan Calvo. Calvo's *Chirurgia universal y particular del cuerpo humano* coincided with Farfán's textbook and included a similar forewarning: 'If he (the surgeon) does not know anatomy, he will be like the blind born carpenter who cuts the wood where it should not be cut.'[35] The opening lines of Farfán's textbook indicated a familiarity with Valverde's *Historia de la composición del cuerpo humano*, whose address to the reader warned of the limited skills of Spanish surgeons with no knowledge of anatomy, and the consequential 'damage, which this causes to the entire Spanish nation.'[36] Yet there are no indications of Farfán's awareness of contemporary medical publications, or of any recently accumulated knowledge in the field, such as the brief references to Vesalius made by both Benavides and Bravo in their earlier medical textbooks.

Apart from a sole reference to Juan Fragoso in the first 1579 edition, and to Francisco Hernández and Amato Lusitano in the 1592 version, Farfán mentioned no other contemporary physicians in his anatomical treatise. He relied instead on classical and medieval authorities, and a considerable number of Arabic authorities, such as Avicenna, Rhazes, Averroes and Albucasis, who were all referenced. This surviving tradition of *galenismo arabizado* was furthermore

---

[33] Jorge Sánchez Silva, 'Augustin Farfán y Alonzo Lopez de Hinojosos. Dos médicos hispanos del Nuevo mundo', *Acta médica*, II(26) (1971), p. 147.

[34] 'Siendo el cuerpo humano (como lo es) el sujeto del medico, y del cirujano, es bien poner aqui una brebe anothomia. Para que sepan y conoscean todos los qualidades y complexiones y officios todas las partes y miembros del. Y porque sepan tambien los sitios y lugares de cada miembro para quando les apliquen las medicinas El cirujano conviene que sepa como y por donde ha de cortar, abrir y cauterizar. Y muchos por no saber la Anothomia, hazen yerros yrremediables cada dia.' Augustin Farfán, *Tractado breve de medicina y de todas las enfermedades...* (Mexico City: Pedro Ocharte, 1592), p. 323.

[35] 'Si (el cirujano) no sabe anatomía será como el carpintero ciego de su natividad, que corta el madero por donde no lo ha de cortar.' Juan Calvo, *La chirurgia universal y particular del cuerpo humano* (Barcelona: J. Cendrat, 1591), p. 168.

[36] 'El daño que desto se sigue a toda la nación Española.' Juan Valverde, *Historia de la composicion del cuerpo humano* (Rome: Antonio Salamanca and Antonio Lafrery, 1556), n.p.

seen in the Arabic terminology and designations of several anatomical categories, which emphasised the swift import of Islamic medical knowledge from Iberia to America. The first chapters of Farfán's treatise were devoted to neuro-anatomical descriptions, which maintained the conception of the imaginary *rete mirabile* as a gateway for the transformation of the vital spirit of the heart to the animal spirit of the brain. This description was another indication that Farfán's work was more a comment on Galenic doctrines than the result of systematic and practical studies of hands-on anatomy. Yet his observations of the large human brain and correct assertions of its unequalled proportional size compared to that of other animals indicate some personal experience with dissection and comparative anatomical studies of this particular organ: 'The head has a great capacity, because the human brain is larger than that of any other animal.'[37]

The author included the brain among three other 'primary organs' – heart, liver and testicles – which were all represented as crucial for the survival and reproduction of the human species. The brain was understood as a container for the soul with three main faculties: an anterior cavity that contained the common sense and the imagination, and a central and a posterior chamber which housed reasoning and memory. This organ, 'no sin misterio', was surrounded and supported by the skull, the cerebral membranes and the *rete mirabile*.[38] In line with this understanding of cerebral anatomy, the sutures of the skull were seen as entrances for the veins, which transported vital spirit from the heart, and as cerebral outlets for the vapours produced during the process of transforming cordial vital spirit into cerebral animal spirit. Farfán's Galenic physiology was furthermore in continuity with a pre-Galenic Aristotelian understanding of this vital and essentially 'cold and humid' organ as a cooling mechanism and a passage for superfluous body heat.[39]

The osteology described in Farfán's treatise was moreover firmly rooted within a Galenic conception of human morphology based on comparative studies of pigs, dogs and apes. The latter category of dissected animals had fostered the erroneous conception of a septipartite human sternum, which was maintained in Farfán's 1579 textbook.[40] Another description entirely rooted in Galenic notions, which had been corrected half a century earlier, was the understanding of the

---

[37] 'La cabeza tiene gran capacidad, por ser el celebro del hombre mayor, que el de ningun animal. La cabeza es de ygual diametro y proporcion. En la delantera esgibosa ò corcobada.' Ibid., p. 326.

[38] 'Las cinco partes contenidas son la dura y pia mater las cuales son así llamadas porque son piadosas como la madre que quiere mucho a su hijo y que quiere más recibir el golpe que no que lo reciba el hijo. El otro es rete mirabile, la substancia del cerebro y otra vez es el hueso basilar donde se funda la olla de la cabeza.' Ibid., p. 329.

[39] 'La sustancia de que es hecho el cerebro, es blanca y blanda, de fria y humida complexion y naturaleza. No fue hecha para que de ella se mantuuiese el casco, como las otras medullas o tutanos, sino para templar el calor del espiritu vital.' Ibid., p. 329.

[40] 'El pecho es compuesto de siete huesos, en cuyas estremidades ay ternillas.' Ibid., p. 336.

five-lobed liver, described by Farfán in his chapter on *Anothomia del higado*.[41] Farfán's account of the same organ as the producer and distributor of blood to the rest of the organism was furthermore a reproduction of Galenic physiology, as emphasised by the author.[42] The uncritical reliance on waning medical doctrines of the Old World was supplemented by curious descriptions of New World phenomena. This was seen in references to the excessive consumption of chocolate among both natives and colonisers, which tended to 'thicken' their blood, and in his appraisals of the fine flint arrowheads of the Chichimeca Indians as recorded in an analysis of arrow wounds.[43] Farfán had a similar objective for his treatise as Hinojosos, intending his textbook to be used primarily in rural districts suffering from a scarcity of healers and medical knowledge: 'Having seen this great misery for those who live without access to any medical remedies, I partook to write this great work, and to describe in it all the diseases and cures, and all that can be done without omitting anything. I do not write for other physicians, but for those who live outside large cities and villages.'[44]

Juan de Barrios, *Verdadera medicina, cirugia y astrologia...*

While almost no trace of contemporary anatomical knowledge was found in the treatises of Hinojosos and Farfán, the situation was markedly different in the encyclopaedic work on anatomy, surgery, medicine and astrology published by Juan de Barrios in the early seventeenth century. As shown in an initial full-page portrait of the author, Barrios was born in the Madrid suburb of Colmenar Viejo around 1562. His textbook referred repeatedly to his own medical experience at the Universities of Salamanca and Alcalá, and later practice in Madrid, Seville and Valencia, before he left Spain in 1589. Unlike earlier Mexican works on anatomy and surgery, Barrios' publication included detailed knowledge of anatomical practitioners and publications of the mid- to late sixteenth century. Contemporary references appeared throughout his 600-page opus, which was entitled *Verdadera medicina, cirugia y astrologia, en tres libros dividida... En el primero se trata de la anatomia del cuerpo humano, y de las heridas de la*

---

[41]   'El higado tiene su asiento y lugar debaxo de las primeras costillas altas del lado derecho, y a lo ultimo de ellas. Tiene cinco partes, que llaman pendolas.' Ibid., p. 341.

[42]   'Dize Galeno que en el higado se hazen tres sustancias por la decocion de la sangre. Dos son de superfluydades, y una natural, como cada dia lo veemos, que se haze en el mosto, quando se cueze.' Ibid., p. 342.

[43]   Ibid., p. 124.

[44]   'Y viendo yo esta miseria, y que con ellos no ha de haber remedio, he tomado este gran trabajo y es ponerles aqui en toda enfermedad y curación, todo lo que debe hacer sin que les falte cosa ninguna ... Ya he dicho que no escribo para los que son médicos sino para los que estuvieren apartados de ciudades y pueblos grandes, donde siempre los suele haber.' Ibid., p. 235.

Figure 9.3   Juan de Barrios' textbook on anatomy, surgery, medicine and astrology: *Verdadera medicina, cirugia y astrologia* (Mexico City, 1607). In the introduction of the book, Barrios was referred to as 'a unique anatomist, and among the best of his age'. The anatomical content of thc textbook did not support this claim, consisting mainly of comments and defences of waning Galenic doctrines. Like Farfán and Hernández, Barrios was educated in medicine at the University of Alcalá de Henares before his transfer to New Spain in 1589. Image reproduced from the original preserved in the Biblioteca Históricomédica 'Vicent Peset Llorca, Universitat de València'. Label: IHMC Fons Antic 0162

cabeza, pecho y vientre ... El libro Segundo trata de que sea calentura ... En el libro tercero se trata de la anatomia de la madre, de la formación de la criatura, de los males de las preñadas, paridas, y de los niños... (Mexico City, 1607).

Two of the three 'books' in Barrios' large medical treatise dealt with anatomy, and the initial approval and recommendation written by his medical colleague at the University of Mexico City, *El Doctor Urieta*, presented Barrios as one of the most skilled and experienced anatomists of his time: 'An exceptional anatomist, and one of the best of his time, both through his knowledge and dissections of the human body, which he has carried out many times with his own hands and with great skill, in order to enhance some of the written claims.'[45] Yet these allegedly numerous corrections of existing

---

[45]   'Unico anathomista y de los mejores de su tiempo, assi en conocer las partes del cuerpo humano, como en la dissecion del, muchas veces hecha por sus manos con grande

anatomical knowledge were not found in the contents of Barrios' textbook, which consisted mainly of comments on previous anatomical publications, with an almost constant emphasis on the infallibility of traditional Galenic doctrines. The author was seemingly well informed of recent anatomical corrections and additions, such as the three auricular ossicles, yet his own purported skills as a dissector had not enabled him to unearth and observe these osteological details, unknown by Galen and discovered by Vesalius and Pedro Jimeno during the mid-sixteenth century: 'I have never seen them in all the bodies I have dissected, of humans, dogs, pigs, rabbits, chickens and other animals, and even after a thorough observation I have not found them; only in some sea breams, and in some fish and in a toad.'[46] The author's awareness of such recent findings was a significant development from Hinojosos and Farfán, who had not revealed in their own works any similar knowledge of contemporary anatomical corrections. Yet Barrios' treatise was in clear continuity with these earlier authors and he included several of their errors in his anatomical and physiological descriptions. He seemingly borrowed from the structure of the second edition of Hinojosos' *Summa y recopilacion de cirugia* (1595) and, like Hinojosos, included concluding chapters on paediatrics, gynaecology and obstetrics. However, whereas Hinojosos had used Avicenna's speculative and invalid description of the human skeleton, Juan de Barrios instead advocated Vesalius' renowned successor in Padua, Gabriele Falloppio, as a reliable authority in questions of osteology.[47] An assembly of renowned anatomists from the previous century were compiled in Barrios' textbook, including Falloppio, Vesalius, Colombo, Estienne and his recently deceased Spanish colleagues Juan Fragoso, Juan Calvo and Bartolomé Hidalgo de Agüero. These latter physicians were all given prominent rank in Barrios' narrative, which credited Fragoso as Barrios' master shortly before the author left for New Spain in 1589.[48] Two years earlier, Barrios had purportedly assisted Juan Calvo's lectures on surgery at the Valencian *Estudi General*, and

---

liberalidad, y con estas condicion es para mas perfecionar, lo que en sus escriptos pretendia.' Juan de Barrios, *Verdadera medicina, cirugia y astrologia, en tres libros dividida ... En el primero se trata de la anatomia del cuerpo humano, y de las heridas de la cabeza, pecho y vientre ... En el libro tercero se trata de la anatomia de la madre, de la formación de la criatura, de los males de las preñadas, paridas, y de los niños...* (Mexico City: Fernando Balli, 1607), *Al discreto lector*, n.p.

[46] 'Yo nunca los he visto en quantos he anatomizado, por que en hombres, perros, puercos, conejos, gallinas y otros animales, he mirado muy bien y nunca lo he visto; solo lo he visto en algunos besugos y en algunos peces y en el escuerzo.' Ibid., f. 2v.

[47] 'Quien mas quisiere ver lea a Gabriel Falopio tratando de anatomia de huesos.' Ibid., f. 17r.

[48] 'En Madrid el año de 1588 adviertiendo los casos estraños con el doctissimo Fragoso, en el hospital de los Ytalianos.' Ibid., f. 16r.

his *Verdadera medicina* recognised Calvo as the most reliable authority on the origin and the pathological nature of syphilis.[49]

Barrios' textbook included no such glowing tribute to Bartolomé Hidalgo de Agüero, whose surgical procedures were instead disparaged unequivocally in an entire chapter *Contra el doctor Hidalgo Seuillano*. Since the early 1580s, Hidalgo, a surgeon at Seville's *Hospital del Cardenal*, had advocated a so-called *via particular* which avoided trepanations and other drastic invasive operations. Hidalgo's reforms provoked a long medical controversy with his colleague Juan Fragoso, which continued even after the death of both parties in 1597. A few years later, Hidalgo's new surgical method was compiled and published by his son-in-law in *Thesoro de la verdadera Cirugía y via particular contra la comun* (Seville, 1604), a work refuted by Barrios as worthless. His chapter 'against' Hidalgo was a vehement defence of Fragoso's maintenance of the *via comun* and a systematic condemnation of Hidalgo's surgical procedure.[50]

Barrios' account included several other references to contemporary Spanish physicians, such as his former teachers at the University of Salamanca between 1581 and 1583. He also mentioned Augustín Vazquez, the second Professor of Anatomy at the renowned institution. The biggest compliments were given to his former teachers at the University of Alcalá de Henares, where he later obtained his doctoral degree under the acclaimed mentor Pedro García Carrero, to whom he paid tribute in a sonnet.[51] Dr Urieta's address to the reader outlined Barrios' medical career and his remarkable circulation between several Spanish universities prior to his subsequent employment in Mexico City: 'It is true that there was not a famous university in Spain, in which he did not heal and appear in person, whether it was in Salamanca, Alcala de Henares, Lleida, Valencia or Seville.'[52] Barrios' textbook did not include detailed references to his prior experience as an anatomist, even though the introduction praised him as 'Unico anothomista, y de los mejores de su tiempo'. Arguably, his experience in this field

---

[49]   'El principio de bubas es Nuevo, y en las Indias antiguo, y de muchos años antes que a España vinese. Juan Calvo Medico Valenciano dize que del demasiado uso de el acto venereo que en los Indios tenian, puede ser que desta manera tubiesen su principio en las Indias.' Ibid., f. 36r.

[50]   'Tratando bien a los doctissimos, nobles y bien nacidos (penso por ventura vuestra arrogancia) que en Nueva España, y en lungas tierras como dizen, no abia de aver discipulos que volviesen por sus maestros, por ventura desterrado de España quisieses hazer prueva de que quedasemos engañados pues con titulo de tesoro, nos le embiais (No como nuestro amigo sevillano).' Ibid., f. 19.

[51]   'Al Doctor Pedro García Carrero, cathedrático de prima, de Alcalá de Henares, su discipulo el doctor Ioan de Barrios le dirige este libro por este soneto ... eres del español Nuevo Esculapio, padre de la vida y Dios de la medicina', Ibid., *Soneto*, n.p.

[52]   'Es cierto, que no dexo Universidad insigne en España, que no curase, y viesse personalmente como fueron Salamanca, Alcala de Henares, Lleida, Valencia y Sevilla.' Ibid., *Al discreto lector*, n.p.

derived mainly from his years as a medical student in Spain. Among his few remarks in this respect was an account of vivisections of the recurrent laryngeal nerves carried out on dogs at the University of Alcalá de Henares: 'I have seen this carried out on dogs several times, and any curious person can do it.'[53] Similar experiments were performed by Francisco Hernández during his employment at the *Hospital de Santa Cruz* in Toledo and both cases were imitations of the spectacular demonstrations carried out Vesalius and Galen.

Barrios occasionally defended this principal authority of ancient medicine against his most renowned modern revisionist: 'Even though I admire Vesalius, I follow Galen.'[54] As another indication of Barrios' Galenic stance, his description of the humoral physiology of the liver, heart and brain, and his maintenance of the *rete mirabile* did not show any significant progress from Hinojosos' and Farfán's anatomical treatises three decades earlier.[55] Throughout Barrios' anatomical treatise, several of the physiological theories advocated by Galen's modern critics were discredited by the author. This was seen in the insistent defence of the basic function of the liver as producer and deliverer of blood to the stomach and intestines: 'The opinion of Realdo Colombo, who says it does not serve this purpose, is worthless.'[56]

While the treatise remained indifferent to modern sceptics such as Vesalius and Colombo, it included appraisals of other contemporary physicians such as Juan de la Fuente, who was described by Barrios as 'el doctissimo Doctor de la Fuente cathedratico desta Universidad'.[57] This first New World professor of medicine had approved Hinojosos' and Farfán's earlier books on anatomy and remained the driving force of Mexico City's medical faculty by the time Barrios arrived in 1589. Fuente died in 1595 and therefore failed to grant the approval of Barrios' work as he had done for other anatomist authors. As with the publications by Hinojosos and Farfán, Barrios' textbook was licensed and approved by both the Viceroy and the Archbishop of New Spain – indicating a firm continuing alliance between the Spanish rulers and the most prominent physicians of the

---

[53]   'En el del neruio recurrente y izquierdo en el qual aunque se dañen solo pierde la voz el hombre, y tambien se ve claro que salen del cerebro, por que en perros, yo lo he hecho muchas vezes, y qualquiera por curiosisdad lo puede hazer. Porque quitado el Corazon anda, ladra, y muerde, la qual experiencia hizimos, yo y el licenciado Garro y Francisco Gomez, y la Villa Real quando estauamos en Alcala de Henares, oyendo al doctissimo y consumadissimo maestro el Doctor Iuan Gomez medico de camera de su Magestad.' Ibid., f. 10.

[54]   'Pero de Besalio, cierto que me admiro que siguio a Galeno.' Ibid., f. 9.

[55]   'Porque en el higado todos los humores se hazen los espiritus animales solo se hazen en el cerebro y rete mirabile, y no los de mas humores, y hazense los espiritus animales con virtud natural, como los humores en el higado, con el calor que ay en el rete mirabile los espiritus vitales, que van hechos del corazon adquiren un nuevo ser de mas perfeccion.' Ibid., f. 3.

[56]   'No vale la razon de Realdo Colombo, que dize que no siruen deste fin.' Ibid., f. 10.

[57]   Ibid., f. 14r.

Mexican capital. Within this seemingly small circle, the author showed a fervent antipathy for his recently deceased colleague Augustin Farfán, who had died in 1604. Barrios' remarks on Farfán were reminiscent of his scorn for Hidalgo and presented the elderly physician and Augustinian friar as a thoroughly incompetent and indifferent medical doctor, whose patients were often saved only through Barrios' own last-minute interventions.[58] This animosity was probably rooted more in personal differences than in any significant divergences in medical principles, since the writings of both Farfán and Barrios rested on solid Galenic foundations. A decade after the publication of *Verdadera medicina*, Barrios, Farfán and Hinojosos themselves became subjects of general scorn in the first Mexican edition of Francisco Hernández's opus, which was based on manuscripts left behind by Hernández before his return to Spain in 1577. In the prologue to this work, entitled *Quatro libros de la naturaleza y virtudes de las plantas, y animales que estan receuidos en el uso de Medicina en la Nueva España* (Mexico City, 1615), the translator and commentator Francisco Ximénez lamented the many cases of plagiarism of Hernández's work found in later Mexican textbooks on medicine: 'Many copies of Doctor Hernández, which are his in name only have been corrupted in every way, both in the terms and medicaments, because many doctors have used and printed pieces of his work. The Doctors friar Augustin Farfán, Ioan de Barrios, Alonzo Lopez de Hinojoso of the Company of Jesus, and many others.'[59] These accusations of plagiarism may be justified in some of the works dedicated to the botanical medicine of New Spain by these authors, but these alleged copyists did not reveal any insights into the numerous recent discoveries and corrections of human anatomy found in Hernández's earlier writings, and instead relied on tradition. Despite their obvious flaws, these publications represented an entirely new genre of medical textbooks on the American continent, which remained unparalleled in the neighbouring European colonies until two centuries later. The creation of a formal chair of anatomy at the University of Mexico City in 1621 further emphasised the uniqueness of the Mexican capital in the history of anatomy in the Americas. This university chair remained the only one of its kind in a New World context until the establishment of a similar professorship in Lima in 1729 – 150 years after Hinojosos' and Farfán's first textbooks on anatomy and surgery.

---

[58] 'Si por esta causa sucede es caso mortal, y el peor de todos, como yo vi en esta Ciudad en el año 1602 en el hijo del Presidente Valerrama, que ya quando yo fui, y el Doctor Farfan tenia los labios secos sin materia ninguna.' Ibid., f. 16v.

[59] 'Muchas copias de el Doctor Francisco Hernández, suyas en el nombre y de todo punto corruptas, assi en los vocablos como en los medicamentos, y para que a pedazos se hayan aprovechado ympresso muchos Doctores. El Doctor fr. Augustin Farfan, Ioan de Barrios, Alonso Lopez de Hinojoso, de la compañia, y otros muchos.' Francisco Ximenez, *Quatro libros de la naturaleza y virtudes de las plantas, y animales que estan receuidos en el uso de Medicina en la Nueva Espana* (Mexico City: En casa de la Viuda de Diego Lopez Davolos, 1615), *Al lector*, n.p.

Chapter 10
Images of Spanish Renaissance Anatomy

One of the most crucial aspects of the 'Vesalian Revolution' of the mid-1500s was the deliberate use of visual aid for the study and correction of anatomical knowledge. With the 1543 publication of *De humani corporis fabrica*, printed anatomical images moved from inferior supplements to aids within the written text, and to a higher status as integrated tools in the presentation and clarification of anatomical data. The meticulous research, execution and printing of these images have maintained their appeal ever since. They represent a successful attempt to support textual evidence with a new iconography which enabled the communication of previously unknown or misrepresented details of the human structure. After *Fabrica*, many later publications on anatomy supported their written content with visual clarifications. For the reader, this analogous appreciation of text and image represented a giant leap forward in the study of human anatomy and a significant advance in the history of scientific and medical illustration. Many of the images represented in *Fabrica* in 1543 became the direct templates for illustrated works by future Spanish anatomists, as seen in Bernardino Montaña's *Libro de la anothomia del hombre* (1551) and Juan Valverde's *Historia de la composición del cuerpo humano* (1556).

Through images, readers became familiar not only with recent corrections of anatomical knowledge, but also with an erudite iconography which openly criticised formerly infallible 'truths'. The 1559 Italian edition of Juan Valverde's anatomical publication exemplifies this development, with its new frontispiece illustration of a pig and a monkey holding human femurs – an explicit visual rejection of Galen's erroneous osteological doctrines. It is no surprise therefore that one of Jacobus Sylvius' most violent attacks on his 'insane' pupil was targeted at Vesalius' deliberate use of images. To the Parisian professor, this represented an ominous novelty that would distort and falsify traditional conceptions of the natural world. In Sylvius' view, images should be avoided, as Galen himself had previously emphasised in his forewarnings of the risks and inevitable pitfalls involved in the artistic representation of nature.[1]

---

[1]   'Superstitiosam, & obscurissimam, & planem inutilem, pro nihilo duxeritis, & tales tum picturas tum characteras remoramento esse magis quam compendio aestimaueritis, praesertim medicis, quorum ars longa, ac vita brevis, & qui partium corporis humani omnium naturam visu, & tactu ex multis anatomis perspectam habere debent, non ex picturis aut libris solis, cum nec nauclerum, nec imperatorem, nec alios quosuis artifices hac umbratili exercitatione formare liceat: sed ne plantis quidem ipsas pingi Galenus permiserit.' Jacobus Dubois Sylvius, *Vaesani cuiusdam*

Even though the study of anatomy was promoted through the use of visual evidence, almost none of the illustrators responsible for its iconic images are known with any degree of certainty. Montaña and Valverde referred only briefly to the content of their illustrations and did not provide any clear references to the artists and craftsmen who produced their anatomical images. This was also the case with the original matrix of their reused representations, Vesalius' *Fabrica*, whose illustrator was not mentioned in the 1543 or 1555 editions. We know that Vesalius' earlier *Tabulae anatomicae* from 1538 were signed and paid for by the Flemish artist Jan Stephan Calcar from Titian's Venetian workshop, who was also credited in the 1539 Venesection Letter, but the illustrator(s) of *Fabrica* remain unidentified. Giorgio Vasari attributed the images of *Fabrica* to Calcar three times in the second 1568 edition of his *Vite*, which also referred to Valverde's later copies of the Vesalian plates.[2] Titian was for centuries rumoured to be the real artist behind *Fabrica*, as suggested in a seventeenth-century publication entitled *Notomie di Titiano* (Bologna, 1670). Several other options have been suggested, but nothing has been determined with any certainty, in spite of numerous articles on the subject.[3] An indication of Vesalius' own lack of appreciation for the artists and xylographers behind his anatomical icons was seen in a short reference in his 'China Root Epistle', a 1546 work notably without illustrations (except for a reused author portrait from Fabrica), in which Vesalius presented prior collaborations with artists as difficult and arduous: 'No longer shall I have to put up with the bad temper of artists and sculptors who made me more miserable than did the bodies I was dissecting.'[4] It seems absurd that Vesalius gave so little credit to his illustrators considering that their artistic mastery was instrumental to the appreciation of his written work. Paradoxically, Vesalius' disparaging remarks coincided with the improved social status of artistic practitioners, which was evident in Vasari's contemporary biographies of Italian painters, sculptors and architects, and was seen in Cellini's notorious autobiography, Pietro Aretino's appraisal of Titian, and several contemporary accounts of the life and works of Michelangelo. In spite of their affiliation to

---

calumniarum in Hippocratis Galenique rem anatomicam depulsio (Paris: Viduam Jacopo Gazelli, 1551), f. 118.r

[2]   'Come furono anco gl'undici pezzi di carte grandi di Notomia, che furono fatte da Andrea Vessalio, e disegnate da Giovanni di Calcare Fiamingo, pittore Eccelentissimo, lequali furuno poi ritratte in minor foglio, & intagliate in Rame dal Valuerde, che scrisse della Notomia dopo di Vesalio.' Cited in Francisco Guerra, 'The Identity of the Artists in Vesalius' *Fabrica*', *Medical History*, 13(1) (1969), p. 41.

[3]   J.B. Schultz, *Art and Anatomy in Renaissance Italy* (Ann Arbor: UMI Research Press, 1982), p. 24.

[4]   Cited in C.D. O'Malley, *Andreas Vesalius of Brussels* (Berkeley: University of California Press, 1964), p. 124 and Tom Jones, 'The Artists of Vesalius' *Fabrica*', *Medical History*, 31(3) (1943), p. 223.

the renowned workshops of Titian, Michelangelo and Vasari, the illustrators of Vesalius' and Valverde's textbooks were still unimportant in the eyes of their employers, who did not bother to mention their names in their own publications.

The approximation of art and anatomy was not an entirely unknown concept in Iberia at the time that the first Vesalian representations appeared in Spanish anatomy books. Between 1538 and 1547, the Portuguese artist Francisco de Holanda lived in Rome, where he studied contemporary and ancient works of art, and was a frequent visitor of such local celebrities as Vittoria Colonna and Michelangelo. Holanda recorded his conversations with Michelangelo in 'Four Dialogues', a section in the second book of his unpublished art treatise *De pintura antiga*, which was completed after his return to Portugal in 1548 and was dedicated to King John III. Holanda presented his manuscript as the first Iberian treatise on the visual arts and as an attempt to elevate the lowly reputation of his nation's art to the superior level of contemporary Italy.[5] A chapter entitled 'What Sciences are Suitable for the Painter' described the author's skills in several branches of complex knowledge and praised the intimate relationship between art and science that he had witnessed in Rome, in particular the artistic experience with human dissection, which was deemed 'as least as profitable for the knowledge (of the artist) as for any surgeon'.[6] Formalised art training including human dissection would soon be introduced in the first Italian art academies, but was already being practised privately by Michelangelo and other prominent painters and sculptors, and was seen by Holanda as crucial for the improved skills and social standing of the artist.[7] Holanda's accounts of the close collaboration between Italian artists and anatomists highlighted the fact that nothing similar existed in Portugal during that period. Juan Valverde de Amusco's 1556 *Historia* would also call on his fellow Spaniards to leave Spain for Italy, where the practice of dissection was far more common and less controversial than in their native land.[8]

---

[5]   'Porque quidam alguns que eu me desprezo de ser pintor, eu entento de mostrar n'este livro quau honrado e nobre causa é ser pintor é quau dificil, e de cuanto serve e vale a ilustre e mui necessária ciéncia da pintura na república.' Cited in Ronald W. Sousa, 'The View of the Artist in Francisco de Holanda's Dialogues', *Luso-Brazilian Review*, 15 (1978), p. 45.

[6]   'Tanto le cumple el conocimiento como a cualquier cirujano.' Francisco de Holanda, *De la pintura antigua y Diálogo de la pintura. Versión castellana de Manuel Denis (1563)* (Madrid: Visor, 2003), p. 43.

[7]   'Requierese que la notomia sea hecha de un cuerpo muerto muy flaco, y proporcionado: quitandole poco a poco de la Piel. Y esto hazese en Italia, de muchas maneras: no solamente para la curugía, per para los grandes debuxadores, y statuaries, que contra hazen de vulto. Y dizen Miguel Angelo que el por si saco, todos los pezes y miollos de un cuerpo finado: y que despues los bacio de cera, para poderlos poner en otro cuerpo como el quisiesse, dela manera que staban en la carne, y que de esso se seruja ensus ymagines.' Cited in Valerià Cortés, *Anatomía, Academía y dibujo clásico* (Madrid: Cátedra, 1994), p. 108.

[8]   'Assi por ser cosa fea entre los Españoles despedacar los cuerpos muertos, como por auer pocos, que venidos á Italia, donde la podrian deprender.' Juan Valverde, n.p.

Later evidence of collaboration between Spanish artists and anatomists was seen in a few documented cases such as Francisco Hernandez's and Nicolas de Vergara's earlier mentioned vivisections of animals carried out in Toledo during the early 1560s. Jusepe Martínez's *Discursos practicables del nobilisimo arte de la pintura* (c. 1675) furthermore referred to the anatomical studies carried out by the painter Juan de Juanes in Valencia a century earlier: 'while they were performing anatomical dissections at the hospitals he went there to draw muscles, nerves, and tendons'.[9] The sculptor Mateo de Vangorla and the anatomist Juan Valero Tabar made anatomical mannequins for the University of Salamanca and for the court of Philip II, of which only the former has survived. These cases were, however, few and far between in Spain, and were significantly removed from the formal cooperation between Italian artists and anatomists in universities, hospitals, artistic guilds and workshops, which built on a long tradition already established in the fifteenth century. This association was institutionalised and formally organised with the 1563 inauguration of the Florentine Accademía del Disegno, founded at the behest of Vasari and Duke Cosimo I de' Medici, and imitated during the three subsequent decades in Perugia, Bologna and Rome.[10] As emphasised in the statutes of the Florentine art academy, the teaching of anatomy should be carried out during the cold winter months in the grand city hospital of Santa Maria Nuova.[11] This institutional education replaced a previous tradition of private anatomy studies carried out by individual artists and spawned a genre of writing on art and anatomy which has been passed down in the form of both manuscripts and published textbooks. The founding members of the Florentine academy, Alessandro Allori and Vincenzo Danti, both wrote treatises on the subject a few years after the academy's inauguration: Allori's richly illustrated manuscript was never published, however, and Danti's first printed volume of 15 books (14 of which are now lost) on human anatomy and proportions was without illustrations. Thus, the first printed *and* fully illustrated book on artistic anatomy appeared in Spain, not Italy. This first work, *Varia commensuracion para la escultura y arquitectura*, was written and illustrated by the goldsmith and sculptor Juan de Arfe y Villafañe and was published in Seville in 1585.

---

[9]   'Se iba, cuando hacían anatomias, a los hospitales para ver y dibujar los músculos, nérvios y tendones, para quedar a su deseo satisfecho.' Jusepe Martínez, *Discursos practicables del nobilisimo arte de la pintura* (Madrid: Real Academia de San Fernando, 1866), p. 150.

[10]   Marinella Pigozzi, 'Arte e scienza a Bologna da Papa Gregorio XIII a Papa Clemente VIII (1572–1605)' in *Rapprasentare il corpo, arte en anatomía da Leonardo all'Illuminismo* (Bologna: Bononia University Press, 2004), p. 133.

[11]   'Vogliamo etiamdio che que' Consoli cha saranno in ufficio nel tempo del Verno siano tenuti e debbano procurare che si faccia in Santa Maria Nuova una anathomia a beneficio de giovani dell'Arte del Dissegno, alla quale debbono tutti esser chiamati per ordine d'essi Consoli.' Zygmunt Wazbinski, *L'Accademia Medicea del disegno a Firenze nel cinquecento* (Florence: Olschki, 1987), p. 424.

This pioneering treatise on artistic anatomy was rooted in Arfe y Villafañe's admiration for the scientific training of artists in Italy and France, and the elevated reputation of the visual arts in these countries, which he wished to emulate in his native Spain.[12] While there is no evidence that the author ever left Iberia, he emphasised that an educational background in Italy was crucial for the training of Spanish artists such as Alonzo Berruguete and Caspar Becerra, who studied with Michelangelo, Daniele da Volterra and Vasari in the early and mid-sixteenth century, and later brought their newly acquired skills back to Iberia: 'These two extraordinary men eradicated the barbarity which existed in Spain and gave light to other talents which have followed since then.'[13] The long-established Italian expertise with the dissection and detailed representation of the human body represented a distinctive union of medical and artistic talent, which would be imitated elsewhere during the mid-sixteenth century. Yet the first post-Vesalian illustrations produced by Spanish authors were still direct copies from already-existing Vesalian images with only few original contributions, as seen in the works of Montaña and Valverde.

Bernardino Montaña: Vesalian Iconography and Pre-Vesalian Anatomy

Bernardino Montaña's *Libro de la anothomia del hombre* did not include written references to Vesalius, but instead revealed its reliance on his work in a number of crude woodcuts which were directly copied from *Fabrica*. The 12 drawings were rather inadequately reworked in the last pages of Montaña's textbook and stand out almost as caricatures of the originals. The images were not mentioned in the body of Montaña's text and seem to have been added in the very last phase of the production in order to serve a largely decorative purpose. They were placed at the end of the book with only a few words of introduction: 'The following figures shown here are necessary for a more clear understanding of the important things that we have dealt with in the past book on human anatomy.'[14]

The introduction established a vague connection to the written content of Montaña's textbook and emphasised the use of illustrations as both important

---

[12]   'He querido tomar este trabajo, y aprovechar à los hombres de mi Arte que quisieren acertar en ella, por ver la falta que hasta ahora ha habido en España de gente curiosa de escribir, habiendo muchos que lo pudieran haber hecho, imitando á otras naciones, principalmente á los Italianos y Franceses que no han sido descuidados de la curiosidad de sus tierras.' Juan de Arfe y Villafañe, *De varia commensuracion, A los lectores* (1587), p. IX.

[13]   'Estos dos singulares hombres desterraron la barbaridad que en España había dando nueva luz á otras habilidades, que despues succedieron y succeden.' Ibid., p. 95.

[14]   'Siguiente algunas figuras necessarias para entender mas claramente algunas cosas muy importantes que hauemos tratado en el libro passado de la Anothomia del hombre.' Bernardino Montaña, *Libro de la anothomia* (1551), f. 129r.

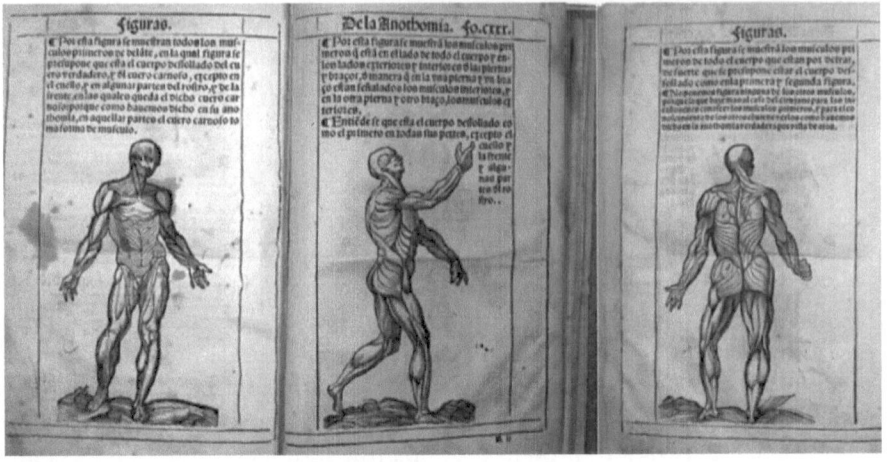

Figure 10.1   Bernadino Montaña's 1551 *Libro de la anothomia del hombre* was an amalgamation of traditional Galenic anatomy and contemporary Vesalian iconography. The first three illustrations were copies of the musclemen from the second book of *Fabrica*, shown in anterior, posterior and lateral views, though without the classical landscape that had provided the background to the original source. The next images represented the venous, arterial, central nervous, and genito-urinary systems, and were taken from Books 3, 4 and 5 of *Fabrica*. Image reproduced from the original preserved in the Biblioteca Históricomédica 'Vicent Peset Llorca, Universitat de València'. Label: IHMC Fons Antic 0617

and necessary, but no other such correlation between text and image is to be found in this work. These first Vesalian plates in a Spanish anatomy book thus failed in their original function as scientific diagrams and visualisations of the written text. With no correspondence between words and images, the plates were reduced to mere visual appendices, seemingly added only to satisfy recent reader appetites for anatomical illustrations whetted by those seen in *Fabrica*. Montaña's text nonetheless recognised crucial aspects of the 'Vesalian revolution', as it openly applauded knowledge obtained through the visual and tactile senses: 'The things that can be seen and touched cannot be understood better than through the sense of sight and touch. For this reason we advise that the physician or surgeon who wants to gain complete knowledge of this science should be trained in observing real and true hands-on anatomy many times.'[15]

---

[15]   'En las cosas que se pueden ver y palpar ninguna manera ay de darlas a entender tan perfectamente como por el sentido de la vista y del tacto y por esta razon es nuestro consejo que el medico o cirujano que quisiere saber cumplidamente esta ciencia, se exercite en ver hazer Anothomia real y verdadera muchas vezes por incision de manos.' Ibid., *Epistola dedicatoria*, n.p.

While this statement recognised the limitations of purely textual evidence, the added illustrations with short written explanations in the last part of Montaña's textbook did not correspond to any of the previous chapters and the Vesalian plates were even organised differently from the written content of *Libro de la anothomia*. The seven books of *Fabrica* presented 'systematic anatomy' – the anatomy of the bones, muscles, veins, arteries, etc. rather than the 'sectional anatomy' of Montaña's own text, which opened with descriptions of the head, thorax and viscera – and examined different regions of the body *a capite ad calcem*. In the process of reworking and copying the original design on new woodblocks, the Vesalian illustrations were not only impoverished and distorted, but mirrored as well, as was also the case in Valverde's subsequent copper prints copied from the Vesalian originals. This process involved an obvious risk of distorting anatomical data in mirror views, which was, however, seemingly acknowledged by both authors. Montaña's first three illustrations were direct copies of the musclemen from the second book of *Fabrica* shown in anterior, posterior and lateral views, but without the classical landscape – arguably the Eugenian hills outside Padua – which provided the background to the original source. The next set of images represented the central nervous, venous, arterial and genito-urinary systems, taken from Books 3, 4 and 5 of *Fabrica*. One of the icons in the history of anatomical representation – the weeping skeleton leaning towards a spade – was shown in anterior and posterior view, in the latter case in a different and even poorer style than the former imitated illustrations. Without a Vesalian matrix at hand, Montaña's illustrator turned to a 'pre-Vesalian' image of a human skeleton seen from behind in another unacknowledged borrowing – this time from Berengario de Carpi's richly illustrated *Isagogae breves* (Bologna, 1522).[16]

In spite of its poor artistic quality and limited medical use, Carpi's original illustration may have been the model for Vesalius' later depiction of a skeleton meditating in front of its own tomb. Montaña's slightly altered representation of this image – depicting the skeleton with a scythe – may also have inspired the alteration of the title page of Vesalius' revised 1555 version of *Fabrica*, where the pictured skeleton had its magisterial cane replaced with a scythe, thereby underlining its *memento mori* symbolism. In the original Vesalian illustrations the musclemen and skeletons from Books 1 and 2 of *Fabrica* were portrayed from many perspectives, and with a confident use of 'cinematographic' techniques which represented the body in different sequences and stages of dissection. The dissected bodies were shown in the foreground of an increasingly ruinous landscape that invited the spectator to contemplate the frailty of man and the brevity of life. This complex imagery linked the human cadaver to the decline of monuments and civilisations, and arguably made further allusions to the seasons, elements and ages

---

[16] Acknowledged by Charles D. O'Malley in his short article 'Bernardino Montaña, Author of the First Anatomy in The Spanish Language', *Journal of the History of Medicine*, 1 (1946), p. 104.

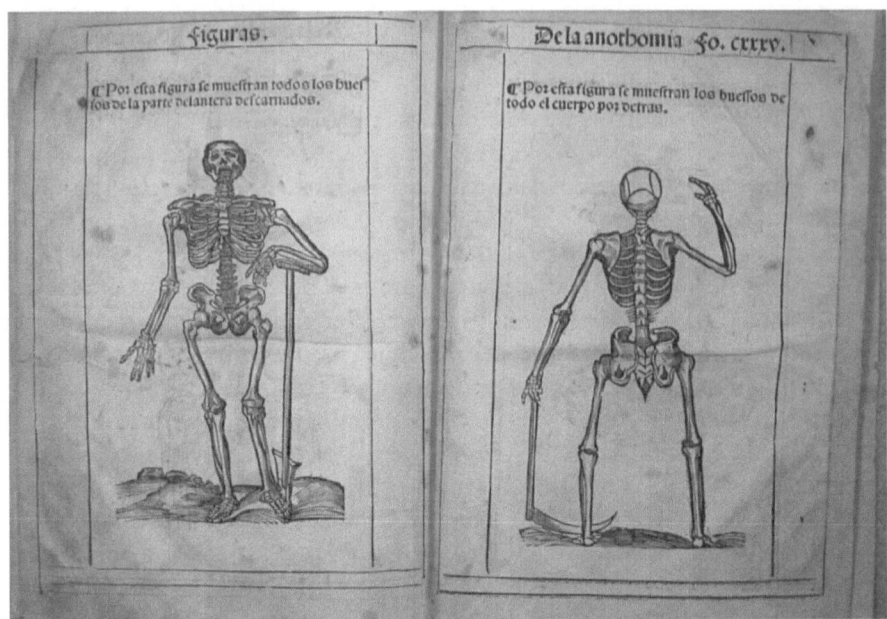

Figure 10.2   One of the 'icons' in the history of anatomical representation – the weeping skeleton leaning on a spade – shown in anterior and posterior view, in the latter case in a different and even poorer style than seen in the former, crudely imitated illustrations. Without a Vesalian matrix, Montaña's illustrator turned to a pre-Vesalian image of a human skeleton seen from behind, borrowed from Berengario de Carpi's richly illustrated *Isagogae breves* (1522). Image reproduced from the original preserved in the Biblioteca Históricomédica 'Vicent Peset Llorca, Universitat de València'. Label: IHMC Fons Antic 0617

of nature and man. Consequently, it can be argued that the qualities of the original plates lay not only in their skilful rendering of anatomical details, but also in a complex iconographical programme intended to stir the reader. Montaña's poor imitations lacked the ability to evoke such strength of feeling, not only because of the copyist's lack of artistic skill, but also due to the absence of the complex allusions created by the iconography of *Fabrica*. In a few brief comments on his replicated illustrations of the Vesalian musclemen, Montaña acknowledged the limitations of his anatomical plates and their restricted usefulness for surgeons with no personal experience in anatomy: 'We will not add images of the other muscles because the surgeon who carries out incisions should first of all know the primary muscles.'[17]

---

[17]   'No ponemos figura ninguna de los otros musculos, porque lo que haze mas al caso del cirujano para las incisiones es conocer los musculos primeros, y para el conocimiento de los

The iconographic and textual complexities of *Fabrica* were almost entirely absent from Montaña small textbook, as the reader was warned in its opening passage: 'In this book we deal only with the things that are helpful and necessary to know for the physicians and surgeon, leaving out everything that does not serve this purpose.'[18] This purely practical approach could be interpreted as a move towards a more restricted style and representation, but in fact the author's style was dictated by his intended target audience of common surgeons unfamiliar with the classical and humanist traditions which Vesalius took pride in imitating. Montaña's work was dedicated to the mighty patron of the arts and sciences, Luis Hurtado de Mendoza, Marquéz de Mondéjar. As in the later 1556 publication by Valverde, the patron's coat of arms dominated the whole title page of *Libro de la anothomia*. The second part of the book, 'the Marquéz's dream' was framed by Renaissance portico richly ornamented with grotesques and was notable as one of the only original illustrations produced for the book. Montaña's textbook on anatomy exists only in the first edition from 1551. It was seemingly never reprinted and appears to have been largely forgotten by the early seventeenth century.

Juan Valverde's *Historia de la Composicion del Cuerpo Humano*: Plagiarism and Modification of Vesalian Iconography

Like Montaña's *Libro de la anothomia*, Valverde's Spanish anatomy from 1556 was not reprinted in later editions, but was instead modified and translated into several Latin and vernacular editions throughout the sixteenth and seventeenth centuries. Similar to Montaña's use of Vesalian imagery, Valverde's illustrations were also reduced versions of the originals from *Fabrica*, and in this case as well the author explicitly aimed his publication at an audience of Spanish surgeons unfamiliar with Latin and humanist learning. According to Valverde, this particular group of medical practitioners had hitherto been kept in the dark; not only were they unfamiliar with dissection and Latin treatises, but they were also ignorant of the 'obscure' and overtly precocious written style of Vesalius. Valverde nonetheless reflected on the visual impact of *Fabrica*, which had proved so successful that alternative anatomical representations could no longer be considered: 'Even though it seemed to some of my friends that I should have made new illustrations, without using those of Vesalius, I decided not to do so, in order to avoid the confusion that could follow from this; from those who do not know

---

otros conuiene verlos como hauemos dicho en la anothomia verdadera por vista de ojos.' Montaña, *Libro de la anothomia*, f. 130v.

[18] 'En este libro trataremos solamente aquello que cumple y es necessario que sepa el médico y cirujano, dejadas todas las otras cosas que no hacen a su propósito.' Ibid., *Epistola dedicatoria*. n.p.

where I agree or disagree with him, and because his figures are so well executed
that it would seem envious and malignant not to wish to make use of them.'[19]
Both Valverde, who modified and sometimes criticised the Vesalian images, and
Montaña, who crudely imitated them, preferred to make use of extant Vesalian
illustrations of human anatomy rather than trying to create their own. Valverde's
book was published by the recently merged Franco-Spanish publishing houses
of Antonio Salamanca and Antonio Lafrery, and was probably illustrated by
Caspar Becerra and/or Pedro de Rubiales, two Spanish painters who were
active in Rome during the 1540s and 1550s under the patronage of the Spanish
Cardinal Juan Álvarez de Toledo. The prints for the book were produced by the
French engraver Nicolas Beatrizet, whose initials can be found in numerous
illustrated publications by Salamanca and Lafrery. Rubiales and Becerra worked
as assistants to Giorgio Vasari, and Becerra also collaborated with Daniele
da Volterra on the frescoes of the Roman Palazzo della Cancelleria and the
church of Trinitá dei Monti.[20] Their Italian masters belonged to Michelangelo's
innermost circle of friends and apprentices, and the two Spaniards were both
followers and admirers of the ageing yet unchallenged artistic authority of mid-
sixteenth-century Rome. Juan Valverde's master Realdo Colombo afforded him a
further connection to Michelangelo. Colombo enjoyed close links to the mature
artist, having served as his personal physician for almost a decade by the time of
Valverde's publication, and from then until Colombo's death in 1559. Before
Valverde's 1555 employment as Cardinal Toledo's personal physician, he worked
as an assistant to Colombo, who succeeded Vesalius as Professor of Anatomy
and Surgery at the University of Padua in 1543. Valverde also accompanied
Colombo to his new appointments at the Universities of Pisa and Rome in
1546 and 1548 respectively. According to Ascanio Condivi's contemporary
biography, *Vita di Michelagnolo Buonarroti* (Rome, 1553), Colombo provided
Michelangelo with bodies for dissection and both men agreed to jointly author
a work on anatomy in the late 1540s. This planned collaboration was confirmed
by a leave of absence from 1547, when Duke Cosimo I granted Colombo
permission to leave his professorship at the University of Pisa in order to execute
this – unfortunately unrealised – project with 'il primo pittor del mondo'.[21] If the
planned work had been carried out, it would arguably have challenged the iconic
status of the *Fabrica* produced by Vesalius and Titian's workshop, and may have
confirmed Michelangelo's prejudice against the Venetian school of painters and

---

[19]   'Aunque a algunos amigos mios parecia que yo deviesse hazer nueuas figuras, sin seruirme
de las de Vesalio, no lo é querido hazer por, euitar la confusión que dello se pudiera seguir, no se
conociendo tan facilmente enlo que conuengo ó desconuengo conel, y porque sus figuras estan
tambien hechas que me pareciera inuida ó malignidad no querer aprovecharme dellas.' Ibid., *Al letor*.

[20]   Albert Frederick Calvert, *Sculpture in Spain* (London: John Lane, 1912), p. 98.

[21]   Schultz, *Art and Anatomy in Renaissance Italy*, p. 102.

their lack of 'disegno' compared to Florentine masters like himself.[22] The only illustration in Realdo Colombo's work appeared on the title page, on which the author was depicted dissecting a male cadaver while a small *putto* offered drawing tools to Michelangelo, who was prominently placed among the spectators in the foreground. Valverde's *Historia de la composicion del cuerpo humano* was in many ways a poor substitute for the unfulfilled collaboration between Colombo and Michelangelo, which was eventually executed by their less renowned apprentices and admirers. While the text included numerous original observations and corrections of human anatomy, the images lacked the originality and the complex iconographical programme of Vesalius' *Fabrica*. Two-thirds of Valverde's images were directly copied or compiled from Vesalius' *Fabrica* and were reversed in the process of re-creating the original woodcut illustrations on new copperplates. Valverde's choice to reuse Vesalius' plates would, in his view, enable the readers to understand where the two authors differed in their anatomical research and observations.

In Book 1, which was dedicated to the bones of the human body, Valverde emphasised that his illustrations were not mere replicas of the Vesalian images, but were significant improvements on the originals, as described in an explanatory text

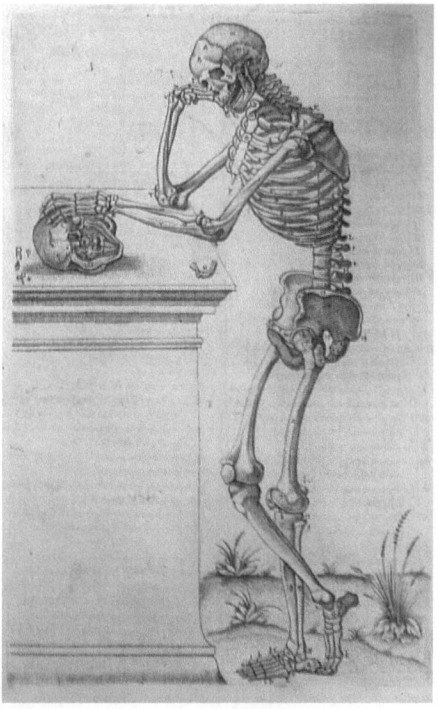

Figure 10.3 Even though Valverde's anatomy criticised some of the images from *Fabrica*, the author still preferred to use established Vesalian illustrations rather than creating his own. Among Valverde's minor modifications of the original images was the addition of the third auricular ossicle to the two already shown on the pedestal bearing of Vesalius' *Fabrica*. Valverde deprived Pedro Jimeno of the honour of being recognised as the discoverer of the stapes in his supporting reference to the ossicle 'of which no one before me has made any mention'. Courtesy of the Library of the Department of Cellular and Molecular Medicine, Panuminstituttet, University of Copenhagen

---

22   Giorgio Vasari, *The Lives of the Great Artists* (Oxford: Oxford University Press, 1991), p. 501.

next to his modified Vesalian image of a meditating skeleton in lateral view: 'I want to make known to the reader that the first figure is different from that of Vesalius, because his was not very well done, as anyone can see.'[23] Among his modifications of the original image was the removal of the inscription on the pedestal, 'Vivitur ingenio, caetera mortis erunt' ('Genius lives on, everything else is mortal'). Valverde furthermore added the third auricular ossicle – the stapes – to the two auricular bones already shown on the pedestal in Vesalius' *Fabrica*. His explanatory note wrongly claimed that he (and not the Valencian anatomist Pedro Jimeno) was the discoverer of 'the third ossicle of which no one before me has made any mention'.[24]

Despite this false attribution, the visual addition of newly accumulated data of anatomical structures presented Valverde's imagery as an integral part of his textbook. It is questionable, however, whether his illustrations were significant improvements on Vesalius' anatomical representations; many of Valverde's revisions of the Vesalian images were distorted and less elegantly executed, and some of his corrections were plainly wrong, as seen in his disproportionate visual enlargement of the human sternum. Still, Valverde's book included significant innovations, such as the most detailed description of the time of 'the smaller circulation', which abandoned the Galenic notion that blood circulated through invisible pores in the septum between the two heart ventricles. Besides the stapes, which he presented visually for the first time, Valverde's anatomy included numerous innovative descriptions of the abdominal and facial muscles, and a correction of Vesalius' inaccurate placement of the crystalline lens in the centre of the human eye.[25]

While Valverde often praised Vesalius, he was also critical, and even chastised Vesalius for his inability to distinguish between human and animal anatomy – an accusation which was remarkably reminiscent of Vesalius' own claims of Galen's misconceptions based on animal dissection: 'This figure is different from Vesalius' because this one shows only the muscles as they appear in humans, and in his some are shown, which exist only in monkeys and other brute animals.'[26]

---

[23] 'Solo quiero advertir al letor, que la primera figura es differente de la de Vesalio, porque la suya no estava bien hecha, como cada uno podra ver, conferiendo las partes en que differimos con el natural.' Ibid., Tab. Segunda del lib. Primero, Fig. II.

[24] 'El tercer ossezuelo del qual ninguno antes de mi ha hecho mención.' Ibid., Libro primero. Fig. IIII.

[25] Further corrections of the Vesalian plates were repeated in the description of the muscles of the face and the body, which were shown in lateral view after superficial dissection. 'Esta figura es la misma que la primera buelta del lado, y differe de la del Vesalio, en que se veen en esta los morzillos de la cara; y la tela del sesto morzillo que mueve la pierna, este alcada, para que se vean major los morzillos.' Ibid., Libro segundo. Tab. Segunda.

[26] 'Esta figura differe de la del Vesalio en que en esta no se veen los morzillos sino como estan en el hombre, y en la suya se veyan algunos que se hallan en las monas, y en otro animales brutos.'

Valverde justified his choice to reproduce the Vesalian images by claiming that his readers could more easily follow his visual corrections if he remained faithful to the existing iconography. His *Historia* was in fact so loyal to the Vesalian matrix that only one-third of its illustrations were original. Of the 42 images in *Historia de la Composición del Cuerpo Humano*, 27 were copied directly – and according to the author improved significantly – from *Fabrica*, while only 15 new illustrations were created for his anatomical textbook. Among the new illustrations was the famous 'Valverdean muscleman', which has since become an icon of anatomical illustration – often juxtaposed with the most celebrated images of Vesalius. It represented a flayed man carrying his own skin in one arm and a dagger in the other. This was arguably a depiction of Marsyas, the ill-fated satyr from Greek mythology who was flayed alive by Apollo, or alternatively a portrayal of St Bartholomew, the Christian martyr who suffered a similar fate in the hands of his captors and who was also depicted grasping a knife. This second suggestion links Valverde's muscleman to the representation of St Bartholomew in Michelangelo's *Last Judgement* fresco, which had been finished a decade earlier. The specific posture of the figure may have been inspired by Benvenuto Cellini's contemporary sculpture *Perseus*, which was completed for Duke Cosimo I de' Medici in Florence in 1554.[27] Valverde took great pride in this new image and emphasised its originality: 'It shows a man who has been flayed of his skin and fat, and all the veins that run between the skin and the flesh, and all the blood vessels, except for the muscular parts. And it should be noted that it is different from that of Vesalius, because in this figure the shadows show the lines and structures of the flesh as they appear in every muscle.'[28]

The first attributions of the Valverdean illustrations to the Spanish painter Caspar Becerra appeared in Spanish art treatises in the mid-seventeenth century, such as Vicente Carducho's *Dialogos de la pintura* (1633) and Francisco Pacheco's *Arte de la pintura, su antiguidad y grandezas* (1649). Both authors agreed that Becerra was the illustrator of *Historia de la composición del cuerpo*

Ibid., Libro segundo. Tab. Quarta.

[27] Valverde and Colombo were closely linked to Tuscany and Florence since Duke Cosimo I de' Medici's employment of Colombo at the University of Pisa – and through the Spanish cardinal Juan Álvarez de Toledo (the patron of Valverde, Rubiales, and Becerra) who was the uncle of Duke Cosimo's wife, Eleonora de Toledo, and arguably the Duke's most powerful ally in Rome. Another link with Valverde's network is seen in a recently discovered 1551 letter of introduction by Toledo, which recommended the services of Valverde's supposed illustrator, Caspar Becerra to Duke Cosimo. Doc. ID 4957 MdP, 1176, inserto 11, f. 26. Cited in Salvador Salort Pons, 'Gaspar Becerra en Florencia', *Archivo Español de Arte*, Vol 78, No 309 (2005), p. 102

[28] 'Muestra un hombre quitado el pellejo y la gordura, y las venas que van entre cuero y carne, y toda la tela carnosa salvo las partes della que se convierten en morzillo. Y es de saber, que esta figura es differente de las del Vesalio, en que en esta las sombras muestran el andar del hilo de la carne, segun que en cada morzillo particularmente caminan.' Ibid., Libro segundo. Tab. Primera.

humano, but Valverde instead credited the artistic – as well as anatomical – skills of the painter Pedro Rubiales, who was even praised in the same sentence as the mighty Michelangelo: 'In our times Michelangelo of Florence and Pedro Rubiales of Extremadura have united anatomy together with painting, and they have therefore become the most excellent and famous painters of all times.'[29] The Spanish medical historian Juan Riera has used this and other circumstantial evidence to attribute the illustrations in Valverde's book to Rubiales.[30] Becerra, however, has been repeatedly credited as the artist responsible for the same illustrations since Carducho's short and somewhat flawed account in his 1633 *Dialogos de la pintura*: 'The anatomical drawings of the French author Vesalius, were carried out excellently by the French artist Juan Calcar, and those of Valverde the Spaniard by the famous Becerra, and both followed Vesalius in one way or another.'[31] Francisco Pacheco's subsequent *Arte de la pintura, su antiguidad y grandezas* made multiple similar references to Valverde's anatomy and its illustrations by Caspar Becerra, who was described as: 'Our brave Spaniard who brilliantly demonstrated what he knew about the muscles (an as imitator of Michelangelo) in Valverde's book of anatomy.'[32] Jusepe Martínez's unpublished art treatise *Discursos practicables del nobilisimo arte de la pintura* (1672) also praised the 'anatomical drawings of our great Becerra'.[33] Enthusiastic references to Becerra as the artist appeared well into the eighteenth century; his anatomical skills were described in detail in Palomino de Castro y Velasco's art treatise *El museo pictorico y escala optica* (1715–24): 'Our Becerra was a truly great anatomist, and today some of his anatomies remain ... and I own them, together with a leg from an anatomical study of baked clay, which is the left, an original of his, about half the natural size, which everyone admires; in my time

---

[29]   'Quando esto sea verdad nos lo han hecho ver en nuestros tiempos Miguel Angel Florentin, y Pedro Rubiales estremeño, los quales por averse dado ala Anatomia juntamente con la pintura an venido á ser los mas excellentes y famosos pintores que grande tiempos à se avisto.' Juan Valverde, *Historia de la composicion de cuerpo humano*, Libro segundo, Tab. tercera.

[30]   Juan Riera, *Valverde y la anatomia del renacimiento* (Valladolid: Ediciones de la Universidad de Valladolid, 1981), p. 18.

[31]   'Para la Natomia, el Vesalio, Autor Francés; cuyus dibujos excelentemente hizo Juan Calksuz Frances, y los del Valverde Español, el insigne Bezerra, ambos siguiendo en todo lo uno, y en lo otro al Vesalio.' Vicente Carcucho, *Dialogos de la pintura* (Madrid: Fr. Martinez, 1622), f. I.

[32]   'Nuestro valiente español Caspar Becerra que hizo tan gran demonstración de lo que sabía de músculos (como singular imitador de Miguel Angel) en el libro de Anatomía de Valverde.' Francisco Pacheco, *Arte de la pintura, su antiguidad y grandezas* (Seville: Simon Faxardo, 1649), p. 276.

[33]   Jusepe Martínez, *Discursos practicables del nobilisimo arte de la pintura*. (Madrid: Real Academia de San Fernando, 1866), p. 12.

it has prevented the amputation of some legs, enlightening the surgeons, who examined its organisation of the muscles, tendons, and nerves.'[34]

The authorship of these illustrations is still disputed in spite of these later claims. Recent stylistic investigation suggests that Becerra may have produced some of the high-quality imagery, such as the flayed St Bartholomew and the elegant title page of the original Spanish edition (where the architectural background is somewhat similar to Becerra's surviving design for the high altar in Las Descalzas Reales kept at the Biblioteca Nacional in Madrid), while the Vesalian copies were probably carried out by a lesser artist.[35] We can be sure that the engraver was Nicolas Beatrizet, whose identity is confirmed by his initials on some of the illustrations of this first Spanish-language book on science and medicine to be illustrated with copperplate engravings.[36] This reproductive technique was not used in an Iberian book publication until 30 years later in Juan de Herrera's *Sumario y breve declaración de los diseños y estampas de la fábrica de San Lorenzo del Escorial* (Madrid, 1589), which was illustrated with coppers by the Flemish engraver Pieter Perret, who is credited with the introduction of the technique of engraving in Spain.[37]

While the complex classical references in Vesalius' original work were downplayed in Valverde's textbook, the iconography of the latter work's few original illustrations also referred to classical antiquity, and in particular to antique Roman sculptures. Some of these, such as the Belvedere Torso (already used in Vesalius' illustrations) and the Standing or Pudic Venus were shown in different stages of dissection, in the latter case representing the uterus and the placenta. The inclusion of these (and arguably other iconic statues from the papal collections such as Laocoön and Apollo Belvedere)[38] may have served as

---

[34] 'Fue nuestro Becerra grandissimo anatomista, y hoy permanecen unas anatomias, una grande como de a vara, y otro como de a sesma, que son suyas, y otra como de un crucifijo, cosa excellente, y yo las tengo, juntamente con una pierna de anatomia de barro cocido, que es izquierda, original suya, como la mitad del natural, que admiran a quantos que ven; y en mi tiempo ha excusado cortar algunas piernas, llevandolas y sirviendoles de luz a los cirujanos, para reconocer por la organisación de sus músculos, tendons y nervios, por donde va y viene la corrupción, y cauterizar o manifestar la parte que convenga a la curación.' Palomino de Castro y Velasco, *El museo pictorico y escala optica* (Madrid: Juan Bernabé, 1795), p. 173.

[35] Diego Suárez Quevedo, 'Arte-ciencia (anatomía) en el Renacimiento español. La obra de Juan Valverde de Amusco y su clasicismo' in *Actas del X Congreso del CEHA. Los Clasicismos en el Arte Español* (Madrid: Departamento de historia del arte de la UNED, 1994), p. 479.

[36] Pilar Rodriguez Marin, *Juan Valverde de Amusco* in *Facultad de medicina de Valladolid. VI Centenario* (Valladolid: Junta de Castilla y León, 2006), p. 83.

[37] José Maria López Piñero, *El grabado en la ciencia hispánica* (Valencia: CSIC, 1987), p. 34.

[38] According to Rose Maria San Juan's article 'Restoration and Translation in Juan de Valverde's *Historia de la composición del cuerpo humano*', the third image of Valverde's Book 3, *De los miembros necessarios ala Digestion y ala Generacion*, 'evokes the Laocoön by the addition

references to the patronage of Cardinal Juan Álvarez de Toledo. Other allusions to Roman sculptures in Valverde's 'intestinal cuirasses' were inspired by the trophies of Marius on the Esquiline, and were reminiscent of images in Antonio Salamanca and Nicolas Beatrizet's earlier illustrated books on Roman antiquity. In another publication, Valverde's Spanish publisher referred to himself as *Antonius Salamanca, Orbis et Urbis Antiquitatum imitator.*[39]

The author, patron, illustrator(s) and publisher of Valverde's anatomy were of Spanish origin, and in spite of its specific Roman context, Valverde's work maintained some Spanish characteristics even in its later impressions and translations. The subsequent Italian translation was carried out by Valverde's friend Antonio Tabo of Albengo, allegedly under the close supervision of the author: 'Making sure that no error was made in the translation and nothing added or taken away.'[40] Like Valverde's original Spanish edition, the *Anatomia del corpo umano* was printed by Salamanca and Lafrery (Rome, 1559) and was also dedicated to a Spanish patron – not to Juan Álvarez de Toledo, who had recently died in 1557, but to the Spanish monarch Philip II. The timing of Valverde's appeal to the mighty sovereign was hardly coincidental. In a royal provision from 1559, Philip II recalled all Spanish students and scholars studying and working outside Spain with an announcement which has since been seen as an ominous event in the cultural and intellectual history of late Renaissance Spain: 'No one can leave these kingdoms to study, teach, learn, or reside at universities, schools, or colleges outside these realms, and those who have gone to stay and reside at these universities, schools, and colleges are required to leave within a period of four months after the announcement of this provision.'[41] In the dedication of Valverde's 1559 edition of *Anatomia del corpo umano*, the author tried to avoid being relocated and so emphasised his own natural affiliation with the Italian

---

of a turned head and an arm that contains the snakelike intertwined intestines'. Valverde's second image in his Book 2 on muscles allegedly shows the 'disruption of the contours of the Apollo Belvedere by stressing the ragged edges of skin left behind by the knife's cut' in Rebecca Zorach (ed.), *The Virtual Tourist in Renaissance Rome* (Chicago: University of Chicago Press, 2008), pp. 57–60.

[39] Andrea Carlino, 'Tre piste per la anatomia di Juan di Valverde', *Mélanges de l'ecole francaise de Rome. Italie et Méditerranée*, 114 (2002), p. 529.

[40] 'Nel tradurre non vi si commettesse errore alcuno; ne vi si aggiungesse, ó levasse nulla.' Giovanni Valverde, *Anatomia del corpo umano* (Rome: Antonio Salamanca et Antonio Lafrery), f. 2r.

[41] 'No pueden ir ni salir de estos reinos a estudiar ni enseñar, ni aprender ni a estar ni a residir en universidades, estudios, ni colecios, fuera de estos reinos, y que los que hasta agora y al presente estuvieren y residieren en tales universidades, estudios y colegios se salgan y no estén más en ellos dentro de cuartro meses, después de la data y prohibición de ésta.' José Maria López Piñero, *Ciencia y tecnica en la sociedad Española de los siglos XVI y XVII* (Barcelona: Labor, 1979), pp. 142–43.

Peninsula and cleverly highlighted the Prudent King's status not only as Spanish monarch, but also as 'patron and protector of all Italy'.[42]

Apart from his attempt to escape the restrictions imposed by the royal decree, the most remarkable information in the revised Italian edition of Valverde's anatomy was his disclosure of the unexpectedly negative consequences of having copied the Vesalian illustrations in his own work. This decision had led to the misconception among non-Spaniards that his work was a mere Spanish translation of Vesalius' *Fabrica*: 'Many of those who did not understand the Spanish language, and who saw that my illustrations were not very different from his began to claim that I had just translated the work of Vesalius.'[43] Valverde wished to prove the inaccuracy of these accusations with the translation of his work into Italian, thereby vindicating his status as an original author and researcher. Yet he also admitted that the new Italian edition was produced in order to make further use of his engraved copperplates, as indicated in both the subtitle and the introduction of his *Anatomia del corpo umano composto per M. Giovan Valuerde da Hamusco & da luy con molte figure di rame ed eruditi discorsi in luce mandata*: 'To satisfy the wishes of many of my Italian gentlemen friends (who on seeing that my work was much shorter than that of Vesalius, and understanding that it was in many ways different from his; and seeing also that some of my illustrations were more graceful and adept than his, wanted to see the book written in their own language), and being in possession of these engraved copperplates, I wanted to translate it into Italian.'[44] This decision seems somewhat contradictory, given that in his dedication to Philip II, Valverde had lamented the fact that the same illustrations had led to confusion and accusations of plagiarism.

---

[42] 'Al che si é aggiunto, l'esser io piu tenuto alla natione Italiana che á niun'altra dalla Spagnuola in fuori. Hor uolendo io mandar questo mio libro sotto un buon appoggio, non ho potuto trouare un'altro piu á proposito che la Maesta vostra, essendo ella comun padrone, & protettore della Italia tutta; Alla quale humilmente supplico che accetti il buon animo, se l'opera non sará degna d'un tanto Prencipe. La cui uita il signore Iddio conserui, & seliciti, secondo che a tutta la Republica Christiana piu bisogna.' Giouanni Valverde, *Anatomia del corpo umano*, Alla S.C.R Maesta el Re Filippo, n.p.

[43] 'Succede dapoi, che molti non intendando la lingua Spagnola, & vedendo le mie figure non molto diverse da quelle, cominciarono á dire ch'io havea tradotta l'historia del Vesalio.' Ibid., f. 2v.

[44] 'Non dimeno per satisfare á prieghi di molti gentilhuomini italiani amici miei (liquali veggendo l'opera mia essere assai piu breue che quella del Vesalio & intendendo che era in molte cose differente dalla sua; & parendo anche loro le mie figure alquanto piu leggiadre & accommodate, che le sue, desiderando di vederla nella lor lingua) & anche trouarmi con li rami intagliati, ho voluto pigliar questa sarica di ridurla in lingua Italiana.' Ibid., Alla S.C.R Maesta el Re Filippo, n.p.

The revised work was given a new title page that depicted a marble niche flanked by two standing skeletons, and a pig and a monkey holding a human femur. This was a clear reference to the new discoveries made by anatomical reformers such as Vesalius, Colombo and Valverde, who had found numerous errors in Galen's anatomy and many obvious cases of discrepancy between human and animal anatomy. At the bottom of this title page were three small images of anatomical practice, representations which have been analysed in Andrea Carlino's *Books of the Body*.[45] These small vignettes depicted both public dissections for the benefit of a large university audience and more intimate anatomy studies performed for a few select pupils, indicating the dual nature of this practice. Outside medical circles, Valverde's Italian anatomy found a broad audience among contemporary artists and scientists, such as Egnazio and Vincenzo Danti, whose treatises on anatomy, optics and perspective praised Valverde's correction of Vesalius' erroneous placement of the crystalline lens in the centre of the human eye.[46] Other Florentine artists and members of Florence's recently founded *Accademia del Disegno* (1563), such as Alessandro Allori and Ludovico Cigoli, frequently referred to Valverde's anatomy in their own written treatises, such as Allori's *Le regole del disegno* (Florence, 1565), and in numerous sketches and artworks. The *Nuttumia di Valverde* was also found in the 1608 book inventory of the *Accademia di San Luca*, the first Roman art academy, which was founded in 1593.[47]

After the publication of the first editions of Valverde's book in Spanish (Rome, 1556) and Italian (Rome, 1559 and 1560), the work appeared again in 1566 in a revised Latin edition entitled *Vivae imagines partium corporis humanae aereis formis expressae*. This new Latin edition was published by Plantin in Antwerp and was soon followed in 1568 by a Dutch version, *Anatomie, oft levende beelden vande deelen des menschelicken lichaems*. While the facts of the artistic collaborations behind the illustrations of the original Roman Salamanca/ Lafrery editions of Valverde's anatomy remain fogged by uncertainty, the reworking of these images for the Plantin editions is documented in remarkable detail. The surviving account books of the Antwerp publishing house reveal

---

[45] Andrea Carlino, *Books of the Body* (Chicago: University of Chicago Press, 1999), pp. 54–56.

[46] 'Questa e la descritione dell'occhio, tratta da libri dell'Annotomia di Vincentio Danti; Doue perche si vede il centro dell humor Christallino fuor del centro della sfera dell occhio per la quinta parte in circa del suo diametro, non lascerò in questo proposito di auuertire che il Vessalio et altri che possero l'humor Cristallino concentrico all'occhio, hanno errato, non pure per quello che ho osseruato nel Valuerde, et in Vincentio Danti, ma anco per la proua, che ne ho da me stesso fatta in molte Annatomia, che feci alter volte in Firenze et in Bologna doue sempre trovai il centro dell'humor Cristallino fuori di quello della palla dell'occhio.' Egnazio Danti, *Le due regole della prospettiva pratica...* (Bologna: Francesco Zannetti, 1583), p. 3.

[47] Archivio di Stato di Roma, TNC, uff. 11, 1608, pt I, vol. 76, f. 818.

that the production and printing costs of the copperplate illustrations for Valverde's book represented a significant investment amounting to 606 florins – more that six times the total cost of the paper used for the first 600 copies of *Vivae imagenes*.[48] These expenses help to explain why fully illustrated books on anatomy were still relatively rare by the mid-sixteenth century and why Valverde was so eager to reuse the copperplate illustrations made for his 1556 *Historia* in later editions, even though these images had led to accusations of plagiarism. The Plantin accounts show some of the economic risks involved in contemporary book production; in Sachiko Kusokawa's estimate, an illustrated quarto book was seven times as expensive as an unillustrated octavo book from the same publishing house.[49] As indicated in the book title, *Vivae imagines partium corporis humanae aereis formis expressae*, the attraction of Valverde's anatomy lay as much in its images as in its text, which was drastically reduced in the Plantin editions and supplemented with the full text of Vesalius' *Epitome* (Basle, 1543).

The 42 illustrations included in the Plantin editions of Valverde's book were executed more poorly even than the 1556 images and were often reversed again in the process of copying Valverde's Vesalian copies. These images for the Plantin editions were reworked on new copperplates which were designed and engraved by the Flemish artist brothers Pieter and Franz Huys. The new title page was designed by Lambert van Noort. All three artists involved were members of the Antwerp painters' Guild of Saint Luke and were frequently employed in the production of illustrated publications by the Plantin publishing house. Van Noort's title page reused some of the architectural background from Salamanca/ Lafrery's 1559 Italian edition, but in a significantly moderated form. The two human skeletons in the original were replaced by the figures of Adam and Eve, and the femur-carrying pig and monkey above the porch were supplanted by the Hapsburg insignia of Philip II, and the Most Catholic King's motto 'Dominus mihi adiutor' ('The Lord is my helper').[50] The same title page was used for the 1568 Dutch edition of Valverde's anatomy *Anatomie, oft levende beelden vande deelen des menschelicken lichaems* and appeared again two centuries later in an extraordinary and unexpected context: this time it formed the cover for the first modern Japanese textbook on anatomy, Sugita Genpaku's *Kaitai Shinsho* (*New Text on Anatomy*), which was published by Suharaya Ichibee (Tokyo, 1774). While the text of *Kaitai Shinsho* was based on works by later seventeenth-

---

[48]   Sachiko Kusokawa, *Picturing the Book of Nature. Image, Text, and Argument in Sixteenth-Century Human Anatomy and Medical Botany* (Chicago: University of Chicago Press, 2012), p. 53.

[49]   Ibid., p. 54.

[50]   The 1566 Plantin edition of Valverde's anatomy occasionally included anatomical broadsides or 'flap-books' bound at the end, as seen in a copy from the Wellcome Library rare books collection (EPB 7341). These fugitive sheets reveal the internal organs of the male and female body and were inspired by the cut-outs made for Vesalius' *Epitome* (Basle: per Ioannem Oporinum, 1543)

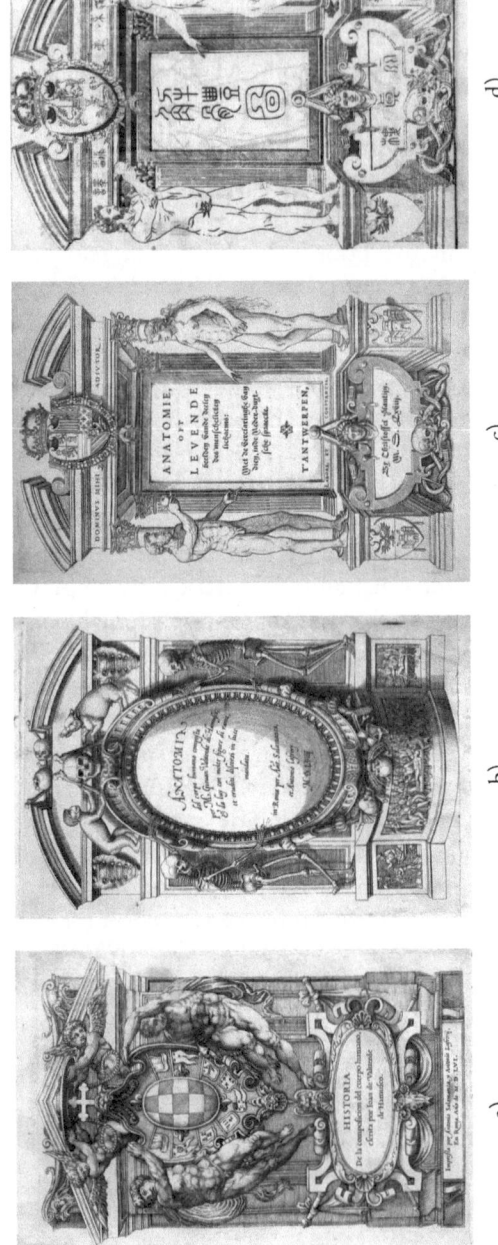

a)   b)   c)   d)

Figure 10.4   Title pages of the Spanish, Italian and Dutch editions of Juan Valverde de Amusco's anatomy. The original Spanish edition from 1556 (a) revealed the influence of Michelangelo's contemporary *Last Judgment* fresco in its exaggerated surface musculature of the two musclemen under the insignia of Valverde's patron, Cardinal Juan Álvarez de Toledo. The second 1559 Italian edition (b) represented two standing skeletons flanked by a monkey and a pig holding a human femur in a reference to recent anatomical evidence which proved the discrepancies between animal and human osteology. Valverde's anatomical textbook was translated into Latin and Dutch editions by Plantin in 1566 and 1568 (c). The Plantin title page appeared again two centuries later as the cover for the first modern Japanese textbook on anatomy, Sugita Genpaku's *Kaitai Shinsho* (*New Text on Anatomy*) (d), published by Suharaya Ichibee (Tokyo, 1774).

Source: (a)–(c) Courtesy of the Archives & Special Collections, Health Sciences Library, Columbia University; (d) courtesy of the C.V. Starr East Asian Library, Columbia University

century Dutch, German and Danish anatomists, the double-headed eagle and the golden fleece of Imperial Hapsburg Spain from the Plantin edition was retained on the title page of this unprecedented Edo period publication.

Italian and Latin versions of Valverde's anatomy were published by Giunti in Venice in 1586 and 1589, using new copperplates and with added illustrations. The author's death occurred at some point between these two publications. The latter publication, *Anatome corporis humani ... nunc primum a Michaele Colombo latine reddita*, was revised by Michele Colombo, son of Realdo, and showed the remarkable continuity of Valverde's Roman network some 30 years after the death of Realdo Colombo in 1559. The continued association of the creators of the original 1556 edition of Valverde's anatomy was also evident in the 1589 Latin edition, which included the only contemporary portrait of Valverde. This portrait was inscribed with the name of the author, 'Ioannes Valverdus Hispanus', and was signed by Nicolas Beatrizet, who had also produced the original engraved illustrations for Valverde's first Spanish and Italian editions three decades earlier. Several Valverdean images appeared in new anatomical textbooks throughout the sixteenth, seventeenth and eighteenth centuries, including David van Mauden's *Bedieninghe der anatomien, dat is maniere ende onderrichtinghe om perfectelijck des menschen lichaem t'anatomizeren* (Antwerp, 1583), André du Laurens' *Historia anatomica humani corporis* (Paris, 1600), Caspar Bauhin's *Theatrum anatomicum* (Frankfurt am Main, 1605) and the first German anatomy to be written in the vernacular, Johann Andreas Schenck's *Anatomia, Das ist: Sjnnreiche, Künstliche, Gegründte Auffschneidung, Theilung, und Zerlegung eines vollkommenen Menschlichen Leibs und Cörpers erkläret* (Frankfurt am Main, 1609).

The increasing amalgamation of the Vesalian and Valverdean images can be seen in a 1617 Dutch edition of Vesalius' work, whose title page was taken not from *Fabrica*, but from the first 1559 Italian edition of Valverde's anatomy textbook.[51] Valverde and Vesalius were even depicted side by side on the title page of a contemporary Venetian edition of Vesalius' anatomy, entitled *Andreae Vesalii Anatomia* (Venice, 1604) (Figure 10.5). Another Venetian edition of Valverde's *Anatomia del corpo humano*, which was published in 1606, eventually reached Ottoman Greece, where it circulated as the first modern Greek text on anatomy in a manuscript translation carried out in 1738 by L. Patousas and the priest-monk superior Demoscenes Petrakes. This *Anatomia of the Human Body by Joannes Valverde* included all the text from the 1606 Venetian edition translated into vernacular Greek on 396 folios. The translation even left room for inserting the (unfortunately unrealised) copies of Valverde's images next to the written

---

51   Andreas Vesalius, *Anatomia viri in hoc genere Princip Andreae Vesalii Bruxellensis; in qua tota humani corporis fabrica, iconibus elegantissimi iuxta genuinam Auctoris delineatio...* (Amsterdam: Ioannes Ianssonius, 1617).

text.[52] A few years earlier, a shortened French version of Valverde's anatomy had been published, including only the illustrations and their explanatory texts. It was intended for use by both physicians and artists, and was published under the title *L'anatomie universelle de toutes les parties du corps humain, Representée en Figures, & exactement expliquée, Ouvrage curieux, & utile aux Etudians en Medicine, Chirurgie, Sages-Femmes, & aux Peintres & Sculpteurs* (Paris, 1730).[53] Valverde's anatomical textbook was originally produced by Spanish residents in mid-sixteenth-century Rome and was intended for a limited audience of Spanish surgeons. However, it has been shown that the book appealed to a much larger readership than expected. We may note that only one impression was made of Valverde's original Spanish anatomy from 1556, while the Italian version from 1559 was reprinted the very next year, implying a demand similar to that for both the Plantin and Giunti editions which were reproduced in numerous new editions during the late sixteenth and early seventeenth centuries. While Valverde's *Historia de la composicion del cuerpo humano* in many ways failed in its original purpose, it instead succeeded as a smaller, cheaper and handier textbook on anatomy than Vesalius' magnum opus. Valverde's anatomy would reappear in translations and copies throughout the seventeenth and eighteenth centuries in leading medical and printing centres of early modern Europe – and occasionally far beyond.

Anatomy for the Artist: Juan de Arfe y Villafañe's *Varia commensuración para la escultura y arquitectura*

Nowhere in an Iberian context was the close, even dependent, relationship between anatomists and artists more evident than in Juan de Arfe y Villafañe's *Varia commensuración para la escultura y arquitectura* (Seville, 1585). This pioneering and richly illustrated treatise on science and theoretical knowledge applicable to the visual arts included four books on geometry, anatomy, zoology and architecture, of which the *Libro segundo* is a rare and unique source produced by an anatomically trained artist of the late Spanish Renaissance.[54] The second book was divided into four subchapters on the proportion, osteology, myology and foreshortening of the human figure, respectively entitled 'Titulo primero de la medida y proporcion del cuerpo humano', 'Titulo segundo de los huesos

[52]   Pan S. Codellas, 'Vesalius-Valverde-Patousas: The Unpublished Manuscript of the First Modern Anatomy in the Greek Language', *Bulletin of the History of Medicine*, 14 (1943).

[53]   *L'anatomie universelle de toutes les partes du corps humain, Representée en Figures, & exactement expliquée, Ouvrage curieux, & utile aux Etudians en Medicine, Chirurgie, Sages-Femmes, & aux Peintres & Sculpteurs* (Paris: Chez F. Gerard Jollain, 1730).

[54]   Books 1 and 2 on geometry and anatomy were published together in 1585. Books 3 and 4 on animals and architecture were added in the 1587 edition of *Varia commensuración para la escultura y arquitectura*.

Figure 10.5   The Latin Giunti version of Valverde's anatomy from 1589 included reworked and new images and the only verified contemporary portrait of Valverde ((a) and (b)). Valverde and Vesalius were depicted side by side on a Venetian edition of Vesalius' anatomy (Venice, 1604) ((c) detail). A contemporary Venetian edition of Valverde's *Anatomia del corpo humano* published in 1606, reached Ottoman Greece, where it circulated in a 1738 manuscript translation as the first modern Greek text on anatomy. Courtesy of the Archives & Special Collections, Health Sciences Library, Columbia University

del cuerpo humano', 'Titulo tercero de los morcillos del cuerpo humano' and 'Titulo quarto de los escorzos'.

Arfe's text included several references to prior sixteenth-century art theorists, such as Albrecht Dürer, Pomponius Gauricus and Sebastian Serlio, as well as classical authors like Pliny and Vitruvius. Yet much of his work was site-specific and deeply rooted in his own particular Castilian setting and context. Arfe's first book on geometry concluded with a chapter on gnomonics, solar clocks and quadrants based on prior astronomical observations carried out in Madrid and Valladolid; Book 4 on *Arquitectura y piezas de iglesia* included descriptions and depictions of monstrances made by Arfe and his relatives for Castilian cathedrals, as well as references to surviving ancient Roman architecture in Mérida, Ciudad Rodrigo and Cáparra. Arfe's 'Titulo tercero de los morcillos del cuerpo humano' ('Third Chapter on the Muscles of the Human Body') furthermore referred to the anatomical teaching of Cosme de Medina, who held the professorship of anatomy at the University of Salamanca between 1552 and 1561. Arfe's account of his own experience with human dissection was somewhat ambivalent, as he considered the practice to be of only limited value for the artistic study of surface anatomy, superficial myology and osteology: 'The other internal and external parts, like the vessels, nerves, and veins do not serve our purpose, and will not be dealt with here.'[55] He therefore dealt exclusively with the studies of bones and muscles, which were of direct use to the artist, who was summoned by Arfe 'to understand the bones and muscles of the human figure, because one would commit a thousand errors by not understanding them'.[56] The anatomical scrutiny of certain structures deemed useless for the artist was disregarded by the author, and the dissection of poor and condemned men and woman, as witnessed in Salamanca, was brushed aside as 'a horrific and cruel practice'.[57] Arfe also stressed this point in his decision to omit the visualisation of the facial muscles, which did not serve the artistic representation of the human face and therefore should be studied 'only by physicians and surgeons, and

---

[55] 'Los demas instrumentos, como las telas, los nervios, y las venas que tienen el cuerpo dentro y fuera no trataremos de ellas por no ser a este proposito.' Juan de Arfe y Villafañe, *Varia Commensuración para la Escultura y Architectura*, p. 141.

[56] 'Para entender los huesos y morcillos de una figura, pues no entendiéndolos, no sabrá hacerse sino con mil errores.' Ibid., Prologo, p. XI.

[57] 'Vimos desollar por las partes del cuerpo algunos hombres y mugeres, justificados y pobres, y demas de ser cosa horenda y cruel, vimos no ser muy decente para el fin que pretendiamos, porque los muscolus del rostro y barriga, nunca se siguen en la Escultura, sino por unos bultos redondos, que diremos adelante, y los de los brazos y piernas en el natural, se ven en los vivos casi determinada y distintamente, y asi los mostraremos, con los terminus altos y baxos que el natural muestra sobre el pellejo.' Ibid., p. 141.

not for sculpture and painting'.[58] Instead of representing the facial sphincters and mimic muscles, Arfe's illustrations divided the human face into a series of *bultos* or defined surface regions which did not correspond to the underlying musculture. By ignoring these muscles, Arfe seemingly followed Valverde, who admitted that the mapping of the facial muscles represented an unresolved problem for contemporary anatomists: 'The muscles of the face are so confusing and intricate that they have caused a lot of contradiction among anatomists, as well as disagreement between Vesalius and Galen, and between Realdo and Vesalius.'[59] Even though Arfe did not make any explicit references to Valverde, several descriptions from Valverde's anatomy were repeated almost verbatim in Arfe's treatise. Another major and unacknowledged source was Giorgio Vasari's *Vite*, which Arfe imitated in his account of the decline of Spanish art following the Gothic invasions, and in his description of the artistic revivals and triumphs spearheaded by sixteenth-century masters such as Alonzo Berruguete, Caspar Becerra and the builders of El Escorial, Juan Bautista de Toledo and Juan de Herrera. In his introduction to Book Two on human proportions and anatomy, Arfe praised a number of Italian painters and sculptors: 'Polayolo, Bacho Brandinel, Raphael de Urbino, Andrea Manteña [sic], Donatelo y Michael Angelo.'[60] All of these artists – except Mantegna – had been praised by Vasari for their direct studies of human anatomy.[61]

The references to Arfe as the 'Spanish Cellini' in the few accounts of his life are rooted in his dual career as a renowned goldsmith and gifted sculptor. This analogy can also be seen in his prior publication of a treatise on the value, weight and manufacturing of gold, silver and precious stones, which was entitled *Quilatador de la plata, oro y piedras*. This work was published in Valladolid in 1572, only a few years after Benvenuto Cellini's 1568 treatise written for goldsmiths and bronze sculptors, which appeared among the 23 known titles in Arfe's book inventory.[62] As in Cellini's earlier writings, Arfe's art theory tried to elevate the goldsmith's craft to the rank of painting, sculpture and architecture,

---

[58] 'No se compone esta parte de la cara con los morzillos que se muestran en los carrillos y frente del rostro dessollado, porque aquella manera es para solo Medicos y Cirurgianos, y no para la Scultura y Pintura.' Ibid., p. 142.

[59] 'Los morcillos de la cara estan tan confusos, e intrincados, que han causado gran contradiccion entre los anatomistas, de manera que ni Vesalio se acuerda con Galeno, ni Realdo con Vesalio.' Valverde de Amusco, *Historia*, f. 32 r.

[60] Arfe y Villafañe, *Varia Commensuración*, p. 95.

[61] This was noted by Boris Röhrl in his recent study of *History and Bibliography of Artistic Anatomy* (Hildesheim: Olms, 2000), p. 98.

[62] José Luis Barrio Moya, 'El platero Juan de Arfe Villafañe y el inventario de sus bienes' in *Anales de Estudios Madrileños* (Madrid: Instituto de Estudios Madrileños, 1982), p. 24.

with which it had traditionally been associated.[63] The parallel agenda of the two artists was also seen in their increasing desire to prove themselves in large-scale sculptural works after an initial career devoted to the decorative arts. Arfe emphasised this aim by titling himself *Escultor de oro y plata* rather than gold/silversmith or *platero* on the title page of *Varia Commensuración*.

We can also note several similarities to the life and career of Albrecht Dürer in Arfe's artistic development. Both Dürer and Arfe were from families who had had a long tradition as goldsmiths and both were of German descent, as evidenced by Arfe's original family name of Harff, a small town near Cologne.[64] Arfe was the third generation of this family dynasty of jewellers who had practised in Spain since the early sixteenth century and whose large monstrances or *custodias* can still be seen in numerous Castilian cathedrals and museums. Dürer was not only praised repeatedly in Arfe's references to other artists, but was also quoted in a number of Arfe's illustrations, which were copied directly from Dürer's prior treatise on human anatomy, symmetry and proportion, *Vier Bucher von menschlicher Proportion* (Nuremberg, 1528). These borrowings were openly admitted by the Spanish author: 'What we write here is most precise because it was shown and described to us previously by the

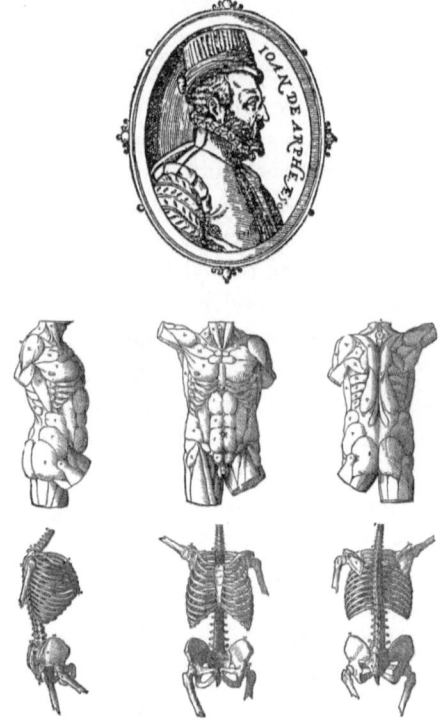

Figure 10.6   Illustrations from Juan Arfe de Villafañe's artistic anatomy, *De varia commensuración para Escultura y Architectura* (Seville, 1585 and 1587). This richly illustrated treatise included four books on geometry, anatomy, zoology and architecture, of which the *Libro segundo* is a rare and unique source produced by an anatomically trained artist from the late Spanish Renaissance. Courtesy of the *Sociedad Española de Historia de la Construcción*

---

[63]   'Antiguamente no había diferencia de los Artífices que ahora llamamos Escultores y Arquitectos á los que ahora son Plateros: por lo qual es cosa cierta que los preceptos de los unos son necesarios á los otros.' Ibid., Prologo, p. XI.

[64]   F.J. Sanchez Canzon, *Los Arfes* (Madrid: Callista, 1929), p. 8.

miraculous inventiveness of Dürer.'[65] The fourth chapter of Book 2, 'Titulo quarto de los escorzos', dealt with the foreshortening of the human figure drawn from different perspectives and acknowledged its repeated use of Dürer: 'We shall here deal with an infallible and precise rule described in more length by Albrecht Dürer, the bright German painter, most skilled in the mathematical sciences, in his fourth book on symmetry, and on the correct shape of the human body.'[66]

The images used in *Libro segundo* of Arfe's work were not direct replicas of the Vesalian illustrations seen in Montaña and Valverde's prior Spanish textbooks on anatomy, but also represented standing skeletons and musclemen placed in open landscapes.

None of these images equalled the mastery of the Vesalian plates in their depictions of human bones and muscles. Their pedagogical advantages nevertheless secured a continued use of Arfe's treatise more than two centuries after the original 1585 version and led to the publication of eight new editions of *Varia commensuración para la escultura y arquitectura* throughout the seventeenth, eighteenth and early nineteenth centuries. While the exact nature of Arfe's own anatomical education remains somewhat unclear, the author repeatedly emphasised his

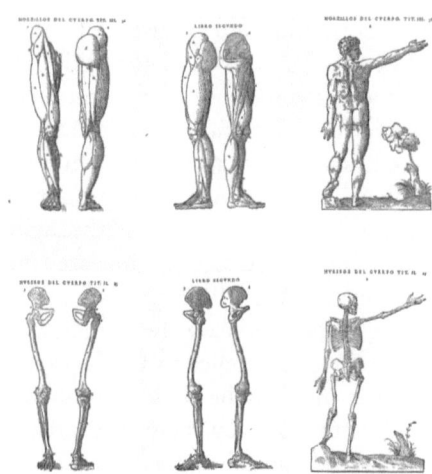

Figure 10.7   Surface musculature and skeleton seen in lateral, anterior and posterior views, from the *Libro Segundo*, the second book from Juan Arfe de Villafañe's *De varia commensuración para Escultura y Architectura*. This second book was divided into four subchapters on the proportion, osteology, myology and foreshortening of the human body, entitled: 'Titulo primero de la medida y proporción del cuerpo humano', 'Titulo Segundo de los huesos del cuerpo humano', 'Titulo tercero de los morcillos del cuerpo humano' and 'Titulo quarto de los escorzos'. These images from Books 2 and 3 depict the surface musculature and skeleton of a male body seen in lateral, anterior and posterior views. Courtesy of the *Sociedad Española de Historia de la Construcción*

---

[65]   'Muy precisa esta que ponemos; lo qual nos escribió y hallo primero el milagroso ingenio de Durero.' Ibid., Titulo IV de los escorzos, p. 172.

[66]   'Trataremos por una regla infallible y precisa que escribió mas largamente Alberto Durero Alemán, clarísimo pintor y muy ejercitado en las ciencias matemáticas en su cuarto libro de Simetría y recta forma del cuerpo humano.' Ibid.

former training and experience in human dissection. He summarised the number of bones and muscles of the body in a series of mnemonics, homespun rhymes and verses, which were often erroneous, as seen in his descriptions of the five-parted human sternum, which had been one of the most contested matters among Vesalius' followers and critics.[67] In spite of the artistic quality of Arfe's illustrations of bones, muscles and human proportions, written and visual misrepresentations of anatomical features appeared throughout his treatise. His allegedly long personal experience with dissection appeared dubious as well, despite his claims: 'In order to demonstrate this I have spent a lot of time and diligence carrying out anatomies of many bodies, and have always benefited from having skeletal bones placed in front of me.'[68] His reference to a collection of human bones as didactic tools for his art did not indicate the possession of an assembled human frame, the indispensable tool for the osteological studies carried out by Vesalius and his Valencian followers. Arfe's study of only loose human bones might account for some of his erroneous osteology, which reduced the total number of human bones to 182 (far fewer than the actual number of 206) and deliberately ignored a number of minor ossicles of no apparent use in the anatomical training of the artist: 'Whoever wants to see them can go to the cemeteries, which are full of them.'[69] Arfe's ambivalence towards dissection and his often conflicting interests in art and anatomy were apparent in his descriptions of the number of bones in the human skull, which approached the correct number of 22. His own estimate indicated a thorough examination of the bones in the cranium, all closely attached by sutures, with the exception of the autonomous mandible: 'The skull has 20 bones, eight cranial and twelve facial.'[70] Given this anatomical insight, it is surprising that Arfe later reduced this number to 'two bones in the head' in one of his rhymes which accounted for all the bones in the human skeleton.[71] The description of only two bones in the human skull complied with his resolve to concentrate only on those anatomical details that were of direct use for the artist. This explained his reduction of the many paired and single bones of the skull into one integrated geometrical structure, and the five

[67]   'El huesso que haze el pecho es como espada y tiene su principio en inas asillas.

Fenece en una punta algo delgada y prendese con el siete costillas.

Partese en cinco partes y pegada està cada una con ternillas.' Ibid., p. 123.

[68]   'Para demonstracion de esta parte hemos gastado mucho tiempo, y puesto toda diligencia, haciendo anatomia de muchos cuerpos, y aprovechandonos de tener los huesos siempre delante.' Ibid., p. 119.

[69]   'Quien quisiere verlos, los cementarios están llenos y podrá.' Ibid., p. 120.

[70]   'Tiene, pues, la cabeza veinte huesos, ocho en el casco, y doce por la cara.' Ibid.

[71]   'Ciento y ochenta y dos sin las ternillas son los huessos de un cuerpo en sus pedazos. En la cabeza dos, dos las asillas costillas veinte y cuartro, y seis los brazos.

Cinco el pecho, las ancas, y espaldillas, sessanta pies y piernas en sus tracos. Los manos veinte y siete un par de veces.

Y el espinazo nueve con dos diezes.' Ibid., p. 138.

bones of the sacrum to one *huesso grande*, and accentuated Arfe's explicit priority of art over anatomy. In another rhyme reference to his anatomical training, Arfe mentioned his detailed studies of a mummified corpse, but unfortunately without revealing the exact nature and objective of this scrutiny: 'The mastery of this science, on which I have spent a great part of my life, was obtained through long discourses and inquiries, and a rare carefulness: in a remote area of my workplace, I studied a mummified corpse for a lengthy period, and also the thickness, length, and shape of all its bones.'[72]

Arfe was born in 1535 and must have been in his early twenties during his alleged anatomical training under Cosme de Medina at the University of Salamanca, which took place three decades before the publication of his artistic treatise. This long interval might explain some of his obvious misrepresentations and flawed memory of human anatomy. The small human figures on his surviving monstrances in Avila, Valladolid and Seville do not clearly reveal this anatomical education, but instead represent a continuity with the tradition instigated by his grandfather, Enrique de Arfe, and continued by his father, Antonio de Arfe. Only Arfe's last sculptural works of life-size human figures, such as the exquisite tomb monument for the Archbishop of Seville, Don Cristóbal de Rojas reveal his development as a leading naturalist sculptor of the late Spanish Renaissance. His last work was a bronze cast of two sculptures of the Duke and Duchess of Lerma, sculpted by Pompeo Leoni, son of the Milanese artist Leone Leoni. The Leoni worked as official court sculptors to the Spanish Habsburg monarchs during the latter half of the sixteenth century, and their artistic careers cover the reigns of Charles V, Philip II and Philip III. The emphasis on anatomical training among cultured sixteenth-century artists was seen in several works by 'los Leoni' and particularly in their perfect mastery of human anatomy, as revealed under the removable imperial armour of *Carlos V y el furor*, which was carried out by Leone between 1550 and 1553, and completed by Pompeo in 1564. Pompeo Leoni later gained a reputation as a leading Spanish antiquarian with his unequalled Madrid collection of books and artworks. The library inventory of this Hispanicised Italian sculptor included not only deluxe bindings of works by Vesalius and Valverde but also the entire collection of original anatomical drawings by Leonardo da Vinci.[73] The cooperation between Arfe and Pompeo Leoni in their later careers would explain the apparent inspiration from Leonardo's anatomy

---

[72]　'Fué con discursos largos inquirida por mi la certidumbre de esta sciencia,
en que gaste gran parte de mi vida, poniendo en esto extraña diligencia:

que de mi propia estancia en ascondida parte, parte mire gran tiempo la presencia.

· De un cuerpo embalsemado, do los gruesos largos, y formas, vi de todos huessos.' Ibid., p. 119.

[73]　Margarita Estella, *Los Leoni, escultores entre Italia y España* in Los Leoni (1509–1608), *Escultores del renacimiento al servicio de la corte de España* (Madrid: Museo del Prado, 1994), pp. 44–45.

in some of Arfe's illustrations. Pompeo Leoni, however, did not purchase the Leonardo drawings until around 1589 – four years after the publication of Arfe's artistic anatomy.[74] Peter Paul Rubens is known to have visited Leoni's Madrid house in 1603, after Pompeo had acquired even more of Leonardo's codices from Italy.[75] Perhaps Arfe had been invited to see Leoni's rich art and manuscript collections during their collaboration two years earlier, which united the owner of Leonardo's anatomical drawings with the author of the first printed and illustrated treatise on artistic anatomy. While Arfe's publication represented an exceptional case in sixteenth-century Spain, numerous art treatises of the next century emphasised that anatomical training was essential for the professional artist, as outlined in Gutiérrez de los Ríos' *Noticia general para la estimación de las artes* (Salamanca, 1600) and its chapter 'Emulacion que estas artes del dibuxo tienen con la Medicina': 'It is already well known that in order to know medicine perfectly one must also know the anatomy of the human body, veins, arteries, bones, and all the other things. In the same way (but for a different purpose) the teachers of the arts should know about the anatomy of the human body so that they can show both inner and outer features in their paintings in a natural and truthful way; and they should not only study the human body, but also the animals of the land, the sky and the sea ... The two arts will profit from one another whenever they can. Painting can use anatomy for its own purpose. Medicine can benefit from drawn images.'[76] Both Vicente Carducho's *Dialogos de la pintura* (Madrid, 1622) and Francisco Pacheco's *Arte de la pintura, su antiguidad y grandezas* (Seville, 1649) praised Arfe's representation of bones and superficially dissected muscles, which were even considered superior to the 'excess knowledge' offered by the illustrated treatises of Vesalius and Valverde.[77]

---

[74] Nicolas García Tapia, 'Los codices de Leonardo en España', *Boletín del seminario de estudios de arte y arqueología*, 63 (1997), pp. 373–74.

[75] Kelley Helmstutler Di Dio, *Leone Leoni and the Status of the Artist at the End of the Renaissance* (Aldershot: Ashgate), p. 134.

[76] 'En la Medicina ya se sabe que para ser uno perfecto, debe tener también particular noticia de la notomía de un cuerpo humano, venas, arterias, huesos, y de todas las demás cosas de él. Pues de esta misma manera (aunque para diferente fin) deben saber la notomía del cuerpo humano los profesores de estas artes, es a saber, para poder significar en las figuras que pintasen lo de adentro por lo de fuera, de manera que parezcan naturales y verdaderas; y no sólo deben saber esto acerca del cuerpo humano, son también en todo género de animales terrestres, celestes, y marítimos. De donde no sin causa se aprovechan estas artes la una de la otra en cuanto pueden. La pintura en coger la notomía para su propósito. La Medicina en aprovecharse del dibujo, como se ve en Dioscorides.' Gutíerrez de los Rios, *Noticia general para la estimación de las artes* (Salamanca: Pedro Madrigal, 1600).

[77] 'La Anatomía que es la tercera parte que pertenece al debuxo, sitio, verdad, numero de músulos convenientes a la pintura, se hallarán en Andrea Vesalio, el qual sobrepuxo a todos sus antepasados. Pero mucho mejor en el doctor Juan Valverde ... Dada esta general noticia de la

Some artists from the emerging Italian art academies developed the same ambivalence towards hands-on anatomical studies, as Arfe's account from the anatomical theatre of Salamanca. Vincenzo Danti's 1567 treatise on proportion claimed to be based on 83 dissections carried out by Danti himself and called for the artist to 'anatomize the human body, in a way little different from that of the physicians'.[78] Pietro Accolti's 1625 *Lo inganno de gl'occhi, prospettiva pratica*, on the other hand, warned against such excessive studies and claimed that only life drawing, not anatomy taught from dissection, should be studied by the young artistic academicians: 'Not those parts studied by the physicians and anatomists, but only the external parts that pertain to the study of painting.'[79] Life drawing classes, henceforth often referred to as 'academies', would become the basis of art education in Europe for centuries thereafter and were generally based on direct life observation rather than systematic dissection of the human body. No formal art academy existed in Spain until 1744, and the illustrations of Vesalius, Valverde and Arfe were still held in great esteem as tools for the artistic study of anatomy in Palomino de Castro y Velasco's 1724 art treatise *El museo pictorico y escala optica*. This ambitious work by the so-called 'Spanish Vasari' included numerous verbatim passages from Arfe, and direct copies of images from Valverde's *Historia* and even maintained the septi-partite sternum in its Valverdean illustration of the human skeleton.[80] Almost two centuries after the publication of *Historia de la composicion del cuerpo humano*, the poorly executed replicas of Valverde's crude Vesalian images were plainly anachronistic as representations of human anatomy. Contemporary and masterly illustrated anatomy books were at that time published elsewhere in Europe and their authors had accumulated significant new knowledge of the human structure since the mannerist representations of Vesalius, Valverde and Arfe y Villafañe.

Anatomía por mayor, a los más valientes y diestros y habiéndoles señalado las Fuentes de donde se han de enriquecer, será justo que los menos escrúpulos y presumidos se aprovechen de lo que trabajó Juan de Arfe y Villafañe, pues no se halla tanto junto en otro autor nuestro: que si bien no tuvo tanta gracia en los perfiles de afuera, como se valió de buenos autores, escribió con verdad en la material de los músculos; trabajo que no se debe despreciar y que los que son aventajados deben seguir seguramente.' Pacheco, *Arte de la pintura, su antiguidad y grandezas*, p. 385. Like Pompeo Leoni, Vicente Carducho was another example of a succesful Italian artist working in Spain who gathered a rich collection of books and artworks, including numerous anatomical figurines and body parts in bronze and plaster. For a complete inventory of Carducho's collection, see Kelley Helmstutler di Dio and Rosario Coppel, *Sculpture Collections in Early Modern Spain* (Farnham: Ashgate, 2013), Appendix, p. 375-381

[78]   Vincenzo Danti, *Trattato delle perfette proporzioni* (Florence, 1567), p. 23.

[79]   Pietro Accolti, *Lo inganno de gl'occhi, prospettiva pratica* (Florence: Pietro Cecconelli, 1625), p. 144.

[80]   Antonio Palomino de Castro y Velazco, El museo pictorico y escala optica. Tomo segundo, Practica de la pintura. (Madrid: Por la Viuda de Juan Garcia Infancon, 1724), Lámina 5, n. p.

While Montaña and Valverde's books represent a remarkably early Spanish acceptance of contemporary Vesalian iconography, the reprints of the same – still uncorrected – illustrations almost two centuries later emphasised the slow development in Spanish anatomy during the long intervening period.

Conclusion

Sebastián de Covarrubias' famous Spanish dictionary (Madrid, 1611) included an entry on Anatomia, which was described as 'the flaying and opening of a human body, which is made in order to study its inner parts and structure. It is a very necessary thing for the physicians and surgeons; and in the universities there are chairs of this discipline, which is carried out on the bodies of executed criminals, and those who die in hospitals, and other particular persons. And in their place it can also be carried out on a monkey and a pig, to show the inner parts, such as the heart, the entrails, and the intestines.'[1] Anatomy was evidently still seen as a thriving branch of learning in Spain, even though the momentum of the previous century had often become more sluggish or even discontinued by the beginning of the 1600s. It has been the aim of this study to shed some light on this neglected area in the history of early modern anatomy. A systematic examination of anatomy in sixteenth-century Spain has focused on its introduction, maintenance and, in many cases, later decay and abandonment. This study has demonstrated both regional variations and general tendencies in the institutionalisation of anatomy as a novel medical discipline taught at, and occasionally even beyond, the principal Iberian universities and hospitals. The creation of new university chairs between 1552 and 1583 represented a significant innovation and expansion of contemporary medical practice and education with no Iberian precedents apart from the University of Valencia.

The consideration of each of these particular university contexts, as well as supplementary investigations of anatomical practice in contemporary hospitals, in overseas colonies and among anatomically trained artists, represent the first detailed synthesis of Spanish Renaissance anatomy. The broad approach has required the integration of a wide variety of sources and regional contexts, which, had they been studied separately, would have offered only a fragmentary and incomplete understanding of the subject. A wide-ranging combination of textbooks, printed images, administrative, literary and legal sources has demonstrated a multifaceted anatomical tradition during Spain's Golden

---

[1] 'Anatomía: Es la descarnadura y abertura que se hace de un cuerpo humano para considerar sus partes interiores y su compostura; cosa necesarisimo a los médicos y cirujanos; y asi en las universidades hay cátedras de esta facultad, y se ejecuta algunas veces en los cuerpos de los ajusticiados y otras en los que mueren en los hospitales y en algunas otras personas particulares. Y en su lugar se suele hacer de una mona, y de un puerco para lo que es lo interior del corazón, asadura y tripas.' Sebastian de Covarrubias (ed.), *Tesoro de la lengua castellana o española* (Madrid: Luis Sanchez, 1611), f. 68v.

Age. Individual regional contexts would not have enabled the reconstruction and general comprehension of the didactic programmes, physical facilities, publications, salaries and transdisciplinary activities of Spanish anatomists during the late Renaissance. The comparative studies of universities and hospitals with institutionalised practices in this field have enabled a deeper understanding of this practice, aided by the textbooks, artworks, statutory reforms and evidence of basic anatomical research which survive from various regional contexts.

In addition to this material, surviving provisions provided specific details regarding how the new academic discipline was instigated and supported by the highest authorities of that period. A comparative geographical rather than thematic approach to this topic has helped us to comprehend both the general patterns and significant regional differences in the earliest dispersal of anatomy studies throughout the Iberian Peninsula. As this study shows, the enduring exclusion of Iberia in accounts of early modern anatomy, and the widely held idea that no activities took place in this area, can no longer be maintained or justified. Prior narratives of scientific forerunners and discoveries have not found room for lesser-known anatomists from the late Spanish Renaissance, whose general and often more unassuming contributions consisted of providing anatomical training for medical students and practising surgeons. While this study may disregard conceptions of inactivity and ignorance of anatomy in Renaissance Spain, it has perhaps not offered significant evidence, which radically changes the history of anatomical knowledge in sixteenth-century Europe. It was, however, never the purpose to trace or introduce new pioneers into the history of early modern anatomy, as previously attempted in various nationalist accounts of the medical and scientific history of Spain. The objective of this study was instead to chart the development of late sixteenth-century anatomy in a particular, and largely ignored, national context and to trace the establishment of this discipline within various institutional and regional settings.

Throughout this undertaking, it has become clear that the state-mandated introduction of anatomy at the principal Castilian universities should be understood in line with previous and later attempts by the Spanish monarchy to improve the education and competence of its physicians and surgeons. In this respect the 'proto-absolutist' state of Charles V and Philip II was far from isolated from contemporary intellectual developments and reforms, and was in fact actively engaged in the incorporation of recently applied knowledge within its own primary universities. This evidence is in direct opposition to most biographies of Vesalius, which portray the Spanish monarchy as neglectful and indifferent to the skills and practices of the renowned anatomist. The conception of Spanish ignorance of Vesalius' new anatomy was seen both in contemporary writings by Juan Valverde and in later accounts of the 'Vesalian Revolution', which presented the Spanish kingdoms of the mid-sixteenth century as isolated from contemporary anatomical learning. Yet during that very period, Vesalius' royal employers were actively engaged in formally establishing anatomy as a university discipline at the foremost medical faculties of Castile. This was most evident in the state-mandated introduction of anatomy, and later of surgery, at the Universities of Salamanca, Valladolid and Alcalá de Henares.

The contemporary initiatives within the Crown of Aragon were in line with developments in Castile and the newly established universities in Barcelona and Zaragoza appointed chairs of anatomy during the initial phase of their foundation. In spite of the significant differences between the crown-regulated universities of Castile and the relatively autonomous universities of Aragon, parallel developments in both kingdoms favoured the expansion of medical faculties with newly appointed professorships of anatomy and surgery. However, despite these analogous developments, the Aragonese Universities of Valencia, Barcelona and Zaragoza presented their own noticeable particularities. The appointment of anatomists to other medical chairs such as botanical medicine or *simples* and surgery encouraged a transdisciplinary association between disciplines that linked both theoretical and practical branches of the medical curriculum. The joint chairs of *anatomia y simples* were engaged in empirical studies of both human cadavers and medical plants, which were seemingly understood as an inextricable whole with overlaps, borrowings and adaptations rather than isolated or loosely united branches of medical learning. This association arguably ensured the survival of anatomy studies in Aragon during the contemporary decline of the practice in Castilian universities.

This study has shown the circulation of scholars, publications and applied knowledge between and beyond the medical faculties of Castile and Aragon. Four focal points have emerged as educational training grounds for the first appointed anatomists of Castile, Aragon, Portugal and New Spain. Among these centres, the *Estudi General* of Valencia was most noteworthy, with an established tradition of anatomy preceding other medical faculties of Iberia by half a century. This head start led to the appointment of Valencian anatomists at the first permanent university chairs of anatomy in sixteenth-century Castile. The University of Salamanca contributed to an influx in the opposite direction, as the Castilian educational base for anatomists such as Francesc Micó and Juan Tomás Porcell, who went on to practice in Barcelona and Zaragoza. The University of Alcalá de Henares stood out as the foremost contributor of New World anatomists, with Francisco Hernández, Augustin Farfán and Juan de Barrios as the most prominent representatives. Finally, the elusive activities of Alonso Rodríguez de Guevara led to short and discontinuous courses of anatomy at the Universities of Valladolid and Coimbra. The different regional contexts have left behind traces of both 'Vesalian' anatomy at the Universities of Valencia, Salamanca and Alcalá de Henares, and 'pre-Vesalian' doctrines at the Universities of Valladolid, Coimbra, Barcelona and Zaragoza. The later consolidation of Galenic anatomy continued throughout the late sixteenth century and was eventually found even in former Vesalian strongholds during the last decades of the 1500s. The gradually reinforced authority of Galen in the late sixteenth- and seventeenth-century university statutes of Salamanca and Alcalá coincided with complaints of decaying facilities and vacant professorships, and showed an apparent waning of the former crown-authorised anatomy studies at these leading Castilian universities.

The differing attitudes to anatomy were seen in the appointment of the notorious opponent of this new practice, Luis Mercado, as *Protomédico General* after the death of Francisco Valles in 1592. The new principal overseer of medical education in the Spanish kingdoms judged anatomical practice to be irrelevant to medical learning, in direct contrast to his immediate predecessor, who had instigated the same practice at the University of Alcalá de Henares four decades earlier. Beyond resistance and fading interest from above, the discrepancies in the development of anatomy in Aragon and Castile may have been increased by the promotion of Castilian anatomists and surgeons to higher university chairs or to the medical staff of the court. Such positions in traditional medicine arguably offered more prestigious and well-compensated careers than the new and minor chairs of anatomy and surgery at the bottom of the medical university hierarchy. While such promotions did not facilitate the continuation of anatomical practice, as seen in the careers of Vesalius and Hernández, the monastery hospital of Guadalupe presented a Castilian refuge beyond the universities, where systematic anatomy studies were carried out by recently qualified physicians. This Renaissance hospital was conspicuous as a setting related to the universities, and with similar application of knowledge in the field of anatomy, which even attracted anatomists from the Universities of Salamanca and Alcalá de Henares where similar practices already existed. Another setting restricted to Castilian physicians was offered by the recently conquered overseas territories, where New Spain – and Mexico City in particular – stood out as a remarkably early centre for studies of anatomy and pathology, and for the publication of the earliest American textbooks on anatomy and surgery. The late-sixteenth-century disputes among Iberian advocates of Vesalian and Galenic doctrines were conspicuously absent from this New World context, which seemed to be an enduring repository of traditional Galenism.

Another interesting development and effect of contemporary anatomy studies was the increasing artistic association with the new medical discipline. This affiliation was seen in the documented collaboration between Spanish anatomists and artists who left behind both anatomical art, as seen in the illustrated textbooks of Juan Valverde and Bernardino Montaña, and Juan de Arfe y Villafañe's treatise on artistic anatomy. The late medieval authority and crude imagery found in Montaña's textbook was far from the advancement in anatomical discoveries and representations produced at the Italian centres of art and science. Juan de Arfe y Villafañe's artistic anatomy, however, was directly related to the contemporary developments in art education, which endorsed and promoted the reputation of anatomically trained artists. Remarkable cases of the affiliation of artists and anatomists were seen in Arfe de Villafañe's anatomical studies under Cosme de Medina, and the vivisections of animals carried out by Francisco Hernández and the Toledan sculptor and architect Nicolas de Vergara.

The Iberian Peninsula was involved in several aspects of anatomical studies and practices during the mid- to late sixteenth century which changed existing university structures and statutes, and led to the creation of new professorships and working spaces devoted to the new discipline. Spanish Renaissance anatomy even reached beyond these academic institutions and entered into new contexts in contemporary hospitals and distant colonies, while artists produced illustrations, prints and mannequins for the visual transfer and representation of anatomical learning. The empirical data gathered from these regional and institutional settings show the active engagement of the Spanish kingdoms both in the initial encouragement and eventual dismissal of some of the most significant developments in early modern anatomy. However, the parallel development of Vesalian anatomy and gradual reinforcement of Galenic doctrines were not limited to this national context, but also appeared outside Spain even in the statutes and curricula of some of the foremost Italian and French centres of contemporary anatomy. The reformation of centuries of medical learning was slow and discontinuous, not only in Spain, but also at the European focal points of this development, and several centuries would have to pass before the study of anatomy obtained its status as the bedrock of medical education.

Bibliography

Sources

Accolti, Pietro, *Lo inganno de gl'occhi, prospettiva pratica* (Florence: Pietro Cecconelli, 1625)

Arfe y Villafañe, Juan de, *Varia Commensuración para la Escultura y Architectura* (Seville: Andrea Pescioni y Juan de León, 1585)

Arias de Benavides, Pedro, *Secretos de Chirurgia especial de las enfermedades de morbo galico y lamparones y mirrarchia* (Valladolid: Francisco Fernandez de Córdoba, 1567)

Barreiro, Caspar, *Chorographia de algums lugares que stam em um caminho que fez G.B ó anno de MDXXXVI* (Coimbra: Joao Alvarez, 1561)

Barrios, Juan de, *Verdadera medicina, cirugia y astrologia, en tres libros dividida* (Mexico City: Fernando Balli, 1607)

Bottoni, Federico, *Evidencia de la circulación de la sangre* (Lima: Ignacio de Luna, 1723)

Bravo, Francisco, *Opera medicinalia* (Mexico City: Pedro Ocharte, 1570)

Calvo, Juan, *La chirurgia universal y particular del cuerpo humano* (Barcelona: J. Cendrat, 1591)

———. *Cirurgia de Guido de Cauliaco con la glosa de Falco : agora nueuamente corregida y emendada y muy añadida con un tratado de los simples, por Iuan Caluo Doctos en Medicina, Lector en la misma facultad en la ciudad de Valencia* (Valencia: Pedro Patricia, 1596)

Carcucho, Vicente, *Dialogos de la pintura* (Madrid: Fr. Martinez, 1633)

Castrillo, Martinez de, *Coloqio breve sobre la dentadura y maravillosa obra de la boca* (Seville: Vassallo de Mumbert, 1577)

Cervantes, Miguel de, *Don Quixote* (Madrid: Juan de la Cuesta, 1604)

Chacón, Daza Dionisio, *Practica y Teorica de Cirurgia* (Valladolid: Ana Velez, 1582)

Collado, Luis, *Cl. Galeni Pergameni liber de ossibus ad tyrones, interprete Ferdinando Balamio, Enarratore Ludovico Collado medico* (Valencia: Typis Ioannis Mey Flandri, 1555)

Covarrubias, Sebastian de (ed.), *Tesoro de la lengua castellana o española* (Madrid: Luis Sanchez, 1611)

Danti, Vincenzo, *Trattato delle perfette proporzioni* (Florence, 1567)

Diaz, Fransisco, *Compendio de chirurgia* (Madrid: Pedro Cosin, 1575)

———. *Tratado nuevemente impresso, de todas las enfermedades de los Riñones, Vexiga, y Carnosidades de la verga, y Urina, diuidido en tres libros* (Madrid: Francisco Sanchez, 1588)

Enriquez, Enriquez Jorge, *Retrato del perfecto medico* (Salamanca: Juan y Andres Renaut, 1595)

Escolano, G., *Decada primera de la historia de la insigne y coronada ciudad y reyno de Valencia* (Valencia: Pedro Patricio May, 1610)

Esteve, Pedro Jaime, *Hippocrates Coi Medicorum omnium principis epidemion liber secundus* (Valencia: Typis Ioannis Mey Flandri, 1551)

Farfán, Augustin, *Tractado breve de Anathomia y Chirurgia y del Conoscimento y Ovra de algunas Enfermedades* (Mexico City: Pedro Ocharte, 1592)

Fragoso, Juan, *Erotemas chirurgicos en los que se enseña todo lo mas necessario del arte de cyrurgia* (Madrid: Pierres Cosin, 1570)

———. *Discurso de las cosas aromaticas, arboles y frutales, y de otras muchas medicina simples que se traen de la India Oriental, y sirven al uso de la medicina* (Madrid: Casa de Fransisco Sánchez, 1572)

Fraylla, Diego, *Lucidario de la Universidad y Estudio General de la Ciudad de Zaragoza* (Manuscript, Zaragoza, 1603. Zaragoza: Institucion 'Fernando el Catolico', 1983)

Guevara, Alonzo Rodriguez de, *In pluribus ex iis quibus Galenus impugnatur ab Andrea Vesalio Bruxelensi in constructione et usu partium corporis humani, defensio* (Coimbra: Juan Baverius, 1559)

Hernández, Francisco, *Obras completas*, German Somolinos d'Ardois (ed.) (Mexico City: Universidad Nacional de Mexico, 1960)

Heseler, Baldasar, *Andreas Vesalius' First Public Anatomy at Bologna, 1540*, translation and commentary by Ruben Erikson (Uppsala: Almqvist & Wiksells boktrykkeri, 1959)

Huarte, Juan de, *Examen de Ingenios* (Baeza: Juan Baptista de Montoya, 1564)

Jimeno, Pedro, *Dialogus de re medica compendia ratione, prater quaedem alia, universam Anatomen humani corporis perstringens, summe necessarius omnibus Medicinae candidatis* (Valencia: Typis Ioannis Mey Flandri, 1549)

León, Andrés de, *De la anathomía* (Granada, 1590)

López de Hinojosos, Alonzo, *Summa y recopilacion de cirugia* (Mexico City: Antonio Ricardo, 1578)

Luis de Granada, *Introducción del símbolo de la Fe* (Salamanca, 1583; Madrid: Fundación Universitaria Española, 1996)

Magarola, Jeroni, *Republica original sacada del cuerpo humano* (Barcelona: Pedro Malo, 1587)

Martínez, Jusepe, *Discursos practicables del nobilisimo arte de la pintura* (Madrid: Real Academia de San Fernando, 1866)

Mercado, Luis, *Institutiones chirurgicae* (Madrid: Luis Sanchez, 1594)

————. *Instituciones para el aprovechamiento y examen de los algebristas* (Madrid: Pedro Madrigal, 1599)

Micón, Fransisco, *Libro del regalo y utilidad de beber frío y refrescado con nieve* (Barcelona: Casa de Diego Galván, 1576)

Montaña de Montserrate, Bernardino, *Libro de la Anathomia del Hombre* (Valladolid: Sebastian Martinez, 1551)

Pacheco, Francisco, *Arte de la pintura, su antiguidad y grandezas* (Seville: Simon Faxardo, 1649)

Padilla, Dávila, *Historia de la fundación y discurso de la Provincia de Santiago de Mexico* (Mexico City, 1596; Mexico City: facsimile, Editorial Academia Literaria, 1955)

Palomino de Castro y Velazco, Antonio, *El museo pictorico y escala optica. Tomo segundo, Practica de la pintura.* (Madrid: Por la Viuda de Juan Garcia Infancon, 1724)

Pascual, Miguel Juan, *Libro e practica de Cirurgia* (Valencia, 1548)

Pinheiro da Veiga, Tomé, *Fastigina. Vida cotidiana en la corte de Valladolid* (Manuscript, 1603; Valladolid: Facsimile, 1973)

Platter, Thomas, *Journal of a Younger Brother: The Life of Thomas Platter as a Medical Student in Montpellier at the Close of the Sixteenth Century* (London: F. Muller, 1963)

Porcell, Juan Tomás, *Información y curación de la peste de Caragoca y praeservación contra la Peste en General* (Zaragoza: Casa de la viuda de Bartholome de Nagera, 1565)

Reina, Francisco de la, *Libro de Albeyteria en el qual se veran todas quantas enfermedades y desastres suelen acaecer a todo genero de bestias* (Zaragoza: Agustin Millan, 1562)

Rios, Gutíerrez de los, *Noticia general para la estimación de las artes* (Salamanca: Pedro Madrigal, 1600)

Robledo, Diego Antonio de, *Compendio chirúrgico útil y provechoso* (Madrid: Vicente Cabrera, 1686)

Salazar, Eugenio de, *Obras festivas*, Alejandro Cioranesu (ed.) (Santa Cruz de Tenerife, 1968)

Servetus, Michael, *Restitución del Cristianismo* (Vienna, 1553; Madrid: Fundación Universitaria Española, 1980)

————. *Obras completas* (Zaragoza: Ángel Alcalá, 2005)

Soto, Lázaro de, *In librum Hippocratis De dieta commentationes* (Madrid: Luis Sanchez, 1594)

Sylvius, Iacobus Dubois, *Vaesani cuiusdam calumniarum in Hippocratis Galenique rem anatomicam depulsio* (Paris: Viduam Jacopo Gazelli, 1551)

Talavera, García de, *Historia de nuestra Señora de Guadalupe* (Toledo: Thomas de Gúzman, 1597)

Torquemada, Antonio, *Colloqios satiricos* (Mondeñado: Augustin de Paz, 1553)

Vadillo, Pedro Gago de, *Discurso de la verdadera cirugia* (Madrid: Juan Micól, 1632)

Valles, Fransisco, *Controuersiarum medicarum & philosophicarum libri decem* (Alcalá de Henares: Juan de Brocar, 1556)

——. *Claudii. Gal. Pergameni de Locis Patientibus Libri Sex, cum Scholiis* (Lyon: Luis Gutierrez, 1559)

Valverde de Amusco, Juan, *Historia de la composicion del cuerpo humano* (Rome: Antonio Salamanca and Antonio Lafrery, 1556)

Vasari, Giorgio, *The Lives of The Great Artists* (Florence: Giunti, 1550/1568; Oxford: Oxford University Press, 1991)

Virués, Geronimo, *Dialogo en qual se trata las heridas de la cabeza* (Valencia: Emprenta de la compañía de los libreros, 1588)

Ximenez, Francisco, *Quatro libros de la naturaleza y virtudes de las plantas, y animales que estan receuidos en el uso de Medicina en la Nueva España* (Mexico City: En casa de la Viuda de Diego Lopez Davolos, 1615)

Secondary Literature

Addy, George M., *The Enlightenment in the University of Salamanca* (Durham, NC: Duke University Press, 1966)

——. 'Alcalá before Reform – The Decadence of a Spanish University', *Hispanic American Historical Review*, 48(4) (1968)

Alejo Montes, F.J., *La Universidad de Salamanca bajo Felipe II 1575–1598* (Salamanca: Estudios de Historia/Castilla y León, 1998

——. *La reforma de la universidad de finales del siglo XVI* (Salamanca: University of Salamanca, 1990)

Alvar Esquerra, Antonio, *La Universidad de Alcalá de Henares a principios de siglo XVI* (Alcalá de Henares: University of Alcalá de Henares, 1996)

Alvarez, Manuel Fernández, *Felipe II y su Tiempo* (Madrid: Espasa-Calpe, 1998)

Álvarez de Morales, A., *Estudios de la historia de la Universidad española* (Madrid: Pegaso, 1993)

Amorós I Gonell, Francesc, *Correspondencia diplomatica de Joan Francesc Rossell, 1616–1617* (Barcelona: Institut d'Estudis Catalans, 1992)

Andretta, Elisa, The Medical Cultures of "the Spaniards of Italy": Scientific Communication, Learned Practices, and Medicine in the Correspondence of Juan Páez de Castro (1545-1552." *Medical Cultures of the Early Modern Spanish Empire* (Farnham: Ashgate Publishing Limited, 2014)

Arana Amurrio, José Ignacio de, *Medicina en Guadalupe* (Badajoz: Diputación Provincial de Badajoz, 1990)

Baquero, Aurelio, *Buquejo historico del Hospital Real y General de Hospital de Nuestra Señora de Gracia* (Zaragoza: Institución Fernando el Católico, 1952)

Barbosa Sueiro, M.B. de, 'Sumula da vida interlope de Alonzo Rodrigues e Guevara. Arquivo de anatomia e antropologia'. *Sep. de A Medicina Contemporânea*, 72(3) (1954)

Baron, J., *Viaje de Vesalio a Terra Santa, Medicina e historia, Vol. 52* (Madrid: Editorial Rocas, 1969)

Barona, Lluis Josep, *Sobre medicina y filosofía natural en el renacimiento* (Valencia: University of Valencia, 1993)

Barral i Altet, Xavier, *Historia del arte de España* (Barcelona: Lunwerg, 1998)

Barreno, Pedro García, *La medicina en El Quijote y en su enterno* in *La Ciencia y El Quijote* (Barcelona: Crítica, 2005)

Barreiro, Agustin Jesús, 'El testamento del Dr. Fransisco Hernández', *Boletín de la Real Academia de la Historia*, 94 (1929)

Bartolome de Jiménez, Natalia, *Historia de la Universidad Española* (Madrid: Alianza, 1971)

Beaujouan, Guy, *La medicina y la cirugía en el Monasterio de Guadalupe* (Paris: Diana, 1966)

Beltrán de Herédia, Vincente, *Cartulario de la Universidad de Salamanca* (Salamanca: University of Salamanca; Secretariado de Publicaciones de la Universidad, 1970–73)

Bennassar, Bartolomé, *La America española y la America portuguesa, siglos XVI–XVIII* (Madrid: AKAL, 2001)

————. *España.* Los siglos de oro (Barcelona: Crítica, 2000)

Bleichmar, Daniela (ed.), *Science in the Spanish and Portuguese Empires* (Stanford: Stanford University Press, 2008)

Blinkhorn, Martin, 'Spain, the "Spanish Problem" and the Imperial Myth', *Journal of Contemporary History*, 15 (1980)

Bouza, Fernando, *Enanos, bufones, monstrous, brujos y hechiceros* (Barcelona: DeBolsillo, 2005)

Bruño, W., 'El primer texto de anatomia publicado en America', *Archivo iberoamericano de historia de la medicina*, 10 (1958)

Bueno, Mar Rey, 'Concordias medicinales de entrambos mundos. El proyecto sobre material médica peruana de Matías de Porres', *Revista de Indias*, LXVI(237) (2006)

————. 'Primeras ediciones en castellano de los libros secretos de Alejo Piamontes', *Boletín de la Biblioteca histórica de la Universidad Complutense de Madrid*, 2 (2005)

Busacchi, Vincenzo, 'Gli studenti spagnoli di medicina e di arte in Bologna dal 1504 al 1575', *Bulletin Hispanique* (1956)

Cabrera, Leoncio López-Ocón, *Breve historia de la ciencia española* (Madrid: Alianza, 2003)

Cajal, Santiago Ramón, *Los tónicos de la voluntad, Reglas y consejos sobre la investigación científica* (Madrid: Edición de Leoncio López-Ocón, 2006)

Camillo, Ottavio, 'Interpretations of the Renaissance in Spanish Historical Thought', *Renaissance Quarterly*, 48(2) (1995)

Canzon, F. J. Sanchez, *Los Arfes* (Madrid: Callista, 1929)

Capelot, Fransisco, 'La obra chirurgica de Juan Fragoso', *Primera Serie*, I(6) (1956)

———. *La obra quirugica de Francisco Diaz* (Salamanca: University of Salamanca, 1957)

Carlino, Andrea, *Books of the Body* (Chicago: University of Chicago Press, 1999)

———. 'Tre piste per la anatomia di Juan di Valverde', *Mélanges de l'ecole francaise de Rome. Italie et Méditerranée*, 114 (2002)

———. *Nel solco di Roma tra filologia e autopsia : note su scienza e antiquaria nel Cinquecento, Rome et la science moderne : entre Renaissance et Lumiéres* (Rome: École française de Rome, 2008)

———. *Vesalio e la cultura delle anatomia e le stampe del rinacimento, Rapprasentare il corpo – Arte e anatomia da Leonardo all' illuminismo* (Bologna: Bononia University Press, 2004)

Carmona Arroyo, Francisco, 'La literatura odontologica en el renacimiento', *Medicina Española*, 68 (1972)

Carvalho, Joaquim Martins Teixeira de, A anatomia em Coimbra no século XVI. *Revista da Universidade de Coimbra*, 4 (1915)

———. *A Universidade de Coimbra no seculo XVI* (Coimbra: Imprensa da Universidade, 1922)

Castellote Cubells, Salvador, 'La anatomia y la fisiologia en la obra de Francisco Suarez', *Archivo Ibero-americano de Historia de la Medicina y Antropología Médica* (1958)

Castro, Americo, *The Structure of Spanish History* (Princeton: Princeton University Press, 1954)

Chico-Ponce de León, Fernando, 'The First Neuroanatomical Text Published on the American Continent: Mexico City, 1579', *Child's Nervous System* (2004)

Chinchilla, Anastasio, *Anales historicos de la medicina general* (Valencia: Lopez y compañia, 1841)

Cleruzio, Antonio, *La macchina del mondo* (Rome: Carocci, 2005)

Clouse, Michele L., *Medicine, Government and Public Health in Philip II's Spain. Shared Interests, Competing Authorities* (Farnham: Ashgate, 2011)

Codellas, Pan S., 'Vesalius-Valverde-Patousas: The Unpublished Manuscript of the First Modern Anatomy in the Greek Language', *Bulletin of the History of Medicine*, 14 (1943)

Conforti, Maria e Rezi, Silvia de, 'Sapere anatomico negli ospedali romani: formazioni dei chirurghi e pratiche sperimentali (1620–1720)' in *Rome et la science moderne: entre Renaissance et Lumiéres* (Rome: École française de Rome, 2008)

Cook, Alexandra Parma and Cook, Noble David, *The Plague Files: Crisis Management in Sixteenth-Century Seville* (Baton Rouge: Louisiana State University Press, 2009)

Cortejoso, Leopoldo, 'Los hospitales de Valladolid en tiempos de Felipe III', *Asclepio*, 12 (1958)

Cortés, Valerià, *Anatomía, academía y dibujo clásico* (Madrid: Cátedra, 1994)

Cowans, Jon, *Early Modern Spain: A Documentary History* (Philadelphia: University of Pennsylvania Press, 2003)

Cruz, Anne J., *Culture and Control in Counter-Reformation Spain* (Minnesota: University of Minnesota Press, 1992)

Cunningham, Andrew, 'The Kinds of Anatomy', *Medical History*, 19 (1975)

———. *The Anatomical Renaissance* (Aldershot: Scholar Press, 1997)

Dandelet, Thomas James, 'Spanish Conquest and Colonization at the Center of the Old World: The Spanish Nation in Rome, 1555–1625', *Journal of Modern History*, 69 (1997)

———. *Spanish Rome* (New Haven: Yale University Press, 2001)

———. (ed.), *Spain in Italy, Politics, Society and Religion 1500–1700* (Leiden: Brill, 2006)

Danon Bretos, José, 'Notas medicas en los libros de "Estudi General" de Barcelona (Siglos XVI–XVII)', *Cuadernos de de historia de la medicina española*, 10 (1971)

———. *Visió histórica de l'Hospital General de Santa Creu de Barcalona* (Barcelona: Fundació Salvador Vives Casajuana, 1978)

David-Peyre, Y., 'La alegoria del cuerpo humano en la peninsula iberica', *Asclepio*, 28 (1976)

Dávila, Joaquin Herrera, *El hospital del Cardenal de Seville y el Doctor Hidalgo de Agüero* (Seville: Edición de la fundación de Cultura Andaluza, 2010)

Eamon, William, *Science and the Secrets of Nature* (Cambridge: Cambridge University Press, 1996)

———. '"Nuestro males no son constitucionales, sino circunstianciales": The Black Legend and the History of Early Modern Spanish Science', *Colorado Review of Hispanic Studies*, 7 (1989)

———. *The Professor of Secrets. Mystery, Medicine, and Alchemy in Renaissance Italy* (Washington DC: National Geographic Society, 2010)

Elliott, John, *The Old World and the New: 1492–1650* (Cambridge: Cambridge University Press, 1992)

———. *History in the Making* (New Haven: Yale University Press, 2012)

———. *Imperial Spain, 1469–1716* (London: Penguin, 2002)

———. *Spain, Europe, and the Wider World* (New Haven: Yale University Press, 2009)

Elorza, Juan Carlos (ed.), *El arte en las cortes de Carlos V y Felipe II* (Madrid: Alpuerto, 1999

Entrambasaguas, Joaquin de, *Grandeza y decadencia de la Universidad Complutense* (Madrid: Editorial Complutense, 1996)

Escribano García, Victor, *La Anatomia y los Anatomistas españoles del siglo XVI* (Granada: Traveset, 1902)

———. *La cirurgia y los cirujanos del siglo XVI* (Granada: Traveset, 1938)

Esperabé de Arteaga, Enrique, *Historia pragmática e interna de la Universidad de Salamanca* (Salamanca: Francisco Nuñez, 1917)

Espuche, Albert García, *Un siglo decisivo. Barcelona y Cataluña, 1550–1640* (Madrid: Alianza, 1998)

Estella, Margarita, *Los Leoni, escultores entre Italia y España* in *Los Leoni (1509–1608) Escultores del renacimiento al servicio de la corte de España* (Madrid: Museo del Prado, 1994)

Felipo Orts, Amparo, *La universidad de Valencia durante el siglo XVI* (Valencia: University of Valencia, 1993)

Fernández Luzón, Antonio, *La Universidad de Barcelona en el siglo XVI* (Barcelona: Edicions Universitat Barcelona, 2003)

Fernández Martin, Luis, 'Origines de la diseccion anatomica en la Universidad de Valladolid', *Cuadernos de historia de medicina española*, 13 (1974)

French, Roger, *Dissection and Vivisection in the European Renaissance* (Aldershot: Ashgate, 1999)

Fernández, Asunción, *Documentos para la historia de las profesiones sanitarias: El colegio de médicos y cirujanos de Zaragoza. (Siglos XV–XVIII)* (Zaragoza: Colegio Oficial de Médicos de Zaragoza, 1997)

———. *El Hospital Real y General de Nuestra Señora de Gracia de Zaragoza en el siglo XVIII* (Zaragoza: Institución Fernando el Católico, 1987)

Fresquet Febrer, José Luis, *El tratado de anatomía de Juan Calvo, Estudios dedicados a Juan Peset Alexandre* (Valencia: University of Valencia, 1982)

———. *La chirurgia universal y particular de Juan Calvo* (Valencia: Tesis de licenciatura, 1979)

———. *La experiencia Americana y la terapéutica en los Secretos de chirurgica (1567) de Pedro Arias Benavides* (Valencia: University of Valencia, 1993

———. *La práctica médica en los textos quirúrgicos españoles en el siglo XVI* (Granada: Dynamis, 2002)

Gachard, Luis Prosper, *Don Carlos et Philippe II* (Brussels: Emm. Devroye, 1863)

Ganizares-Esguerra, Jorge, 'Iberian Science in the Renaissance: Ignored How Much Longer?', *Perspectives on Science*, 12 (2004)

Garcia Ballester, Luis, 'La cirugía en la Valencia del siglo XV, El privilegio para disecar cadaveres en 1478', *Cuadernos de historia de la medicina española*, 7 (1967)

———. *Las obras medicas de Luis Collado. Nota a proposito de un manuscrito del British Museum* (MS Sloane, 2489) (Madrid: Gráficas Orbe, 1971)

———. *Historia social de la medicina* (Madrid: Akal, 1971)

———. *La medicina a la Valencia medieval* (Valencia: Edicions Alfons el Magnanim, 1988)

———. *Los moriscos y la medicina. Un capítulo de la medicina y la ciencia marginada de la España del siglo XVI* (Barcelona: Labor, 1984)

———. *La busqueda de la salud. Sanadores y enfermos en la España medieval* (Barcelona: Ediciones Península, 2001)

García Cárcel, Ricardo, *Historia de Cataluña. Siglo XVI–XVIII* (Barcelona: Ariel, 1985)

———. *La leyenda negra, Historia y opinion* (Madrid: Anaya, 1992)

García de la Concha, Víctor, *Nebrija y la introduccion del renacimiento en Espana* (Salamanca: University of Salamanca, 1983)

García Hourcade, Juan Luis, *Andrés Laguna. Humanismo, ciencia y política en la Europa renacentista* (Segovia: Junta de Castilla y León, Consejería de Educación y Cultura, 1999)

Garcia Oro, Jose, *Cisneros, el Cardenal de España* (Barcelona: Ariel, 2002)

———. *Felipe II y la Universidad de Salamanca* (Santiago de Compostela: Imprenta Aldecoa, 1998)

———. *La Universidad de Alcalá de Henares en la etapa fundacional* (Santiago de Compostela: Independencia Editorial, 1992)

García Tapia, Nicolas, 'Los codices de Leonardo en España', *Boletín del seminario de estudios de arte y arqueología*, 63 (1997)

Gonzalez, Enrique Gonzalez, *La enseñanza médica en la Ciudad de Mexico durante el siglo XVI*, *El mestizaje cultural y la medicina novohispana del siglo XVI* (Valencia: Instituto de Estudios Documentales e Históricos sobre la Ciencia, 1995)

Gonzales Navarro, Ramon, *Universidad Complutense: Constituciones originales Cisnerianas* (Alcalá de Henares: University of Alcalá de Henares, 1984)

González, Nazario, *Una historia abierta* (Barcelona: Edicions Universitat Barcelona, 1998)

Goodman, David C., *Power and Penury* (Cambridge: Cambridge University Press, 1988)

Goyanes, Joan José Barcía, *El mito de Vesalio* (Valencia: University of València, 1994)

Granjel, Luis S., *Discurso sobre el pasado de la enseñanza del saber y el arte medicos, La Universidad de Salamanca* (Salamanca: University of Salamanca, 1953)

———. *Historia y Medicina en Espana* (Valladolid: Junta de Castilla y Leon, 1994)

———. *Medicina española renascentista* (Salamanca: University of Salamanca, 1980)

Grell, Ole Peter and Elmer, Peter (eds), *The Healing Arts: Health, Disease and Society in Europe 1500–1800* (Manchester: Manchester University Press, 2004)

Guerra, Fransisco, 'Juan Valverde de Amusco', *Clio medica*, 2 (1967)

Harvey, L.P., *Muslims in Spain, 1500–1614* (Chicago: University of Chicago Press, 2005)

Haskell, Francis, *Taste and the Antique. The Lure of Classical Sculpture, 1500–1900* (New Haven: Yale University Press, 1991)

Helmstutler di Dio, Kelley, *Leone Leoni and the Status of the Artist at the End of the Renaissance* (Farnham: Ashgate, 2011)

———. *Sculpture Collections in Early Modern Spain* (Farnham: Ashgate, 2011)

———. *Sculptures in Spanish Collections from Philip II to Philip IV, Collecting Sculpture in Early Modern Europe* (New Haven: Yale University Press, 2003)

Huguet-Termes, Teresa (ed.), *Health and Medicine in Hapsburg Spain: Agents, Practices, Representations* (London: The Wellcome Trust, 2009)

Huisman, Tim, *The Finger of God. Anatomical Practice in 17th-Century Leiden* (Leiden: Primavera Pers, 2009)

Iborra, Pascual, *Historia del protomedicato en España (1477–1822)* (Valladolid: University of Valladolid, 1987)

Jimenez, Catalan, M., *Memorias para la Historia de la Universidad Literaria de Zaragoza. Tomo II* (Zaragoza: Typ. La Académica, 1926)

Kamen, Henry, 'The Decline of Spain: A Historical Myth?', *Past and Present*, 91 (1981)

———. *La Inquisición española* (Barcelona: Crítica, 1999)

———. *Spain in the Later Seventeenth Century* (London: Longman, 1980)

Kagan, Richard L., *Clio and the Crown: The Politics of History in Medieval and Early Modern Spain* (Baltimore: Johns Hopkins University Press, 2009)

———. *Students and Society in Early Modern Spain* (Baltimore: Johns Hopkins University Press, 1975)

———. 'Universities in Castile, 1500–1700', *Past and Present*, 48 (1970)

Keen, Benjamin, 'The Black Legend Revisited: Assumptions and Realities', *Hispanic American Historical Review*, 49(4) (1969)

Kusukawa, Sachiko, *Picturing the Book of Nature. Image, Text, and Argument in Sixteenth-Century Human Anatomy and Medical Botany* (Chicago: University of Chicago Press, 2012)

Lain Entralgo, Pedro, 'La anatomia de Vesalio', *Archivo iberoamericano de historia de la medicina* (1951)

———. 'La ciencia española' in *Diccionario de historia de España*, vol. 1 (Madrid, 1952)

———. 'The Spanish Contribution to World Science', *Cahiers d'historie mondiale*, 6 (1961)

Leite, Serafim (ed.): *Estatutos da Universidade de Coimbra (1559)*. (Coimbra: Por ordem da Universidade, 1963)

Lejeune, Fritz, *Zur spanischen Anatomie vor und um Vesal* (Cologne, 1926)

Lopez, Enrique Alvarez, 'El Dr. Fransisco Hernández y sus comentarios a Plinio', *Revista de Indias*, 3 (1942)

Lopez, Luis Alberti, *La anatomia y los anatomistas del renacimiento. Collección de monografias de historia de medicina* (Madrid: Consejo Superior de Investigaciones Científicas, 1948)

López Piñero, Jose Maria, *Antologia de la Escuela Valenciana del siglo XVI* (Valencia: Cátedra e Instituto de Historia de la Medicina, 1962)

——. *Ciencia y tecnica en la sociedad Española de los siglos XVI y XVII* (Barcelona: Labor, 1979)

——. *Diccionario historico de la ciencia moderna en España. 2 Volumes* (Barcelona: Península, 1983)

——. *El Codice Pomar (ca 1590), El interes de Felipe II por la historia natural y la expedición Hernández a América* (Valencia, 1991)

——. *El grabado en la ciencia hispánica* (Valencia: CSIC, 1987)

——. 'El saber anatomico y la dissección de cadaveres humanos en la España en la primera mitad del siglo XVI', *Cuadernos de la historia de la medicina Española*, 13 (1974)

——. *El triunfo de las ciencias aplicadas, Arte y saber: La cultura en tiempos de Felipe III y Felipe IV* (Valladolid: Ministerio de Cultura, 1999)

——. *Historia de la Ciencia y de la Tecnica en la Corona de* Castilla, vol. III (Madrid: Junta de Castilla y León, 2002)

——. 'La obra de Juan Tomas Porcell', *Medicina Española*, 52 (1965)

——. *Los saberes medicos y su enseñanza en el siglo XVI, Historia de la medicina valenciana* (Valencia: Vicent Carcía, 1988)

——. *Los temas polemicas de la medicina renacentista – Los controversias de Frensisco Valles* (Madrid: Consejo Superior de Investigaciones Cientificas, 1988)

——. *Medicina moderna y sociedad Española (siglos XVI–XIX)* (Valencia: Catedra e instituto de la historià de medicina, 1976)

——. *Santiago Ramón y Cajal* (Valencia: University of Valencia, 2006)

——. 'The Vesalian Movement in Sixteenth-Century Spain', *Journal of the History of Biology*, 12 (1979)

López Terrada, Maria Luiz, *Medicos, Cirujanos, Boticarios y Albéitares*, Historia de la Ciencia y de la Tecnica en la Corona de Castilla, vol. III (Madrid: Junta de Castilla y León, Consejería de Educación y Cultura, 2002)

Lovci, Radovan, *Michael Servetus, Heretic or Saint?* (Prague: Sharpless House, 2011)

Lozoya, Xavier, *El preguntador del rey. Francisco Hernández* (Mexico City: Consejo Nacional para la Cultura y las Artes, 1991)

Mandressi, Rafael, *Disecciones y anatomía, Historia del cuerpo del renacimiento a la illustración* (Madrid: Taurus, 2005)

———. *Le regard de l'anatomiste. Dissection et invention du corps en Occident* (Paris: Seuil, 2003)

Marañon, Gregorio, *La literatura científica en los siglos XVI y XVII, Historia general de las literaturas hispanicas, vol. 3, Renacimiento y Barrocco* (Madrid: Editorial Vergara, 1953)

Maravall, José Antonio, *Estudios de la historia del pensamiento español* (Madrid: Centro de estudios constitucionales, 1984)

Martinez Ruiz, Enrique, *Felipe II, la Ciencia y la Tecnica* (Madrid: Actas Editorial, 1999)

Martínez-Vidal, Àlvar, 'Harvey. Dal vecchio al nuovo mondo' in *Harvey e Padova. Atti del convegno celebrativo del 4o centenario della laurea di William Harvey* (Padua: Antilia, 2002)

Martínez-Vidal, Àlvar and Pardo-Tomás, Jose, 'Anatomical Theatres in Early Modern Spain', *Medical History*, 49 (2005)

Meyer, A.W., 'The Amuscan Illustrations', *Bulletin of the History of Medicine*, 14 (1943)

Morejón, Hernández, *Historia bibliográfica de la medicina española* (Madrid: La viuda de Jordan e hijos, 1845–52)

Moreno Espinosa, Gerardo, *Don Carlos, El Príncipe de la leyenda negra* (Madrid: Marcial Pons Historia, 2006)

Muñoyerro, Alonso, *La facultad de medicina de Alcalá de Henares* (Madrid: Consejo Superior de Investigaciones Cientificas, 1944)

———. *Libro de Actas* (Madrid: Instituto Jerónimo Zurita, 1935)

Nader, Helen, *The Mendoza Family in the Spanish Renaissance, 1350–1550* (New Brunswick: Rutgers University Press, 1979)

Navarro, Rafael, 'El doctor Juan Valverde', *Bol. del col. med.*, 191 (1927)

Navarro, Ramón Gonzalez, *Felipe II y las reformas constitucionales de la Universidad de Henares* (Madrid: Sociedad Estatal para la Commemoriacion de los Centenarios de Felipe II y Carlos V, 1999)

Navarro Brotons, Victor, *Espanya i la revolució científica: Aspectes historiogràfics, reflexions i perpectives, Actes de la VII trobada d'història de la ciencia i de la tècnica* (Barcelona: SCHCT, 2003)

———. 'La actividad astronomica en la España del siglo XVI: perspectivas historiograficas', *Arbor*, 142 (1992)

———. 'The Reception of Copernicus in Sixteenth-Century Spain – The Case of Diego de Zuñiga', *Isis*, 86(1) (1995)

———. (ed.) *Más allá de la Leyenda Negra. España y la Revolución Científica* (Valencia: Consejo Superior de Investigaciones Cientificas, 2007)

Naya, Juan, *Servetus Michael, Heartfelt: Proceedings of the International Servetus Congress, Barcelona, 20–21 October, 2006* (Lanham: University Press of America, 2006)

Noreña, Carlos G., *Studies in Spanish Renaissance Thought* (The Hague: Springer, 1975)

Olmedilla y Puig, Joaquin, *Francisco Arceo, illustre medico y escritor español del siglo XVI* (Madrid: Real Academia de Medicina, 1913)

O'Malley, Charles D., 'Andres Laguna and his Anatomica Methodus', *Physis*, 5 (1963)

————. *Andreas Vesalius of Brussels, 1514–1564* (Berkeley: University of California Press, 1964)

————. 'Bernardino Montaña, Author of the First Anatomy in The Spanish Language', *Journal of the History of Medicine*, 1 (1946)

————. *Los saberes morfologicos en el renacimiento. La anatomía. Historia universal de la medicina* (Barcelona: Salvat, 1973)

————. *Miguel Servetus* (Philadelphia: American Philosophical Society, 1959)

————. *Pedro Jimeno, Valencian Anatomist of the Mid-Sixteenth Century, Science, Medicine and Society. Essays to Honor Walter Pagel* (London: Heinemann, 1972)

Otero Carvajal, Luis Enrique, *La destrucción de la ciencia en España: las consecuencias del triunfo militar de la España franquista* (Madrid: Complutense, 2006)

Parrado del Olmo, Jesus Maria, *Grandes hitos del renacimiento español* (Madrid: Actas, 2003)

Pardo, Jose Bernia, *La differencia entre el animal y el hombre en el Antoniana Margarita* (Valencia: University of Valencia, 1975)

Pardo-Tomás, José, *Ciencia y Censura – La Inquisicion Española y los Libros scientificos en los siglos XVI y XVII* (Madrid: CSCI, 1991)

Pardo-Tomás, José, and Piñero, José Maria Lopez, *La influencia de Fransisco Hernández (1515–87) en la constitución de la botánica y la material médica modernas* (Valencia: University of València, 1996)

Park, Katharine, 'The Criminal and the Saintly Bodies', *Renaissance Quarterly*, 47 (1994)

Pedretti, Carlo, *L'anatomia di Leonardo da Vinci* (Florence: Grantour, 2005)

Persaud, T.N.V., *A History of Anatomy – The Post-Vesalian Era* (Springfield: Charles C. Thomas, 1997)

Peset Llorca, V., 'La Universidad de Valencia y la renovacion cientifica española', *Asclepios*, 17 (1965)

Pi-Sumyer, Jaume, 'Joan d'Alos and the Doctrine of the Circulation of the Blood', *Yale Journal of Biology and Medicine*, 28 (1955–56)

Pigozzi, Marinella, 'Arte e scienza a Bologna da Papa Gregorio XIII a Papa Clemente VIII (1572–1605)' in *Rapprasentare il corpo, arte en anatomía da Leonardo all'Illuminismo* (Bologna: Bononia University Press, 2004)

Poole, Stafford, 'Juan de Ovando's Reform of the University of Alcala de Henares, 1564–1566', *Sixteenth Century Journal*, 21(4) (1990)

Porter, Roy (ed.), *The Scientific Revolution in a National Context* (Cambridge: Cambridge University Press, 1992)

Prieto Carrasco, Casto, *La enseñanza de la anatomía en la Universidad de Salamanca*(Salamanca: University of Salamanca, 1935)

Puerto, Javier, *La fuerza de fierabrás, Medicina, ciencia y terapeutica en los tiempos del Quixote* (Madrid: Just in Time, 2005)

———. *La leyenda verde* (Salamanca: Junta de Castilla y León, 2003)

Redondo, Augustin, *Le corps dans la société espagnole des XVI et XVII siécles* (Sorbonne: Publications de la Sorbonne, 1990)

Rekers, B., *Benito Arias Montano, 1527–1598* (London: Studies of the Warburg Institute, 1972)

Riera, Juan, *Cirujanos, urulogos y algebristas del renacimiento y barocco* (Valladolid: University of Valladolid, 1990)

———. *Valverde y la anatomia del renacimiento* (Valladolid: Ediciones de la Universidad de Valladolid, 1981)

———. 'Vida y obra de Luis Mercado, Seminario de historia de la medicina española', *Cuadernos de Historia de la Médicina Española* (1968)

Rifkin, Benjamin A., *Human Anatomy* (London: Harry N. Abrams, 2007)

River, Elias S. (ed.), *Poesía lírica del Siglo de Oro* (Madrid: Cátedra, 2003)

Rivero, Manuel, *Felipe II y el gobierno de Italia* (Madrid: Sociedad Estatal para la Conmemoración de los Centenarios de Felipe II y Carlos V, 1998)

Rodriguez-San Pedro Bezares, Luis E., *Historia de la Universidad de Salamanca. Volumen III: Saberes y confluencias* (Salamanca: University of Salamanca, 2002)

———. *Historiografía de la Universidad de Salamanca en la edad moderna: siglo XV-XVIII, Actas I congreso Historia de Salamanca* (Salamanca: University of Salamanca, 1992)

———. *La Universidad Salmantina del Barrocco, periodo 1598–1625* (Salamanca: Ediciones University of Salamanca, 1986)

———. *Las universidades hispánicas de la monarquia de los Austrias al centralismo liberal* (Salamanca: Ediciones University of Salamanca, 1998)

Rojo Vega, Anastasio, *Enfermos y sanadores en la Castilla del siglo XVI* (Valladolid:, Secretariado de Publicaciones, University of Valladolid, 1993)

———. *Medicina barroca Vallisoletana* (Valladolid:, Secretariado de Publicaciones, University of Valladolid, 1984)

Rossi, Paolo, *The Birth of Modern Science* (Oxford: Blackwell, 2001)

Rujula, José de, *Indice de los colegiales del mayor de San Ildefonso y menores de Alcalá* (Madrid: Consejo Superior de Investigaciones Cientificas, 1946)

Saliba, George, *Islamic Science and the Making of the European Renaissance* (Cambridge, MA: MIT Press, 2007)

Salort Pons, Salvador, 'Gaspar Becerra en Florencia', *Archivo Español de Arte*, Vol 78, No 309 (2005)

Sanchez Ron, José Manuel, *La ciencia y el Quixote* (Barcelona: Crítica, 2005)

Sánchez Silva, Jorge, 'Augustin Farfán y Alonzo Lopez de Hinojosos. Dos medicos hispanos del Nuevo mundo', *Acta médica*, II(26) (1971)

Santander Rodriguez, Teresa, *El Doctor Cosme de Medina y su biblioteca. (1551–1591)* (Salamanca: Centro de Estudios Salmantinos, 1999)

———. 'La creacion de la catedra de chirurgia en la Universidad de Salamanca', *Cuadernos de la historia de la medicina Española*, 43 (1965)

———. 'La iglesia de San Nicolas y el antiguo teatro anatómico de la Universidad de Salamanca', *Revista española de teología* (1983)

San Vicente, Angel, *Monumentos diplomaticos sobre los edificios fundacionales de la Universidad de Zaragoza y sus constructores* (Zaragoza: Diputacion Provincial, Institucion Fernando el Católico, 1981)

Sawday, Jonathan, *The Body Emblazoned* (London: Psychology Press, 1996)

Schultz, J.B., *Art and Anatomy in Renaissance Italy* (Ann Arbor: UMI Research Press, 1982)

Shapin, Steven, *The Scientific Revolution* (Chicago: University of Chicago Press, 1998)

Sobrequés i Gallicó, Jaume, *Historia de Barcelona* (Barcelona: Plaza & Janes Editores, 2008)

Solano Costa, Fernando, *Historia de Zaragoza* (Zaragoza: Ayuntamiento de Zaragoza, 1976)

———. 'Pedro Cerbuna y el funcionamiento de la nueva Universidad' in *Historia de la Universidad de Zaragoza* (Madrid: Editora Nacional, 1983)

Somolinos d'Ardois, Germán, 'Capítulos de historia médica mexicana', *Soc. Mexicana de Historia* (1979)

———. *Vida y obra de Francisco Hernandez* (Mexico City: Universidad Nacional de Mexico, 1960)

Fransisco Hernández, *Los impresos medicos mexicanos (1553–1618), El mestizaje cultural y la medicina novohispana del siglo XVI* (Valencia: University of Valencia, 1995)

———. *Obras completas* (Mexico City: Universidad Nacional de Mexico, 1958–84)

Sousa, Ronald W., 'The View of the Artist in Francisco de Holanda's Dialogues', *Luso-Brazilian Review*, 15 (1978)

Subiza, E., 'Los medicos de Felipe II, Aportación a su studio', *Archivo Iberoamericano de Historia de la medicina y la antropología médica*, 6 (1954)

Tapia Garcia, Nicolas, *La leyenda negra: Historia y opinion* (Madrid: Alianza, 1981)

Torre y del Cerro, Antonio de la, *Provision de cátedras en la Universidad de Barcelona* (Barcelona: Publicacions de la Universitat de Barcelona, 1926)

Urriza, Juan, *La preclara facultad de artes y filosofía de la Universidad de Alcalá de Henares en el Siglo de Oro, 1509–1621* (Madrid: Diana, 1942)

Valero Garcia, Pilar, *La Universidad de Salamanca en la época de Carlos V* (Salamanca: University of Salamanca, 1988)

Varey, Simon (ed.), *The Mexican Treasury: The Writings of Francisco Hernández* (Stanford: Stanford University Press, 2000)

Vicente, Jose Vazquez, *Los anatomicos de la epoca del renacimiento* in *Trabajos de la cátedra de historia critica de medicina* (Madrid: Minuesa de los Ríos, 1935)

Villalon, L. J. Andrew, 'Putting Don Carlos Together Again: Treatment of a Head Injury in Sixteenth-Century Spain', *Sixteenth Century Journal*, 26(2) (1995)

Zubiri Vidal, Fernando, and Zubiri de Salinas, Ramon, *Las epidemias de peste y cólera morbo-asiático en Aragón* (Zaragoza: Diputación Provincial, Institución Fernando el Católico, 1980)

Index